U.S. Marine Corps Physical Readiness Training for Combat

Plus

Mountain Operations

Department of Defense

Pentagon
Publishing

UNITED STATES MARINE CORPS
Marine Corps Combat Development Command
Quantico, Virginia 22134-5001

29 January 1988

FLEET MARINE FORCE REFERENCE PUBLICATION............0-1B

MARINE PHYSICAL READINESS TRAINING FOR COMBAT

1. PURPOSE

Fleet Marine Force Reference Publication (FMFRP) 0-1Marine Physical Readiness Training for Combat is intended for use by all Marines. It provides the information and references necessary to establish and conduct physical conditioning programs to prepare Marines for the physical demands of combat.

2. APPLICABILITY

Although not directive, this handbook is intended for use as a reference by all Marines in developing unit physical readiness training for combat. This FMFRP does not establish policy for the conduct of physical fitness training. Refer to Marine Corps Order 6300.3_ for applicable Marine Corps policy on physical fitness. It may also be used as a basis for instruction conducted by unit leaders or as a reference for instruction or training.

3. RECOMMENDATIONS

To improve this manual, the MAGTF Warfighting Center seeks comments and recommendations from the user. Officers and noncommissioned officers with constructive comments are encouraged to respond. Lack of response will be considered as concurrence. The User Suggestion Form which follows can be detached and forwarded to - -

Commanding General
Marine Corps Combat Development Command (WF 06)
Quantico, Virginia 22134-5001

4. DISTRIBUTION

Initial distribution of this FMFRP 0-1B is shown below. Addressees on this list should ensure that appropriate subordinate organizations are provided copies.

JOHN I. HOPKINS
Brigadier General, U.S. Marine Corps
Assistant Chief of Staff, Warfighting
Marine Corps Combat Development Command
Quantico, Virginia

DISTRIBUTION: TDJ

From:

To: Commanding General, Marine Corps Combat Development Command (WF 06), Quantico, Virginia 22134-5001

Subj: FMFRP 0-1B, MARINE PHYSICAL READINESS TRAINING FOR COMBAT, RECOMMENDATIONS CONCERNING

1. In accordance with the Promulgation Page to FMFRP 0-1B, which invites individuals to submit suggestions concerning this FMFRP directly to the above addressee, the following unclassified recommendation is forwarded:

Page________Article/Para No.________Line No.________Figure/Table No.________

Nature of Change:_____Add_____Delete_____Change_____Correct

2. Proposed New Verbatim Text: (Verbatim, double spaced; continue on additional pages as necessary.)

3. Justification/Source: (Need not be double spaced.)

NOTE: Only one recommendation per page.

PREFACE

MARINE PHYSICAL READINESS TRAINING FOR COMBAT

Chapter 1
PHYSICAL READINESS LEADERSHIP
Section I. Role of Physical Fitness in Combat

1101. PURPOSE OF PHYSICAL FITNESS TRAINING

Physical fitness training in the Marine Corps has one purpose: to prepare Marines to physically withstand the rigors of combat. All other goals of physical fitness training are subordinate to and must support attainment of this goal. The idea that only infantry or reconnaissance units and their attachments normally face physically demanding combat is wrong. This error must not influence the priority commanders of combat support, combat service support, aviation, and headquarters units give to physical fitness for combat. Physical fitness for combat has a high priority for all Marines. A sound, effective unit program, requiring limited time and material, offers a greater payoff in combat than many more expensive and time-consuming training programs. This handbook describes the concept and provides the ingredients of a program for commanders to use in carrying out one of their most serious responsibilities: ensuring that their Marines are physically ready for combat.

1102. SCOPE OF THE MANUAL

This handbook provides guidance for all leaders, trainers, and planners of physical training programs. It describes unit physical fitness for combat training in the following chapters:

a. Physical Readiness Leadership. Chapter 1 provides guidance to leaders in the conduct of physical combat readiness training.

b. Physical Readiness Training Programs. Chapter 2 discusses how to structure programs to meet physical readiness goals in a variety of combat training situations.

c. Physical Conditioning Activities. Chapter 3 discusses the primary physical conditioning activities which commanders may use to attain readiness for combat goals. Sections in this chapter address foot marches under load, strength-building activities, and activities which build confidence and the aggressive spirit needed in combat.

d. Combat Water Survival. As an amphibious force, Marines cannot ignore the importance of combat water survival training. Chapter 4 describes a program for commanders to conduct this training.

e. Competitive Conditioning Activities. Chapter 5 describes competitive activities which stress the relationship to combat actions, featuring team-building types of competition.

f. Evaluation of Performance During Training. Chapter 6 describes tests which measure individual and unit physical fitness for combat.

g. The Human Body. Chapter 7 covers the structure and functioning of the body.

1103. COMMANDER'S ROLE

Major General Commandant John A. Lejeune in the 1921 edition of Marine Corps Manual expressed our philosophy of leadership. Among other things, General Lejeune stated that "it will be necessary for officers ... to endeavor to enlist the interest of their men in building up and maintaining their bodies in the finest physical condition." It is significant that the Commandant prefaced these writings in 1921 with the statement that his thoughts were greatly influenced by the experience of World War I. We have never varied as a Corps from this belief that one of a commander's most serious responsibilities is to have Marines physically prepared for combat. Senior commanders and their staffs carry out this role primarily by providing command emphasis, including planning, support, and supervision. Commanders at the battalion/squadron and company/battery level execute a program of physical fitness for combat by close supervision and, most importantly, leadership by example.

1104. HISTORY OF MILITARY PHYSICAL READINESS

Every war has revealed our military physical deficiencies during the initial periods of mobilization. This realization followed the Civil War and has recurred regularly with each national emergency.

a. Training programs in each war were geared to the physical need of the era. Success was dependent upon the amount of time available during training to physically prepare Marines for battle conditions. Frequently, casualties in initial engagements were attributed to the inability of our Marines to physically withstand the rigors of combat over rugged terrain and under unfavorable climatic conditions. With adequate preparation, Marines have always handled the test of battle.

b. During World War II, the first physical conditioning doctrine that could be scientifically justified by testing was introduced. As the war progressed, this program was effective in the physical conditioning of millions of Marines for combat.

c. Postwar periods have traditionally been a time of consolidation. Unfortunately, some leaders considered the conditioning phase of training to be a wartime tool. With such a philosophy prevalent between wars, physical readiness was relegated to secondary importance resulting in a serious lowering of combat effectiveness. The initial commitment of Marines in Korea dramatically displayed this failure to recognize the extreme physical demands of warfare. Call-up of large numbers of reserve Marines and pressing them into the war within weeks after call-up mainly caused the loss in effectiveness.

d. Costly lessons learned from our military experiences over a period of years and the course of several wars led to an increasing interest in the physical conditioning of the individual Marine. No longer can we afford to emphasize physical fitness during wartime and de-emphasize it during peacetime. It is evident that, in spite of increased mechanization and modern weapons, physical readiness retains a vital place in the life of each individual Marine and in every unit within the Marine Corps regardless of mission.

e. Commanders are well aware of the need for rugged and well conditioned Marines. The daily demands of housekeeping, maintenance, support, training, operations, and other time-consuming tasks make it necessary for commanders to set aside time for frequent, regular, and vigorous exercise periods.

1105. PHYSICAL DEMANDS OF COMBAT

There are three primary elements of effective physical fitness for combat: lower body strength and stamina; upper body strength and stamina; and a competitive, combative spirit. Cardiovascular functioning is not treated as a separate element since it is a necessary foundation for the other three. Exercises and activities designed to improve Marines' condition in the three primary elements will foster good cardiovascular and respiratory condition.

a. Lower Body Strength and Stamina. Some common demands that combat places on lower body strength and endurance are marching long distances under load and functioning effectively at the destination; moving quickly and evasively under fire; and carrying wounded Marines to safety.

b. Upper Body Strength and Stamina. Some common demands that combat places on upper body strength and stamina are rapidly emplacing crew-served weapons; handling large-caliber ammunition for extended periods; climbing walls, cliffs, and other high obstacles; and performing field maintenance on air craft or heavy machinery.

c. Competitive, Combative Spirit. While part of this critical aspect of combat readiness is mental and emotional, robust physical condition and a training program which promotes physical aggressiveness greatly enhance a competitive and combative spirit. Activities which force Marines to overcome natural physical fear, which directly leads to fatigue, are particularly valuable.

1106. PHYSICAL EFFECTS OF COMBAT STRESS

It has been thoroughly documented that the added uncertainty and stress of combat have a major physical effect on Marines as well as the acknowledged psychological effect. In other words, in combat, fear equals fatigue. In training, we cannot easily reproduce this combat stress which reduces the effectiveness of individuals and units. However, we can produce fatigue and psychological doubt in training situations by developing a challenging physical training program which stresses the participants. This compels Marines to exercise their ability to continue to function under difficult and trying conditions. Such activities as long foot marches under load and difficult confidence courses train Marines to overcome their own fear and fatigue. Rugged competitive activities such as martial arts and pugil stick fighting prepare Marines psychologically to overcome an opponent as well as their own fear and fatigue. This sort of physically demanding training builds self-confidence and unit morale. The adage that "the unit that sweats more in peacetime will bleed less in battle" certainly applies to its physical training program. Athletic coaches have long followed this approach by such practices as having teams shoot foul shots after wind sprints in order to strengthen their mental conditioning and toughness. Although pushing Marines to their limit is necessary, commanders often set lower standards initially than ultimately desired. The program should then call for steadily increasing the difficulty. A good guide for a starting point is a standard that will be physically and psychologically demanding for no less than three-fourths of the unit. Expressed another way, no more than one-fourth of the unit should be working at less than their limit. The program should progress quickly, augmented, if needed, by remedial training for those unable to meet the unit standards, until the training is challenging for all but the most exceptional Marines in the command. Commanders must never forget that Marines expect to be challenged and thrive on being pushed to their limit.

1107. LIMITS OF PHYSICAL READINESS

A well-conditioned Marine unit has a significant advantage in combat, but this does not mean that physical conditioning can substitute for technical knowledge or good planning. For instance, units must, as discussed in chapter 3, pursue a continuing and challenging program of marches under loads. This does not disprove the axiom that fighting men, if required to carry over one-third of body weight, will generally become ineffective in battle. Leaders and planners must decide what is essential to be carried and must use an effective logistic distribution system, such as unit trains, to make remaining items available when they are needed. In a similar vein, the valid requirement to foster a competitive and combative spirit in Marines does not excuse leaders from the responsibility to practice tactics which will not needlessly risk lives. Physical readiness for combat is a vital goal, but it is only one part of a unit's overall readiness for the lest of combat.

Section II. Fundamentals of Physical Fitness

1201. FITNESS FOR MARINES

Total combat readiness includes technical proficiency and mental, emotional, and physical fitness. If any one of these attributes is lacking, combat effectiveness suffers proportionately. Without technical fitness, Marines lack the knowledge and skill to fight; without mental and emotional fitness, they lack the incentive and will to fight; and without physical fitness, they lack the physical ability and confidence to fight. Physical fitness in a Marine means a whole and healthy body, the capacity for skillful and sustained performance, the ability to recover from exertion rapidly, the desire to complete a designated task, and the confidence to face any eventuality.

1202. COMPONENTS OF PHYSICAL FITNESS

A sound body, free of disease and defect, does not in itself constitute physical fitness. Before an untrained Marine can be considered physically fit for combat, the following traits must be developed:

a. Strength. Every Marine needs enough strength to perform the heaviest task encountered in routine and emergency activities. The basic areas where strength is required are in the arm and shoulder girdle, abdomen, back, and legs. Muscles increase in size, strength, and firmness with regular and strenuous exercises. Without work, they grow flabby and weak.

b. Endurance. Each Marine needs sustaining power to maintain the maximum ability without undue fatigue. There are two types of endurance:

(1) Muscular Endurance. The Marine needs muscular endurance to fight the enemy under the most tiring combat conditions. Muscular endurance is the ability to perform continuous work over long periods of time. Endurance depends on the bloodstream's ability to deliver large amounts of oxygen and nutrition to the muscle masses and then carry away the waste products quickly.

(2) Cardiovascular-Respiratory Endurance. The development of cardiovascular-respiratory endurance ("wind") is necessary to maintain muscular endurance. Cardiovascular-respiratory endurance depends on the efficiency of the blood vessels, lungs, and heart. The maximum effort a Marine can exert over a period of time is limited by the capacity to absorb oxygen and expel carbon dioxide. The average Marine's cardiovascular-respiratory capacity can be greatly increased by exercise.

c. Agility. A Marine must be able to change direction quickly and as faultlessly as possible. The ability to react instantly and to maintain orientation during rapid changes of body position is important to survival. This important ability may be developed by conditioning exercises that require varied and rapid changes of body position on the ground and in the air, such as obstacle courses.

d. Coordination. Coordination is the ability to move all parts of the body in a smooth, efficient, and concerted effort (commonly called timing). A well-coordinated individual does not make useless movements. An individual who moves with precision and accuracy saves energy. Coordination is best developed by practicing diversified muscular activities and skills affecting all body parts.

1203. TYPES OF EXERCISE

Basically, there are two forms of exercise: isotonic and isometric. Both forms are contained in the activities found in this manual.

a. Isotonic. Isotonic exercises are those in which the expenditure of energy is regulated and released during concentrated efforts. The regulated expenditure of energy is controlled by both the mode of exercise and the individual's effort. This type of effort is common to most exercise and sports. To develop endurance, coordination, and flexibility, isotonic exercise must be used. Strength can also be developed or increased through this type of exercise.

b. Isometric. Isometric exercises are those in which maximum effort is applied and held until the engaged muscle energy is depleted during a single contraction effort. The individual exerts full force against an immovable object for a relatively short period of time (6 to 10 seconds) and then repeats the contraction several times with a short period of rest between each contraction. This type of exercise develops only strength; therefore, it has limited application.

1204. PRINCIPLES OF PHYSICAL CONDITIONING

Men and women vary in their physical makeup. Physiological (body) function and reaction also vary in proportion to heavy demands placed upon the body. To attain the maximum benefit without sacrificing Marines' welfare, fitness must grow with a careful program of conditioning. To allow for adjustments in body functioning as the conditioning program progresses and to ensure attainment of objectives, the following principles of physical conditioning must be applied:

a. Overload. As strength and endurance increase, the physical load must be increased until the desired level of fitness is reached.

b. Progression. In beginning stages, the load must be moderate. Gradual progression from this low state of fitness to a higher state is possible through application of a progressive program.

c. Balance. An effective program utilizes various types of activities and provides for the concurrent development of strength, endurance, and coordination as well as basic physical skills.

d. Variety. Some programs fail because the routine becomes boring. The most successful programs always include conditioning activities, competitive events, and military physical skill development.

e. Regularity. There is no easy or occasional way to develop physical fitness. Regularity of exercise is a must, with daily exercise preferred.

1205. THREE STAGES OF PHYSICAL CONDITIONING

Unconditioned or poorly conditioned Marines pass through the following stages in reaching the desired state of physical condition.

a. Toughening Stage. This stage is approximately 2 weeks long and is usually characterized by muscular stiffness and soreness followed by recovery.

b. Slow Improvement Stage. This stage is approximately 6 to 10 weeks long and is characterized by slow and steady improvement until the desired level, or a high level, of fitness is attained.

c. Sustaining Stage. This stage goes on indefinitely in order to maintain the level of conditioning achieved by passage through the previous stages.

1206. EFFECTS OF CLIMATIC CONDITIONS

Temperature, both atmospheric and body, affects the physical performance of Marines. Proper maintenance of body temperature through warm up exercise, proper dress in cold weather, and removal or adjustment of clothing in hot weather is necessary for effective performance and health. Climatic factors to be considered are the following:

a. Exercise In High Temperatures and High Humidity. Marines can endure strenuous physical activity in extremely hot temperatures if they are given an opportunity to become acclimated and if they take enough salt and water. It is essential to continue physical training programs in hot climates. Marines can better withstand high temperatures when they are well-conditioned. High humidity combined with high temperatures presents a serious danger. These conditions prevent the natural cooling of the body by the evaporation of perspiration. Training schedules should conform to the provisions of the base commander in respect to wet bulb conditions. Those who conduct training under these conditions should monitor weight loss and be careful to make fluid replacement adjustments.

b. Exercise at High Altitudes. Certain problems are encountered in conditioning Marines stationed in high altitudes because the heart undergoes greater exertion during exercise. It is particularly important that only light exercise be given initially at such altitudes. A Marine's body gradually adjusts to high altitudes within a few weeks. After this adjustment, progressively greater amounts of exercise are possible.

c. Exercise in Arctic Regions. Military duty in the arctic is so arduous that a high level of physical conditioning is essential. Because of the difficulties of carrying on physical conditioning exercises in extreme cold, Marines should be conditioned to the highest level possible before they arrive. A sustaining program will then maintain that level. When exercising in cold weather, Marines should be required to remove excess clothing to prevent them from becoming damp with perspiration.

1207. WARMING UP AND COOLING OFF

It is a fundamental physiological principle that Marines should warm up gradually before taking strenuous exercise. Such action speeds up the circulation to prepare the body to take an overload and helps to prevent injury to muscles and joints after exercising. Marines should be kept mildly active, walking, stretching, or performing some other mild muscular activity until breathing and temperature have returned to normal. Marines should never be allowed to cool off too rapidly; in cool or cold weather, they should put on additional clothing during the cooling-off period.

1208. PHYSICAL ACTIVITY AS AGE INCREASES

Combat makes severe physical demands on Marines. All Marines, regardless of age, must be physically ready to meet these demands. There is no physiological reason to cease exercise or exertion with age. Increased age usually brings increased responsibility which, in many instances, leads to a routine that can become almost devoid of physical activity. The key to fitness with increased age is to continue exercising at a reasonable level and to include exercise of a vigorous type in the daily routine. Older persons who have not regularly maintained a reasonable state of physical fitness require more time than younger persons to become fit. Such individuals usually require a longer period of time to recover from physical effort than younger Marines. If general health is good, evidence shows that older Marines can develop and maintain a degree of fitness which permits vigorous activity and proper performance of duties. It is both the individual's and the commander's responsibility that all Marines continue a daily sustaining exercise program.

Section III. Goals of Physical Readiness Training

1301. NECESSITY FOR PHYSICAL READINESS TRAINING

A very important objective of training is attainment and maintenance of operational readiness. Marines must be physically ready for operations at any time and under all conditions of climate and environment. A combination of training to develop proficiency in physical skills and conditioning to improve strength and endurance results in physical readiness for combat. The degree of physical fitness required of Marines can be acquired through physical exercises in a normal well-coordinated and closely monitored training routine. The performance of purely military duties, such as drills and marching, is not enough to build all the desired areas of fitness. Few recruits are physically fit for the arduous duties ahead of them. The softening influences of our mechanized civilization and the difficulties of conditioning Marines make physical fitness more important than ever before. If Marines are to be fully developed to and maintained at the desired standard of physical fitness, a well-conceived plan of physical readiness training must be a basic part of every training program. Marines cannot be adequately prepared in any other way for the hard work and arduous demands required on the battlefield.

1302. OBJECTIVE OF PHYSICAL READINESS TRAINING

The overall objective of the physical readiness training program is to develop individuals and units who are physically able and ready to perform their duty assignments or missions during training and in combat. To attain the objective of physical readiness, exercise activities must develop the following areas:

a. Strength and Endurance. Developing adequate strength to perform required duties and adequate endurance to sustain activity over a long period of time.

b. Muscle Tone. Developing muscle tone adequate to maintain proper posture and reasonable weight control.

c. Skills. Developing military physical skills which are essential to personal safety and effective combat performance. As skill is developed, agility and coordination are attained. The essential skills are-

(1) **Marching Under Load.** Marching with individual and unit weapons and equipment.

(2) **Running.** Distance and sprint running on roads and cross-country.

(3) **Jumping.** Broad jumping and vertical jumping downward from a height.

(4) **Dodging.** Changing body direction rapidly while running.

(5) **Climbing and Traversing.** Vertical climbing of ropes, poles, walls, and cargo nets. Traversing horizontal objects such as ropes, pipes, and ladders.

(6) **Crawling.** High crawling and low crawling for speed and stealth.

(7) **Throwing.** Propelling objects, such as grenades, for a distance with accuracy.

(8) **Vaulting.** Surmounting low objects, such as fences and barriers, by use of hand assists.

(9) **Carrying.** Carrying objects and employing man-carries.

(10) **Balancing.** Maintaining proper body balance on narrow walkways and at heights above normal.

(11) **Falling.** Contacting the ground from standing, running, and jumping postures.

(12) Surviving in Water (Or Other Specialized Situations) Using water survival techniques.

d. Character Traits. Jnstilling character traits which help accomplish military missions to include-

(1) Confidence. Confidence develops through achieving progressively more difficult tasks as physical ability grows.

(2) Aggressiveness. Participation in combative activities and contests develops desire and willingness to overcome an opponent.

(3) Reaction Under Pressure. Competitive contests and game situations train Marines to think and to act quickly while under pressure.

(4) Teamwork. Teamwork develops through competitive events in which a number of Marines must coordinate their efforts to accomplish a physical task.

1303. BENEFITS OF EXERCISE

The benefits of exercise are not always understood. Some of the more important results of exercise are listed below:

a. Improved Muscle Tone. Muscular tone improves and, at the same time, muscular strength and endurance are built up.

b. Cardiovascular-respiratory Endurance. Cardiovascular-respiratory endurance, or wind, improves through a process of opening up dormant lung capacity to absorb greater amounts of oxygen.

c. Circulation. Circulation of the blood speeds up and extends to a greater portion of the body as exercise forces the blood to service all parts of the body. The efficiency and effectiveness of the heart, lungs, and blood vessels improve.

d. Flexibility. A wider range of muscular movement is possible and rapidity in physical skills grows.

e. Elimination of Body Waste. Bending and twisting the body and the general speedup of body processes caused by exercise regulate and help eliminate body wastes.

f. Tension. Working off excess nervous energy and relief from daily worries and cares relieve tension. Participation in exercises leaves little time for worry.

g. Sleep. Sleep improves because muscles are healthfully tired after about of exercise. A by-product of sound sleep is relief of tension.

h. Obesity Control.. Control of obesity (fat) is made possible by using up excessive amounts of fat-producing food elements.

i. Injury Susceptibility. Susceptibility to injury is reduced through exercise. Muscles, tendons, and joints are strengthened. Injuries such as hernia, back strain, and joint sprains are less likely to occur if muscles are maintained in proper tone.

Section IV. Leadership Roles

1401. PSYCHOLOGICAL LEADERSHIP

The full development of a Marine's resources is not all physical. To be effective in developing physical readiness, leaders must realize that mind and attitude are also important to success. The more important psychological considerations are to—

a. Promote Understanding of the Value of Physical Readiness. A desire to be physically ready should be created in all Marines. Motivation is increased and Marines take greater interest in their individual physical fitness if they understand the value and benefits of vigorous exercise. When Marines realize their efforts are an investment in their own personal welfare, it should not be difficult to obtain their cooperation. Marines should understand the objectives, the benefits, and the value of each type of exercise activity in their program. They should also understand the relation of physical readiness to survival in combat.

b. Maintain a Positive Approach. Physical readiness training for combat is strenuous and demanding. It is a responsibility of leadership to create an atmosphere where all desire to participate fully. This attitude should be fostered. A negative approach must not be identified with physical readiness training, even with those having difficulty. Only in unusual cases should fear of punishment be the motivating factor behind good performance. For those few who cannot keep up or attempt to malinger, an effective remedial program is essential.

c. Seek Cooperation and Develop Morale. In a program placing maximum physical stress upon individuals, it is necessary to gain their cooperation. Favorable reaction is enhanced by proper planning and organization, challenging requirements, use of competition, and application of a progressive program resulting in physical fitness. As physical fitness grows, morale also grows.

1402. COMMAND AND SUPERVISORY FUNCTIONS

a. Command Functions. Commanders should take the following actions to support physical readiness training:

(1) Lead by personal example.

(2) Instill command interest and indicate to subordinate personnel the importance of this training to the welfare of the organization.

(3) Allot sufficient time for the achievement of objectives and monitor the use of such allotted time. The substitution of other training or routine duties for scheduled physical readiness training is unsound and unwise.

(4) Assign and properly utilize qualified personnel to supervise and conduct physical readiness training. If leaders are not competent, take action to ensure they become competent quickly. Reassign those who do not meet this standard to other duties.

(5) Make necessary facilities and funds available to support a program to develop physical readiness within all personnel.

(6) Measure the physical fitness of individuals and units in order to evaluate progress and to determine if the program is successful.

b. Supervisory Functions. Leaders responsible for planning, conducting, and supervising physical readiness training should take the following actions:

(1) Prepare physical readiness training schedules which apply the principles of physical conditioning and which aim for a particular type of program plan.

(2) Provide for wide participation of as many Marines as possible. All Marines, regardless of position or age, will benefit from regular exercise. In some instances, special efforts are necessary to overcome obstacles to regular and frequent training. Special effort is also necessary to ensure remedial conditioning. Such conditioning should occur for those who are physically substandard and after extended absence due to leave, sickness, injury, and travel.

(3) Prevent waste or unwise use of time allotted for physical readiness training. Time-wasters include unprepared instructors; assignment of one instructor to a group larger than a platoon; progression which does not keep pace with the physical development of the Marines; extreme formality; inadequate equipment or facilities which require waiting turns to exercise; and lengthy rest periods between exercises which interfere with the application of overload.

(4) Ensure that the program contains vigorous physical activity. Such activity places progressively greater demands upon the body during each exercise session and also over the duration of the training program. To be of benefit, exercise must tire the muscles and cause the heart to increase its rate of beat.

(5) Set an overall objective for each physical fitness program.Observe the training as necessary to ensure that the established objectives are being achieved.

(6) Observe physical readiness training to insure the use of a positive approach. To implement a positive attitude, small-unit leaders and instructors should personally set the example; have an understanding, fair, and sympathetic attitude; recognize individual differences; and motivate Marines toward their best effort.

(7) Guide and inform small-unit leaders and instructors concerning approved techniques, directives, and literature. As necessary, arrange for local training of instructors to include clinics, conferences, schools, and demonstrations.

(8) Determine the effectiveness of physical readiness training by personal participation in and observation of training, analysis of field inspection reports, and analysis of individual physical fitness test scores. Scores may be combined to reflect the fitness of the unit.

1403. SMALL-UNIT LEADERS AND INSTRUCTORS

a. Responsibility. The instruction and conduct of physical readiness activities are the function of company/battery commanders, platoonleaders and persons assigned as instructors. Experience has proved the effectiveness of physical fitness development when conducted in company-and platoon-size units under direct control of the leader with overall supervision by the parent-unit commander. For example, all the platoons of a company may exercise at the same time under the general supervision of the company commander, with each platoon conducting the assignment separately and under its own leadership.

b. Leader's Assignment. A small-unit leader or an instructor in a school or training activity is assigned to a combat unit or to a support unit. In this assignment, the leader is responsible for all training to include physical readiness training. In a different situation, the noncommissioned officer (NCO) or officer may beassigned as a full-time physical readiness instructor. This contrasts to a unit leader assignment where only part of the time is devoted to such training. In either case, the leader will hold an important and vital position for the physical fitness of Marines.

c. Leader's Training. Leaders may come to the assignment either fully or partially trained. It may be their first responsibility for the development of physical fitness. If they have had previous training through experience, make certain that their information is supplemented with study of this manual. If they have had professional training in physical education during civilian life, but no military experience, they should also use this manual to learn the methods used by the Marine Corps. A new leader should take advantage of various ways to learn including attendance at leader training courses, self-study, practice, and discussion with more experienced leaders.

d.Leader's Objective. As a physical readiness training instructor, the leader has two general objectives. The first is to motivate Marines to want to be physically fit. The second is to conduct a program that will develop a high degree of physical fitnss. Motivated Marines will react enthusiastically to such a program. It aids greatly in achieving local program objectives.

e. Leader's Personal Fitnes. A unit leader who must instruct and demonstrate physical activities must be in physical condition to do the job without undue physical stress.The leader should be able to do those things that must be demonstrated. The leader's strength, endurance, posture, and skill should set the example. This does not mean that the leader must excel, as other Marines do not expect championship performance. However, they do expect, and deserve, a creditable showing of fitness for the job.

f. Leader's Knowledge. The leader must have three types of knowledge to properly administer physical readiness training. Theyare- -

(1) Knowledge of Marines. The leader must understand Marines, know how to lead and motivate them, understand how they learn, and apply this knowledge wisely in the day-to-day training situation.

(2) Understanding of Body Functioning. A more intelligent exercise program results from understanding and applying the principles which govern physical conditioning of the body. The leader with such knowledge can better prescribe, adjust, and regulate exercise types, amounts, and progression to attain fitness.

(3) Understanding Exercise Activities. The leader needs to understand the contribution each type of physical activity makes to physical fitness, and how to use each activity to develop fitness. Skill to demonstrate and lead the various activities is a necessary part of technique and is invaluable to the instructor or small-unit leader.

Chapter 2
PHYSICAL READINESS TRAINING PROGRAMS
Section I. Development of a Program

2101.GUIDANCE FOR PLANNERS

This chapter instructs planners on physical readiness training procedures. It contains program planning guidance including factors to consider when developing programs; steps in assembling a program; definitions of activity packages and systems of exercise; selection of activity packages; and selection of systems used in implementing those packages.

2102. MARINE CORPS MISSION

The mission of the Marine Corps is to seize and defend advanced naval bases and to perform other missions as the President of the United States may direct. Traditionally, to perform this mission, Marines have been projected into the area of operations by Navy vessels and have attacked hostile beaches over the shore in landing craft, assault amphibian vehicles, and helicopters. Recently, performance of this task has been somewhat complicated by the introduction of the concepts of maritime prepositioning and the air-landed Marine expeditionary-brigade. Under these concepts, Marines, landed at airfields in foreign countries, will join their equipment at commercial ports. The distance between these airfields and ports is often 10 or more miles. In time of international crisis, it can be expected that transportation assets will not be adequate. Marines from all elements of a Marine Air-Ground Task Force may have to march from the airfield to the site of their equipment. All Marines must be physically conditioned to rigors of conducting foot marches with individual weapons and equipment.

2103. EVALUATION OF FITNESS

The Marine Corps mission is such that all Marines must constantly and consistently achieve a high level of fitness that prepares them for the demands of combat. While Marines have many different MOS's, all Marines must be prepared for the demands of marching under load and performing basic infantry tasks, such as rear security and patrolling. While the physical fitness test remains our universal measure of individual fitness, it is the commander's responsibility to observe and evaluate the unit's ability to perform effectively in combat. Often the commander's most effective evaluation tool is weekly physical training sessions. The commander should participate in these sessions and see that they are fully integrated into the unit physical conditioning program.

2104. INTEGRATION OF TRAINING

Training time must be used efficiently and wisely. Every opportunity to integrate physical training into other training activities should be seized. For example, foot marches under load can be integrated into the normal daily activities as a way to move from one training or work site to another. This integration reflects realism in training and should be used to maintain overall proficiency. To accomplish this requires imagination and enthusiasm on the part of the commander. The area of operations in which the training is conducted determines what can be integrated and how. If possible, training should also consider the terrain and climate of the area in which the unit will subsequently conduct training or operations. Training should include familiarization with special equipment and the application of specialized techniques to tactical principles. Activities such as gun drills and command post emplacement drills also have physical conditioning value, particularly if supervised and done competitively or measured against time standards.

2105. ACTIVITIES AT THE UNIT LEVEL

Physical readiness training is a command responsibility and is generally conducted in Marine units at the company and platoon levels. Battalion/squadron commanders must constantly supervise, making sure that all Marines are physically prepared for combat.

a. Physical readiness training should usually be conducted weekly by each battalion/squadron-level unit, using events designed for all personnel (e.g., marches under load, battalion/squadron runs, military field days, etc.).These events enable the commanders to demonstrate personal leadership, and observe and evaluate the physical condition and combat physical readiness of their Marines. This exercise period should be designed to elevate the unit's morale and emphasize unit identity through the wearing of uniforms and the carrying of battalion and company guidons. The battalion/squadron commander should make every effort to be a visible participant in the unit's physical readiness program on a weekly basis. These activities could be conditioning marches under load or a weekly calisthenics/unit run.The suggested times for these activities are early on Monday morning or late on Friday afternoon after the training week has been completed. The uniform may be physical training gear orutilities and boots.

b. Commanders should be cautious of long runs in boots and utilities. This is not to suggest that units should not run in utilities, but to remind commanders that boots are designed for marching, not running. If commanders choose boots and utilities, they should select off-road routes which will provide a cushioning effect.

c. Units should be made well aware of the commander's intent before the unit falls out for exercise. If the exercise period is conducted early in the morning, it need not necessitate an extensive shower/clean-up period afterward, so long as all Marines from the commander on down remain dirty. Commanders should inform their Marines not to fall out for these sessions in clean, pressed uniforms.

d. The physical readiness training of headquarters units is often difficult to manage. Often physical training is left for individuals or sections to develop and manage on their own. This approach can be effective, but if not supervised, it can result in a "paper" program. Headquarters unit commanders should endeavor to conduct weekly integrated training which will enable them to evaluate the physical readiness of their personnel. Often this training has to be scheduled before or after normal working hours to maximize participation of the entire unit.

2106. STEPS IN PLANNING

To implement workable and effective programs (as directed by CMC ALMAR 261/87) the planner must-

a. Determine the Type of Program Needed. Marine units are inherently different in organization and mission. The physical readiness program must be tailored to the mission and to the current physical condition of most unit personnel. Programs to meet this need are of the following types:

(1) **Developmental Programs.** Marines in a beginning or poor state of physical readiness need a program which will develop strength, endurance, physical skills, and character traits which are beneficial to successful accomplishment of military missions. Such programs should be applied progressively to rise gradually to a peak of fitness and skill.

(2) Maintenance Programs. Once Marines reach the sustaining stage of conditioning, their goal is then to maintain this level by participation in a maintenance program.

(3) Remedial Programs. The term "remedial" is usually applied to those individuals or groups who possess substandard physical fitness. For example, a remedial physical conditioning program could be applied to persons who are overweight, who fail to reach physical fitness test standards, or who have missed extended periods of conditioning due to illness, injury, extended hospitalization, or other absence. Saturday training sessions are designed as remedial sessions for all individuals who either miss the daily training or have failed to meet adequate standards. Note that personnel in light duty or no-duty status should be expectedto attend all training sessions, observing, supervising, or assisting as needed.

b. Determine the Time Required. The amount of time for training operations varies considerably. However, every unit can find time to conduct physical readiness training. Frequent (i.e.,daily) physical training of short duration-30 to 60 minutes-is preferred over occasional longer periods. Other demands for training time are so urgent that every minute of time allotted for physical readinesstraining should be used. Determine the time required per week anddivide it into daily blocks.

c. Organize for Various Group Size. It is essential to stress exercise rather than formality. Marines must complete the program where they are- -on the training field, in the motor pool, on the range, next to the classroom, in the office area, in the shop, aboard ship, or elsewhere. Although desirable, it is not always possible to assemble company-size units for physical training. Platoon-size groups are appropriate for the proper conduct of physical conditioning activities. Certain situations may require exercise programs for section- or squad-size units. This manual outlines programs for all situations and types of organizations.

d. Allow for Weather and Exercise Area. In programing and scheduling, the climate and terrain often govern the selection of activities.

(1) Weather changes cause differences in temperature, rainfall, wind chill, and snow. These changes should be anticipated as they dictate the type of program. Alternate plans should be part of the schedule. Nevertheless, normal weather changes and rain should not drive training inside.

(2) Local terrain and available exercise areas may also influence the selection of activities and the type of program which it is possible to support. Some activities can be completed in nearly any area.

e. Plan for Seasonal Change. As most physical readiness training is conducted outdoors, it is necessary to recognize seasonal change. A program should be divided into fall, winter, spring, and summer parts. Seasonal change also causes change in light. For example, an early morning program started in the summer will have ideal lightconditions, yet in the fall or winter, darkness will occur at that same hour and interfere with the conduct of the program. Develop programs in seasonal blocks, and make provisions for anticipated changes.

f. Consider Needed Facilities. An excellent program can be conducted with practically no facilities since there are exercises which require no equipment. However, a better program can be developed when supported by certain facilities and items of equipment. Proper command support, plus ingenuity, will solve this problem. (Items of equipment, when necessary to support the recommended exercises, are included in the chapters on exercise activities.)

g. Specify Appropriate Uniform. The uniform worn for exercising depends upon the season of the year, the state of the weather, and local regulations. Whenever possible, Marines should be dressed alike. Undershirts are preferred as the upper garment when the weather permits. A uniform that restricts the free movement of the body should not be worn when exercising.

h. Consider Availability of Instructors. Leaders who can lead and direct the scheduled activity must be available. Organizational units should train junior officers and noncommissioned officers down to squad or section leaders to instruct and lead the various activities.

i. Select Activity and System. With the type of program needed and the objectives in mind, select an activity package or an integrated training package for each day's scheduled physical training. At this same time, the system to be used in employment of the selected activity or activities must be determined. This selection affects equipment, areas, instructors, transportation, and other support requirements.

j. Secure Command Participation and Support. Prepare and brief the commander to assure full understanding of the objectives and administration of the program. The full participation and support of the commander will greatly improve the level of success.

k. Supervise Execution. Determine needs, publish the program, and supervise its execution as a necessary part of developing the unit's training schedule.

2107. SELECTION OF ACTIVITIES AND SYSTEMS

a. Exercise Activities. The planner must determine exercise activities which will be appropriate to include in the program. An exercise activity is a single means of exercise usually identified by the name applied; for example, running, log exercises, and obstacle course. Many exercise activi ties can become part of activity packages.

b. Activity Packages. Many of the physical activities described in this manual are arranged in prescribed sequences and are known as activity packages. An activity package is a number of exercises of the same type, assembled as a group or a set, and arranged in a specific sequence. Exercise packages are organized in such manner that not more than 15 minutes will be required to complete the execution of any package. Each type of activity is explained in laterchapters. The number of available drills, tables, or circuits; the manner of organization; and the contribution each makes to the total program are covered. Full understanding of this information will greatly assist in developing effective programs. Various designations are used to identify exercise packages; for example, conditioning exercises when arranged in a set order are known as drills, and other packaged activities are designated as tables or circuits. The following are activity packages:

(1) Activity packages for groups:

o Conditioning Drill Two
o Conditioning Drill Three
o Rifle Drill
o Log Drill
o Grass Drill
o Running Tables
o Guerrilla Tables
o Circuit-Interval Table
o Combatives Tables
o Relay Tables

(2) Strength Circuits:

o Fixed Circuit
o Movable Circuit
o Simplified Circuit (Circuit Interval Table)

(3) Activity packages for individuals:

o The 6-12 Plan
o Weight Training
o Isometric Exercise

c. Advantages of Using Packages. These of exercise packages simplifies scheduling and conducting of exercise and results in the following benefits:

(1) Schedule development is simplified as the planner assembles packages which will satisfy the training objective. There is no need to deal with selection of individual activities or to be concerned about the amount of time to be expended on each.

(2) Any 15-minute period, and in some cases less time, can be scheduled or used to perform an activity.

(3) If longer periods of time are available or if the objective demands, several packages can be assembled to provide a more complete period of activity.

(4) Marines are assured a balanced set of exercises or activities as each package is carefully arranged to reach all muscle groups.

(5) The instructor can concentrate on the conduct of a vigorous workout as the type and duration of the activity have already been determined.

d. Nonpackaged Activities. Several types of activity are not packaged. Activities in this category are conditioning marches, unit runs, obstacle courses, combat water survival swimming, teamcontests, and team sports. These activities can be scheduled in combination with packaged activities, or they may be scheduled separately. Often they require a longer period of time. Most nonpackaged activities require a 50-minute period to satisfactorily complete their objective. For conditioning marches, a few hours are required. The benefits of these activities should not be overlooked as some desirable objectives cannot be attained without them.

e. Systems of Exercise. Several methods or systems of organizing exercise and activity packages can be used. Each system is based upon a specific organization as follows:

(1) Single Activity System. The unit leader immediately assumes command of the unit at the beginning of the exercise period. The leader moves the unit to a predesignated exercise site at double time, forms the unit in a circle around the leader, groundsclothing and equipment as appropriate, and quickly moves into the exercise routine. There is usually no time to teach; therefore,the Marines must know the activity to be used. At the conclusion of a 5- to 15-minute period, the unit leader returns the unit to the instructional area at double time and releases the unit for the next scheduled activity.

(2) Progressive Activity System. All Marines (company or platoon) complete activities in the same order during the period. For example, Drill One is followed by dual combatives, and finally a 1-mile run. This system is usually progressive from a warm up activity, such as Drill One, to an activity which contributes in a major way to one of the objectives, such as aggressiveness development through combatives or cardiovascular development, such as running.

(3) Rotating Activity System. Set up the same number of activities or stations as there are platoons in the company. Each platoon rotates through each station in turn. With four platoons in a 50-minute period, about 10 minutes can be devoted to each station. With three platoons, approximately 15 minutes may be spent at each station. Activities must be of a type that can be covered in the time allotted. For example, with three platoons, Station 1 could be Conditioning Drill One; Station 2, running; and Station 3, a team contest.

(4) Circuit System. Set up a number of stations to provide various types of exercise equipment and/or items of apparatus. The idea is to keep all Marines busy and exercising vigorously for a short period of time at each station. The fixed strength circuits and the movable strength circuits are examples of this system. Station changes must be rapid, and the exercise must be started quickly after each change. Since the objective is to exercise at top speed, the motivation comes from frequent change of activity by moving to another station. Rotation by station groups continues until all Marines have covered all stations.

(5) Interval System. This system stresses the development of strength and endurance. It involves heavy work for a given distance within a specified time, alternated with lighter work and recovery, but never stopping during the workout. This procedure is repeated, and the intensity is increased gradually as physical condition improves, but always with adequate recovery. The important concept is stress, recover, stress, recover, and so on. This system is often applied through running, but other activities of a continuous nature may also be used. An example is the Circuit-Interval Table.

f. Steps in Selecting Activities and Systems. The planner should follow these steps.

(1) Evaluate the needs of one unit with emphasis upon unit mission, objective, and time available.

(2) For each day, select an activity package or a combination of activities which will contribute to the objective.

(3) Then determine the system to be used in implementing the selected activities. There are several possible choices and much opportunity for flexibility in program development. For example, the weekly program may contain various exercise packages, systems, and time periods. Additional guidance is contained in following chapters.

2108. SAMPLE WEEKLY PROGRAMS

The variety of weekly programs given shows available alternatives. Programs and schedules other than those illustrated may be assembled to provide schedules to fit any situation. Many other scheduling combinations are possible through use of the packaged activities, varied time periods, and different types of assigned duty.

a. Sample A. Figure 2-1 illustrates a 60-minute time allotment in which the single activity system is used with activity packages that vary from day to day. In this program, 5 hours of training are scheduled for the week. This program needs some type of integrated unit training to supplement it each week. A bimonthly conditioning march of 10 miles in 3 hours is suggested to round out this program. The Saturday session is designed for remedial physical training.

NOTE: A physical training activity must be conducted at a minimum of once every 72 hours. After 72 hours, muscles begin to atrophy.

Figure 2-1 is available in the illustration section at the end of the book

b. Sample B. A more comprehensive weekly program (fig. 2-2) illustrates the scheduling of various-length periods, a variety of activities, and the use of three systems during the week. In this schedule, 3.5 hours are included for the week. The following should be noted concerning this schedule:

Figure 2-2 is available in the illustration section at the end of the book

(1) On Monday, a 60-minute period is available for physical readiness training. Three 20-minute packages are scheduled. Each platoon, under its own leadership, will progress through each scheduled activity in turn.

(2) On Tuesday, Wednesday, and Thursday, only 30-minute periods are available. Here each platoon leader supervises each platoon in a single activity system.

(3) On Friday, the company commander directs training to exhibit personal leadership example and supervision. In this example, only one strength circuit and one cross-country course are available, and each will accommodate only one platoon. For this reason, Conditioning Drill One station is included. A platoon rotates to each station within the 60-minute period.

(4) The Saturday session once again is designed for remedial physical training.
(5) Bimonthly, a 10-mile/3-hour march under load will be necessary to round out this program.
NOTE: Cardiovascular/respiratory exercise to improve endurance must be of long enough duration (20 minutes or more) to induce and maintain oxygen debt.

c. Sample C. A third sample schedule (fig. 2-3) illustrates the use of the single activity system for 5.25 hours of training. This schedule includes sustaining-type activities for a unit that has passed through both the toughening and slow improvement stages of conditioning. The use of competitive activities is featured to holdinterest and provide self-motivation. A bimonthly 10-mile/3-hour march under load will be necessary to round out this program. The Saturday session once again is designed forremedial physical training.

Figure 2-3 is available in the illustration section at the end of the book

NOTE: Thc body will respond to hard work or physical training without soreness and stiffness provided the work or physical training is conducted regularly.
NOTE: Excrciscs of short duration increase strength and bulk but do little to effect endurance.
NOTE: Thc times shown in the charts arc participation times.

Section II. Remedial Physical Training

2201. NEED FOR ATTENTION

Remedial physical training is a process by which physically substandard individuals are conditioned to meet standard levels of performance. To achieve a full degree of operational readiness, it is necessary to bring all Marines up to the prescribed standard. Experience has demonstrated that some Marines have difficulty due to a poor state of fitness, obesity, or lack of motivation. Attention to these individual deficiencies will help improve unit combat readiness.

2202. IDENTIFICATION OF INDIVIDUALS

The company or battery commander identifies persons who cannot achieve the prescribed level of proficiency through the physical fitness test or as a result of observation during training. The commander notes particular weakness of body parts as indicated by failure of certain test events. These Marines are then placed in a special remedial program either at platoon, company/battery, or battalion/squadron level. Extra time is devoted to overcoming the weakness. These extra periods of conditioning may be during or after duty hours, as determined by the commanders. Other persons need reconditioning as result of hospitalization. Restoring physical fitness to damaged areas of the body is accomplished through progressively graded physical activities under professional supervision, not by the physical fitness trainers.

a. Types of Deficiencies. Physical deficiencies which can be corrected by exercise fall into several categories.
(1) Lack of Strength in One or More Body Parts. The major muscle areas concerned are the arms and shoulder girdle, back, abdomen, and legs.
(2) Lack of Overall Endurance. Usually, there is deficient muscular and cardiovascular endurance.
(3) Deficiency in Coordination and Agility. In these cases, physical skill is not developed to a satisfactory degree in activities such as crawling, running, jumping, climbing, traversing, vaulting, pushing, pulling, lifting, and carrying.
(4) Overweight or Underweight. Either condition may interfere with physical fitness and mission accomplishment. Lack of exercise is not always the cause. The cause may be malfunction of normal physiological functions or it may be poor health habits such as over- or under-eating, lack of adequate rest, or overconsumption of alcohol.
(5) Lack of Motivation. Not all persons are motivated to attain or maintain **a** desirable state of fitness. Some Marines do not understand the importance of physical fitness, some find proper exercise too difficult, and others find it inconvenient.

b. Causes of Deficiencies. There are several causes for Marines being deficient in physical fitness:
(1) Absence of exercise.
(2) Exercise which fails to develop all muscle groups and components of fitness.

(3) Exercise which is not vigorous enough or which lacks progression.
(4) Injury or illness which depletes fitness.
(5) Inadequate amounts of sleep or rest.

c. Methods of Detection. The commander has several means by which to detect physical deficiencies:
(1)
from those who pass. Make an analysis to determine the cause of failure as revealed by test scores.
(2) Observe Marines as they perform physical tasks-
o Marines who have difficulty during training or in physical types of work,
o Marines who have become obese and therefore experience difficulty.
(3) Be alert to those Marines who are often on sick call, returnees from hospitalization, or newly assigned.

2203. ADMINISTRATION OF REMEDIAL ACTION

a. Group. Attitude Marines who are singled out as being physically deficient are self-conscious and are not always convinced they need extra help. Within a deficient group, motivation may be low. These Marines must be convinced that a remedial program which is tailored to their needs will help them and will eliminate their deficiencies. Encouragement is often needed and desirable.

b. Leadership. The leader of this group must study individuals and know their deficiencies. The unit leader should counsel Marines individually, observe them closely as they progress through the remedial program, maintain records, and adjust the program as required.

c. Measurement.. Whenever improvement in performance is noted, Marines should be measured by the physical fitness test either individually or by group. When testing reveals an individual to be satisfactory, that individual should be released from the remedial group. There may be exceptions to this policy in the case of Marines who are overweight or in the execution of an exercise program prescribed for some special purpose.

d. Organization of a Remedial Group
(1) A remedial group is usually a small group of Marines within a company or battalion. In some situations, it may be a larger group numbering 50 to 75 Marines. In the case of a larger group, divide the Marines initially into subgroups according to ability. Prescribe exercise loads commensurate with their ability. General conditioning is sufficient in the early part of the program to qualify Marines who are on the borderline. Qualification will reduce the size of the group, permitting more individual and specialized attention to those who remain.
(2) As the program progresses, regroup individuals who have like deficiencies in order that they can concentrate on their weakness. For example—

(a) One group may be weak in the arms and shoulders as revealed by failure in the chin up event. This group, with an assistant instruction charge, can work on pull-ups, rope climb, pushups, rifle or logexercises, horizontal ladder, and similar types of exercise.

(b) Another group may be weak in general endurance as revealed by the3-mile run. This group could profit from participation in conditioning drills, running, grass drill, and strength circuit.

(3) It may be necessary to form some groups to overcome weaknesses in skills, such as an inability to throw, to quickly change direction while running, to crawl rapidly, or to carry a load. Lack of coordination or lack of practice may cause these deficiencies. In this instance, instructors must provide an opportunity to practice and correct poor form and other errors as they are noted.

Chapter 3
PHYSICAL CONDITIONING ACTIVITIES
Section I. Marching Under Load

3101. GENERAL

Few physical fitness activities are as directly related to readiness for combat as foot marches under load. In addition to the obvious physical conditioning and unit cohesiveness benefits, marches under load prepare most Marines for numerous foreseeable tasks in combat. The idea that only infantry and reconnaissance units actually require training marches to prepare them for combat is clearly a mistake. In addition to the very real possibility of Marines from all types of units being used to fill combat-depleted infantry units, it is likely in maritime prepositioning force operations that aviation, combat service support, and command element personnel will have to march with weapons and equipment, from the arrival airfield to the equipment marshalling site. In American coastal cities, the harbor is nearly always 10 or more miles from the airport. It is difficult to imagine any different condition in a lesser developed country .A further combat-related benefit is that foot marching under load is the most effective way to develop leg strength in the context of a unit training program. In his classic study of troop performance in World War II, Men Against Fire, S.L.A. Marshall noted that the greatest single weakness of replacements of all occupational fields was leg strength. Truck drivers must push trucks which are stuck in mud, and headquarters communicators must climb hills to emplace antennas. The demands on the lower body in combat apply to all Marines, and foot marching is one activity which can help build these muscles while being conveniently integrated with the rest of the training program.

3102.TRAINING GOALS

The standard for success of a foot march is very simple to measure: did the unit arrive at the destination at the prescribed time with Marines in condition and required equipment present to accomplish the mission?A progressive program can increase a unit's readiness. It can instill pride by increasing distances and rates of march and by selecting routes over increasingly challenging terrain as the Marines become better conditioned. However, the ability to execute the mission at the conclusion of the march must remain the standard of success.

3103.MOVEMENT PLANNING

The basic considerations in planning a foot march are the mission, tactical situation, terrain and weather, and the units to participate. The success of the march will depend largely upon the thoroughness with which it is planned. A successful march is characterized by adherence to prescribed routes and time schedules, the efficient employment of the means available, and the ability of the unit to accomplish its assigned mission upon arrival at the destination.

a. Movement Orders. Movement planning culminates in the preparation and issuance of an operation order prepared in the standard, five-paragraph format. Necessary annexes are attached to furnish detailed information required for the movement. Written movement orders are rarely prepared at company level. A discussion of the annexes is contained in this chapter.

b.March Planning. March planning, as discussed here, is the planning conducted at battalion level. March planning may be organized into the following steps:

(1) **Preparation and Issuance of the Warning Order.** In order to afford subordinate units the maximum possible time to prepare for a pending move, a warning order containing all available information about the march is issued. The amount of planning time available will determine the time of issuance and the content of the warning order.

(2) Estimate of the Situation. In his estimate, the commander considers the mission, terrain, weather, time, and space factors, available routes, available transportation for the movement of equipment and/or shuttling of Marines, enemy capabilities, disposition of own forces, physical condition and training of Marines, and courses of action available to the command.

(3) Organization and Dispatch of a Reconnaissance. Party very march plan is based on as thorough and complete a ground reconnaissance as time and the situation will permit. Map and aerial reconnaissance are valuable in formulating a plan, but are not a substitute for ground reconnaissance. Route reconnaissance is accomplished by a reconnaissance party which usually consists of a reconnaissance element, an engineer element from the attached or supporting engineer unit, and a traffic control element. Unit standing operating procedures generally establish the basic composition of the reconnaissance party. It is modified as necessary to meet the requirements of a particular march. A recommended method of reporting information obtained by the route reconnaissance party is shown in figure 3-1.The minimum information required from the reconnaissance party is—

(a) Available routes and conditions.(Routes may be specified by higher headquarters.)

(b) Recommended rate of march.

(c) Selection of start point and release point, or confirmation of the suitability of start point and release point previously selected by map reconnaissance.

(d) Confirmation of location of the assembly or bivouac area.

(e) Location of critical points on the route.

(f) Distance between critical points on the route and total distance from start point to release point.

Figure 3-1 is available in the illustration section at the end of the book

(g) Location of obstacles and estimation of necessary Marines and equip- ment needed to repair and maintain routes.

(h) Number of guides required and their location on the route.

(4) Development of Detailed Movement Plan. Some of the elements included in a detailed movement plan are-

(a) Organizing the Column. To facilitate control and scheduling, units will be organized into serials and march units and given an order of march. In determining the order of march, the march planner must consider the enemy situation and the desirable order of arrival of the units at the destination. Where dispersion is required, a unit may be organized into two or more columns, each assigned a different route.

(b) Using Reconnaissance Information. Results of the route reconnaissance will be used to select the route(s); determine the start point, critical points along the route(s), and the release point; and select the rate of march.

(c) Determining March Computations. March unit pass time is based on the strength, formation, and rate of march. The pass time of the marching columns, plus necessary time distance computations, will be used to determine the completion time of the march.

(d) Drafting of Road Movement Table. Using the completed march computations, a draft road movement tabic is compiled.

(e) Checking the Plan. Using the draft road movement table and a road movement graph, the movement plan is checked to ensure that it conforms to the directive of the higher headquarters and the battalion commander's instructions.

(5) Preparation and Issuance of the Road Movement Order. After the plan has been checked and approved by the commander, an operation order is prepared and issued. The operation order may be in written form or issued orally, and is accompanied by a road movement table, overlay and/or strip map, and appropriate administrative details.

(6) Road Movement Table. A road movement table, prepared as an annex to the operation order, provides serial commanders with arrival and clearance times at critical points along the route of march. It also provides the column commander with information as to the proposed location of elements of the column at various times.

(7) Overlay/Strip Map. An overlay serves the normal purpose and should show, as a minimum, the present location of units, route of march, critical points, and the new location of units at the destination. A strip map is a schematic diagram of the route of march and shows landmarks and critical points with the distances between them. A strip map may be issued as an annex to the road movement order, in addition to or in lieu of an overlay.

c. March Computations. Prior to issuance of the operation order, the S-3 must verify time and space computations as they provide him with the necessary data for the preparation of a road movement table.

(1) Time-Distance. Time distance (TD) is determined by dividing the distance to be traveled (D) by the rate of inarch (R):

TD = D/RWhere TD = hours D = distance in kilometers R = kilometers per hour

(2) Length of ColumnThe length of column (LC) is used to determine the pass time of a column. The sum total of the following two parts determines the length of column. (See pars, (a) and (b)).The two parts are the space occupied by the Marines alone (including the distance between Marines) and the sum of the distances between the units of the column (column gap).

(a) The length of column of Marines alone is determined by multiplying the number of Marines by the appropriate factor selected from the table below. The length of column does not include distances between units. LC Marines = No. of Marines x factor.

(b) The total distance in meters between units is determined as follows:

Step One. Determine the number of gaps between serials (total serials minus one).

Step Two. Multiply the number of serial gaps from Step One by the length (in meters) between respective units.

Step Three. Determine the number of gaps between march units (total march units minus one, minus the number of serial distances).

Step Four. Multiply the number of march unit gaps from Step Three by the length (in meters) between respective units.

Step Five. Add the totals from Step Two and Step Four to get the total meters for the column gap.

EXAMPLE:A battalion foot column is organized into 12 platoon-sized march units and 3 company-sized serials. REQUIRED: total column gap distances when there are 100 meters between serials and 50 meters between march units.

(3) Pass-Time.Pass-time (PT) is the time a unit takes to pass a specified point. For foot columns, the pass-time is determined by applying the following formula:PT (minutes) = LC x FACTOR (for appropriate rate of march).

PASS-TIME FACTORS - FOOT TROOPS

.0150 for 4.0 km/h
.0187 for 3.2 km/h
.0250 for 2.4 km/h
.0375 for 1.6 km/h

EXAMPLE: Determine the pass-time of a unit whose length of column is 1,500meters and is marching at a rate of 4 km/h.PT (min) = 1,500 x .0150 (the factor for 4.0 km/h) = 22.5 mm.

(4) Completion Time. Completion time is the time of day that a march will be completed. Completion time is determined by using the following formula: Completion Time = SP time + TD + PT + Scheduled Halts (other than normal hourly halts).

EXAMPLE: A column's starting point (SP) time is 0700.The time-distance is6 hours and 40 minutes. Pass-time of the column is 30 minutes. A 35-minutelunch halt has been scheduled. What is the completion time of the march?Employing the 24-hour clock system, the formula can be applied as follows for simplified addition of the times:

HrMin SP time................0700
TD......................0640
PT......................0030
Lunch halt..............0035

Completion time.........13 105or 1445

The march will be completed atl445.

(5) Experience Tables. Based on previous movements made by a unit, data is accumulated to facilitate march planning. Such data includes approximate pass-times lor various elements of the battalion. The S-3 can utilize these data rather than computing them each time a march is scheduled. Such experience tables serve to reduce the time required to complete the computation phase of march planning. Matter appropriate to the unit standing operating procedure should be integrated therein.

d. Road Movement Graph

(1) A road movement graph is a time-distance diagram used in planning, preparing, or checking road movement tables, and for controlling marches. The graph provides a visual representation of a march plan so that conflicts and discrepancies may be prevented in the planning stage, before congestion occurs on the route. It is not usually issued as a part of the order. Road movement graphs may be applied to small units, to movements of a single column, or to a large organization scheduling separate elements, moving by various means, with different rates of march, over one or more routes.

(2) To construct a road movement graph, use the following steps: (See fig. 3-2.)

Figure 3-2 is available in the illustration section at the end of the book

Step One. Determine the number of hours available for the march. Designate the lower left corner of a sheet of graph paper as the SP time or the earliest even hour before the march is to begin. Select a scale and plot the hours available in sequence from the left on the horizontal axis.

Step Two. Determine the distance to be moved in kilometers. Indicate the SP at the lower left corner of the graph sheet and, using an appropriate scale, plot the number of kilometers on the vertical scale from the SP to the release point. Indicate the location of critical points on the vertical scale.

Step Three. At the proper distance from the SP, draw a horizontal line indicating the location of the RP. Indicate by a vertical line the hour when the movement must be completed. Plot lines representing route restriction, if any, at the proper distances and times on the graph.

Step Four. Determine the pass-times of foot and motor elements of the column.

Step Five. Starting at the SP at the specified hour, plot the movement of the head of the leading element. If the rate of march is 3 kilometers per hour, the lead element will have moved 1.5 kilometers in 30 minutes, 3 kilometers in 1 hour, etc. Plot the trace of the lead man to the RP. The last Marine will cross the SP in pass-times after the lead man. Measure this time on the graph and plot the trace of the last Marine of the column. The lines describing the head and tail of a march serial are parallel. Indicate the time subsequent serials reach the SP and plot the trace of the head and tail. Check to see that the plan complies with all restrictions and orders. If there are any violations or conflicts, the plan may be made to conform by changing the order of march, the starting time of the serials, the planned rate of march, or the organization of the column.

e. Conversion to the Metric System. The adoption of the metric system for expressing linear distances requires a simple means of conversion.

(1) To accomplish this, use the following tables:

o Multiply inches x 2.54 to obtain centimeters. Multiply centimeters x .39 to obtain inches,

o Multiply yards x .91 to obtain meters. Multiply meters x 1.1 to obtain yards,

o Multiply miles x 1.6 (or 8/5) to obtain kilometers. Multiply kilometers x .62 (or 5/8) to obtain miles.

(2) Linear distances can be accurately measured on maps which have been converted to the metric system by using a straightedge graduated in centimeters (cm).The centimeter scale can be used in lieu of the graphic scale when the scale of the map is shown, as follows:

Map Scale:	One cm Equals:
1:500,000	5,000 meters
1:250,000	2,500 meters
1:100,000	1,000 meters
1:50,000	500 meters
1:25,000	250 meters

3104. INDIVIDUAL LOAD

Since the backbone of Marine ground combat forces is the infantry, tactical mobility is largely dependent on the marching ability of the individual Marine. The load that each Marine must carry is the greatest impediment to mobility. In general, the commander must make every effort to reduce the individual load of his Marines to the absolute minimum. While the load has the greatest effect on Marines marching on foot, the value of carrying the minimum essential supplies and equipment applies for all Marine forces.

a. Fighting Load. The fighting load consists of items of clothing, equipment, weapons, and ammunition that are carried by, and essential to, the effectiveness of the combat Marine and the accomplishment of the immediate mission. The fighting load should not weigh more than 40 pounds. The commander must ensure that his Marines' loads- -and his own- -be stripped to the bare minimum. The addition of small, light-weight items in excess of the limit should not be tolerated, since cumulatively, these items will take a toll in energy. Every extra pound a Marine must carry decreases his combat effectiveness.

b. Existence Load. The existence load consists of items other than those in the fighting load that are required to sustain or protect the combat Marine, that may be necessary for increased personal and environmental protection, and that are not normally carried by the individual. The existence load is normally carried by the Marine's unit. Marching, but not engaged in combat, a Marine may be expected to carry a maximum of 50 pounds of supplies and equipment.

c. Considerations

(1) The primary consideration is not how much a Marine can carry, but how much he can carry without impaired combat effectiveness-moral or physical. The combat strength of a unit is not counted simply in numbers of Marines, but in the numbers of willing and physically able Marines.

(2) It is generally better to risk temporary inconvenience from lack of health and comfort items than to exhaust Marines due to overloading.

(3) It is a fundamental truth that men become physically exhausted more quickly when under the stress of combat. Marines must be required to carry less into battle than they are conditioned to carry in training. Marines should be conditioned for carrying weight, but should be equipped in combat for fleetness of foot.

(4) A common mistake is to base the fighting load on the gear and supplies necessary to meet every contingency. The commander cannot reasonably expect the Marines to carry enough gear for every possible eventuality. The items to be carried must be based on reasonable expectations.

(5) It is the commander's responsibility to procure the transport to carry any additional gear. As a rule of thumb, a rifle company, or unit of similar size, requires one 5-ton truck and trailer in support to carry additional gear. In cold weather, or other conditions in which the necessary personal gear increases, this requirement will increase.

(6) The commander must ensure that the supply system provides, on a dependable and timely basis, the balance of essential supplies and equipment not carried by the unit. Marines must feel confident that they will be supported with the necessary supplies and equipment.

(7) In training, commanders must implant in their Marines the pride in operating under austere conditions. To effectively demonstrate the need for this spartan mentality, the commander must personally set the standard. Marines should be trained in field craft, foraging techniques, and the use of caches and field expedients. Maximum use should be made of captured stores.

3105. MARCH TECHNIQUES AND PROCEDURES

a. Organization. A command executing a march is organized into march units, march serials, and march columns, as necessary for control.

(1) March Units. A march unit is a unit of command which moves and halts at the command of a single commander. The march unit normally corresponds to one of the smaller Marine units such as a platoon or company.

(2) March Serials. A march serial consists of one or more march units organized under the senior officer and given a numerical or alphabetical designation to facilitate control. A serial is usually a battalion, but may be a company if the battalion is marching alone.

(3) March Columns. A march column is composed of elements of a command moving over the same route. It may be composed of one or more serials. To facilitate control, a column commander is designated. A column is normally a regiment or larger unit, but may be a battalion if marching alone.

b. Order of March. The order of march is determined by the mission, terrain, and the commander's desired order of commitment of units into action.

c. Control and Coordination Measures. The commander establishes initial control of the march by designating control measures in his march order. The most frequently used measures include:

o Start point and release point.

o Time at which head or tail of column passes the start point.

o Rate of march.

o Order of march.

o Route of march.

o Location of command post.

o March communications.

d. The Pacesetter. The pace-setter is an experienced individual carrying the same load as the majority of the Marines and marching from 4 to 10 meters at the head of the column. (See fig. 3-3.)The pace-setter's primary duty is to maintain the rate of march ordered by the column commander.

(l) Setting the Pace. The pacesetter does this by establishing the pace (length of step) and cadence (steps per minute) to obtain the prescribed rate of march. He should be of medium height so normal strides will be taken. Overstriding or understriding tends to tire leg muscles quickly and affects the combat efficiency of marching Marines. The officer marching at the head of the column supervises the pacesetter to ensure that he takes normal strides and maintains a uniform cadence.

Figure 3-3 is available in the illustration section at the end of the book

(2) Reducing Accordion Effect. Accordion effect occurs when the column alternately lengthens and shortens, causing the rear elements of the column to double time to maintain distance. It is caused by changes in the rate of march by the lead elements, usually after passing a slowing obstacle. To reduce this effect as much as possible, lead elements should slow their rate of march for a sufficient time after crossing obstacles to allow the rear elements to maintain the prescribed distance without running. The rear elements can signal the head when the last man has cleared the obstacle and the rate can be resumed. Since some accordion effect is nearly inevitable, the order of march should be periodically rotated to prevent units in the rear from becoming physically exhausted at the conclusion of the march.

e. Length, Rate and Pace

(1) **Length of March.** The length of march varies depending on terrain and weather, enemy situation, and physical and mental condition of Marines. The normal length of march for a 24-hour period is from 20 to 32 kilometers (12 to 20 miles) marching from 5 to 8 hours at a rate of 4 kilometers (2.5 miles) per hour (km/hmi/h).A march in excess of 32 kilometers in a 24-hour period is considered a forced march. Well-trained units that have been progressively trained will be able to achieve a unit goal of 60 miles in a 3-day period.

(2) Rate of March. The same factors of terrain, weather, load to be carried, and condition of Marines affect the rate of march. The commander considers these and selects a rate which will place his unit at the destination in the shortest possible time in condition to accomplish the mission. Unit standing operating procedure should state the standard rate of march for that unit over normal terrain for both road and cross-country movement and in both day and night. The commander then modifies this rate if the situation requires. Normally prescribed standards are as follows:

Figure is available in the illustration section at the end of the book

(3) Pace and Cadence. The normal pace is 30 inches. A pace of 30 inches and a cadence of 106 steps per minute result in a speed of 4.8 kilometers per hour or 3 miles per hour and a rate of 4 kilometers per hour or 2.5 miles per hour if a 10-minute rest halt per hour is taken. Since the pace of each man may vary, the cadence may have to be adjusted to maintain the prescribed rate. Slope of ground and surface footing affect pace. A normal cadence is relatively simple to maintain on level or gently rolling terrain on firm footing. Mud, sand, loose gravel, and hills all greatly affect pace.

f. Halts. Halts during a day's march are taken at regular intervals to rest personnel and adjust loads.Halts are specified by standing operating procedures or march order. Such factors as midday heat or enemy action may force the adoption of long midday halts or night marches. At long halts, each unit disperses to a previously selected location near the route of march.

(1) Normal Time and Frequency. Under normal conditions, a 15-minute halt is made after the first 45 minutes of marching. Following the first halt, a 10-minute halt is made after each 50 minutes of marching. Variations of this schedule are made when a scheduled halt time occurs when passing through a built-up area or when cover and concealment are required by the tactical situation and none is available. Observation posts may be established, if required for security of the unit during halts.

(2) Actions at Halts. All units in the column should be halted at the same time. At the halt signal, Marines should move to the side of the road, staying within the immediate vicinity of their unit. Marines should remove or loosen gear and sit or lie with feet elevated. Commanders inspect Marines and equipment, and corpsmen administer medical treatment, as required.

g. Communications. The four primary means of communications used in controlling foot marches are radio, visual, sound, and messenger.

(1) Radio. Radios are normally used for communications between platoons and higher headquarters in the march column. Radio transmissions should be held to the absolute minimum necessary for control and should be at minimum power required.

(2) Visual.. Visual communications may include flashlights, luminous markers, panel sets, pyrotechnics, and hand and arm signals. When contact with the enemy is not expected, the loss of light discipline from some of these signals may be a lesser security risk than the risk of disclosure from radio transmissions.

(3) Sound. Sound communications include voice and such signaling devices as horns and whistles. Some of their uses include assembling Marines at the conclusions of halts and providing warnings of air or chemical attacks.

(4) Messengers. Messengers are particularly effective during periods of radio silence and reduced visibility. Messages should be simple and brief to preclude mistakes by the messenger.

h. Security. Both active and passive measures are used against attack by enemy aircraft and indirect fire weapons during movement. Active measures include the use of air sentries and organic and attached weapons in accordance with the unit air defense standing operating procedure. Passive measures include the use of concealed routes and assembly areas, night marches, and extended interval between elements of the column. With imagination and planning, all of these measures can be practiced in training without detracting from the physical conditioning benefit of the march. In fact, such drills as rapidly taking cover during simulated attacks and then quickly resuming the march can prove physically challenging.

i. Reconnaissance. Reconnaissance determines in advance critical points along the march route such as bridges, fords, and obstacles in order that congestion or delay may be held to a minimum and local security provided. Paragraph 3103.C describes organization of the reconnaissance party.

3106.TACTICAL MARCHES

These marches are movements of Marines and equipment not in direct ground contact with the enemy but expecting early ground contact either en route or upon arrival at the objective. Some characteristics of this type of movement include security elements to the front, flanks, and, if necessary, to the rear of the march column. Additional considerations include the selection of concealed areas on defendable terrain for start and release points and rest stops. In training, infantry units of regimental size or smaller will normally derive the greatest benefit from tactical marches. The security requirements usually dictate a slower rate of march and, therefore, reduce the physical conditioning value for other types of units which would seldom move in combat with the expectation of early ground contact with the enemy. Two key points should be remembered regarding tactical marches in a training program.

a. Security. Although security requirements often slow the rate of march, it is a valid and important training objective for infantry units to work toward reducing march times while still practicing sound tactical security. March standing operating procedures, good land navigation skills, and simple but complete march orders contribute as much to this goal as physical conditioning. The idea that proceeding slowly enhances security is seldom true. Keeping the enemy off balance by rapid movement frequently is the best guarantee of security.

b. Training. In peacetime training, it is critical that support personnel who normally are located with infantry units in combat, participate completely with infantry units in forced march training. Not only must they not slow the infantry rate of march because of inferior conditioning, but they must be thoroughly familiar with the infantry unit's standing operating procedure for security on the march. Personnel from within the ground combat element such as artillery forward observer teams and combat engineer units, can be easily identified for this training and should be habitually associated with the same infantry units. Personnel from other elements, such as low-altitude air defense teams and radio battalion detachments, are more difficult to identify and earmark for support of the same infantry unit. Their normal mode of employment is in vehicles and their units frequently provide general support vice direct support. This does not, however, lessen their requirement to be prepared to provide support to dismounted infantry over terrain or in a tactical situation not conducive to vehicles. Such training needs to take place, and planning and cooperation among the elements of the MAGTF can overcome the obstacles.

3107.ADMINISTRATIVE MARCHES

These marches are movements of Marines and equipment made when no enemy interference is expected except from aircraft or indirect fire. During these marches, units are administratively grouped for ease of control and speed of movement. The term administrative foot marches should not be taken to imply that tactical considerations are disregarded. There are simply less stringent security measures used than when ground contact with the enemy is a distinct possibility. The normal formation is the route column with one file moving on either side of the road and with negligible security. Air sentries are used and dispersion between individuals and units is practiced. Although roads usually provide the fastest route, there may be reasons to conduct these marches cross-country (particularly as part of a training program). Administrative marches are the type of foot marches most practical for training programs for non-infantry units. These marches may also play a key role in the physical conditioning program of infantry units, because of the greater speeds possible due to lessened security requirements.

3108. FORCED MARCHES

A forced march is a march which covers a greater distance than 32 kilometer (20 miles) in a 24-hour period. Normally the extra required distance should be achieved by increasing the number of hours marched in a day beyond the norm (8 hours) rather than increasing the rate of march beyond the norm (4 kilometers per hour/2.5 miles per hour). However, there will be occasions when the situation or mission demands an increase in the rate of march.

a. Maximum Recommended Distances. The maximum recommended distances for forced marches are-56 kilometers (35 miles) in 24 hours; 96 kilometers (60 miles) in 48 hours; or 128 kilometers (80 miles) in 72 hours.

b. Sample Time Breakdown. A sample time breakdown for a forced march of 52 kilometers, beginning at first light is as follows:

Figure is available in the illustration section at the end of the book

3109. NIGHT MARCHES

Night marches are characterized by closed formations, more difficult control and reconnaissance, and a slower rate of march than day marches.

a. Control. Control is increased by reducing the distance between individuals and units, and by using connecting files to maintain contact between platoons and companies. Connecting files are normally constituted from personnel from the rear most march unit.

b. Safety on Roads. Because of reduced visibility, night marches on roads used by vehicles require attention to safety procedures to prevent accidents. Commanders should not think these safety measures apply only to peacetime training. In combat, nearly all vehicles operate with blackout lights at night, making visibility even more difficult than in garrison driving. If consistent with the tactical situation, the following measures can reduce the possibility of marchers being struck by vehicles:

(1) Use off-road trails and routes as much as possible.

(2) Guards to the front and rear of march units should be marked with strips of luminous tape and carry red-filtered flashlights. These measures are generally consistent with light discipline, while still affording reasonable warning to approaching drivers.

Section II. Conditioning Drills (One, Two, and Three)

3201. GENERAL

a. Description and Objective. Conditioning drills are calisthenic exercises. Each drill contains seven exercises organized and numbered in a set pattern. Each drill takes 15 minutes to complete. The objective of conditioning drills is to exercise all major muscle areas in order to develop strength, endurance, coordination, and flexibility.

b. Area and Equipment. Any level area is satisfactory for conduct of the drills. Drills One and Two contain ground exercises. If ground conditions are unsatisfactory, Drill Three should be used as it contains no ground positions. Usually, no equipment is required;however, if the group exceeds a platoon in size, an instructor's stand is necessary.

c. Formation. The extended rectangular formation is used in this drill. (See app. A.)

d. Starting Level and Progression. The starting level is six repetitions of each exercise. An increase of one repetition for each three periods of exercise in which the drill is performed is an acceptable rate of progression. This rate is continued until 12 repetitions can be completed. To maintain, continue the drill at 12 repetitions. To progress, move to 6 repetitions at a more difficult drill which exercises the same muscle group. Progression can also be gained by moving from Drill One to Drill Two, as Drill Two is more demanding.

e. Starting Positions. Starting positions vary with the exercise and are explained in each exercise.

f. Leadership. A principal instructor demonstrates and leads the drills. The instructor must be familiar with leadership techniques peculiar to conditioning drills to include the exercises, commands, counting cadence, cumulative count, formation, method of teaching the exercises, and utilization of assistant leaders.

g. Use With Other Programs. Conditioning Drills One, Two, and Three reach all major muscles of the body. They are easy to learn and to perform, and they are simple to administer and supervise. These features, coupled with the short time required for completion, the fact that no equipment is necessary, and adaptability to most areas of execution, make these drills possible in any programs.

3202. CONDITIONING DRILL ONE

This conditioning drill is similar to the calisthenics drill commonly known throughout the Marine Corps as the "Daily Seven."The "Daily Seven" may be used as a substitute for Conditioning Drill One.

a. Exercise l:High Jumper

(1) Starting Position. Feet separated shoulder width, knees flexed, body bent forward at the waist, arms aligned with the trunk and hips, elbows locked, palms facing, fingers extended and joined, head and eyes to the front.(See fig. 3-4, A.)(Elbows remain locked throughout the exercise.)

Figure 3-4 is available in the illustration section at the end of the book.

(2) Cadence. Moderate.

(3) Movement. A four-count exercise: at the count of—

(a) ONE -- Take a slight jump into the air, swinging the arms forward and up to shoulder level.

(b) TWO -- Take a slight jump into the air and swing the arms downward and back, returning to the starting position.

(c) THREE -- Take a vigorous leap into the air, swinging the arms forward and up to an overhead position, momentarily looking skyward, on returning to the ground the knees are flexed, head and eyes return to the front.

(d) FOUR--Repeat the action of count two.

b. Exercise 2: Bend and Reach

(1) Starting Position. Feet spread more than shoulder width, arms overhead, elbows locked, palms facing, fingers extended and joined, head and eyes to the front.(See fig. 3-4, B.)

(2) Cadence. Moderate.

(3) Movement. A four-count exercise: at the count of—

(a) ONE-Bend at the knees and waist, swing the arms straight downward and reach between the legs. Touch the ground as far to the rear as possible and look to the rear. (Elbows remain locked throughout the exercise).

(b) TWO-- Recover sharply to the starting position.

(c) THREE--Repeat the action of count ONE.

(d) FOUR-- Repeat the action of count TWO.

c. Exercise 3: Pushup

(1) Starting Position. Front leaning rest position. To assume this position there is a silent one-two count:on the silent count of one, assume the squatting position, heels together, elbows locked inside the knees, hands flat on the ground directly beneath the shoulders; on the silent count of two, thrust the legs to the rear, toes and heels together, body straight from head to heels.(Seefig. 3-4, C.)

(2) Cadence. Moderate.

(3) Movement. A four-count exercise: at the count of—

(a) ONE -- Flex the elbows lowering the body until the thick portion of chest touches the ground.

(b) TWO -- Raise the body until elbows are straight and locked.

(c) THREE-- Repeat the action of count ONE.

(d) FOUR -- Repeat the action of count TWO. (On returning to position of attention, the silent one-two count is used in reverse).

d. Exercise 4: Trunk Twister

(1) Starting Position. Feet are spread more than shoulder width apart, fingers placed behind neck, thumbs pointing downward, elbows back. (See fig. 3-4, D.)(Elbows remain well back throughout the exercise).

(2) Cadence. Slow.

(3) Movement. A four-count exercise: at the count of-

(a) ONE-- Keeping the knees locked and the back straight, bend forward at the waist sharply, with a slight recovery.

(b) TWO-- Twist the trunk to the left vigorously at the waist, keeping the elbow back. The left elbow is higher than the right.

(c) THREE-- Twist vigorously to the right, so the left elbow comes under the right.

(d) FOUR-- Straighten sharply to the starting position. NOTE: Do not attempt to touch the elbows to the knees on counts two and three.

e. Exercise 5: Squat Bender

(1) Starting Position. Feet are spread less than shoulder width apart, hands on hips, thumbs in small of back, elbows back. (See fig. 3-4, E.)

(2) Cadence. Moderate.

(3) Movement. A four-count exercise: at the count of—

(a) ONE-- Assume the squatting position, maintain balance on the balls of the feet, with trunk erect thrust arms forward to shoulder level, elbows locked, palms down.

(b) TWO-- Recover to starting position. Elbows are well back.

(c) THREE-- Keeping the knees locked, bend forward at the waist, touching the ground in front of the toes.

(d) FOUR-- Vigorously recover to the starting position.

f. Exercise 6: Body Twist

(1) Starting Position. On the back, arms extended sideward on the ground, palms down. The legs are raised to a near vertical position, feet together, knees locked.(See fig. 3-4, F.)

(2) Cadence. Slow-fast.

(3) Movement. A four-count exercise: at the count of-

(a) ONE-- Lower legs slowly to the left until they touch the ground near the left hand, keeping the knees straight and shoulders on the ground.

(b) TWO-- Recover the starting position by quickly raising the legs, keep knees straight and feet together.

(c) THREE-- Repeat movement of count ONE, except the movement is to the right side.

(d) FOUR-- Recover sharply to the starting position.

g. Exercise 7: Stationary Run

(1) Starting Position. Position of attention. (See fig, 3-4, G.)

(2) Cadence. Fast.

(3) Movement.

(a) At the command of execution, start running in place at double-time cadence, lifting the left foot first time cadence. Follow the instructor as he counts two repetitions of cadence;

e.g., 1, 2, 3, 4--1, 2, 3, 4. The instructor then gives informal commands such as FOLLOW ME, running on the toes and balls of the feet, keeping the back straight, speeding up the cadence to a sprint, raising the knees high, leaning forward at the waist, and pumping the arms vigorously.

(b) To halt the exercise, the instructor will count two repetitions of cadence as the left foot strikes the ground: 1, 2, 3, 4 - 1 , 2, 3, HALT.

NOTE: When counting cadence, the instructor counts only as the left foot strikes the ground. The duration of the exercise is approximately 1 1/2 minutes.

3203. CONDITIONING DRILL TWO

a. Exercise 1: Jumping Jack

(1) Starting Position. Feet separated more than shoulder width, arms overhead.(See fig.3-5, A.)

(2) Cadence. Moderate.

(3) Movement. A four-count exercise: at the count of~

(a) ONE-- Jump to position with the feet together and assume the squatting position, swinging the arms sideward and downward, placing the hands palms down on the ground, elbows locked inside the knees.

(b) TWO-- Recover to the starting position by jumping to the side straddle and swinging the arms sideward overhead.

(c) THREE-- Repeat the action of count ONE.

(d) FOUR-- Recover to the starting position.

b. Exercise 2: Turn and Bend

(1) Starting Position. Side straddle, arms overhead.(See fig. 3-5, B.)

(2) Cadence. Moderate.

(3) Movement. A four-count exercise: at the count of-

(a) ONE-- Turn the trunk to the left and bend forward over the left thigh, attempting to touch the fingertips to the ground outside the left foot. Keep the left knee straight. On successive repetitions attempt to touch farther and farther to the side.

(b) TWO-- Recover to the starting position.

(c) THREE-- Turn the trunk to the right and bend forward over the right thigh, trying to touch the hands to the ground outside the right foot. Keep the right knee straight.

Figure 3-5 is available in the illustration section at the end of the book.

(d) FOUR-- Recover to the starting position.

c. Exercise 3: Eight-Count Pushup

(1) Starting Position. Position of attention. (See fig. 3-5, C.)

(2) Cadence. Moderate.

(3) Movement. An eight-count exercise: at the count of—

(a) ONE-- Assume the squatting position, palms on the ground directly beneath the shoulders, elbows locked inside the knees.

(b) TWO-- Thrust the legs to the rear assuming the front leaning rest position.

(c) **THREE**-- Flex the elbows until the thick portion of the chest touches the ground.

(d) **FOUR**-- Raise the body on a straight plane until the elbows are locked.

(e) **FIVE**-- Repeat the action of count THREE.

(f) **SIX**-- Repeat the action of count FOUR.

(g) **SEVEN**-- Recover to the squatting position as in count ONE (elbows locked inside the knees).

(h) **EIGHT**—Return sharply to the position of attention.

d. Exercise 4: Turn and Bounce

(1) Starting Position. Feet spread more than shoulder width apart, arms sideward at shoulder level, palms up. (See fig. 3-5, D.)

(2) Cadence. Slow.

(3) Movement. An eight-count exercise: at the count of—

(a) **ONE**-- Turn sharply to the left as far as possible, then recover slightly.

(b) **TWO**-- Again turn to the left as far as possible and recover as in ONE.

(c) **THREE**-- Repeat the action of count TWO.

(d) **FOUR**-- Recover sharply to the starting position.

(e) **FIVE**—Turn sharply to the right as far as possible, then recover slightly.

(f) **SIX**—Again turn to the right as far as possible and recover as in FIVE.

(g) **SEVEN**-Repeat the action of count SIX.

(h) **EIGHT**-Return to the starting position.

NOTE: The head and hips remain to the front throughout the exercise and the knees and elbows are locked at all times.

e. Exercise 5: Squat Stretch

(1) Starting Position. Attention. (See fig. 3-5, E.)

(2) Cadence. Moderate.

(3) Movement. A four-count exercise: at the count of—

(a) **ONE**-- Squat, placing the hands on the ground about 12 inches in front of the feet.

(b) **TWO**— Keeping the fingertips on the ground, straighten the knees completely and raise the hips.

(c) **THREE**-- Recover to position ONE.

(d) **FOUR**-- Recover to the starting position.

f. Exercise 6: Leg Circular

(1) Starting Position. On the back, arms stretched sideward, palms down, feet raised foot from ground, knees straight. (See fig 3-5,F.)

(2) Cadence. Slow.

(3) Movement A four-count exercise: at the count of—

(a) **ONE**-- Swing the legs as far as possible to the left, keeping the knees straight and the legs together.

(b) **TWO**-- Swing the extended legs over head with the thighs as close as possible to the trunk.

(c) **THREE**-- Swing the legs as far as possible to the right.

(d) **FOUR**-- Recover to the starting position.

g. Exercise 7: Stationary Run

(1) Starting Position. Position of attention. (See fig. 3-5, G.)

(2) Cadence. Fast.

(3) Movement

(a) At the command of execution, start running in place at double-time cadence, lifting the left foot first. Follow the instructor as he counts two repetitions of cadence; for example: 1, 2, 3, 4--1, 2, 3, 4. The instructor then gives informal commands such as FOLLOW ME. Running on the toes and balls of the feet, keeping the back straight, speeding up the cadence to a sprint, raising the knees high, leaning forward at the waist, and pumping the arm vigorously.

(b) To halt the exercise, the instructor will count two repetitions of cadence as the foot strikes the ground: 1, 2, 3, 4-1,2, 3, HALT.

NOTE: When counting cadence, the instructor counts only as the left foot strikes the ground. The duration of the exercise is approximately 1 1/2 minutes.

3204. CONDITIONING DRILL THREE

a. Exercise 1: Side Straddle Hop

(1) Starting Position. Position of attention. (See fig. 3-6, A.)

(2) Cadence. Moderate.

(3) Movement. A four-count exercise: at count of-

(a) ONE-- Take a slight jump into the air, moving the legs sideward (more than shoulder width apart); at the same time, swing the arms overhead (to an overhead position) clapping the palms together.

Figure 3-6 is available in the illustration section at the end of the book.

(b) TWO-- Take a slight jump into the air, swing the arms sideward and downward returning to the starting position.

(c) THREE-- Repeat the action of count ONE.

(d) FOUR-- Repeat the action of count TWO.

b. Exercise 2: Back Bender

(1) Starting Position. Standing, feet 12 inches apart, fingers placed behind the head. (See fig. 3-6, B.)

(2)Cadence. Slow.

(3) Movement. A four-count exercise: at the count of-

(a) ONE-- Bend the upper trunk backward, raising the chest high, pulling the elbows back, and looking upward. Keep the knees straight.

(b) TWO-- Recover to the starting position.

(c) THREE-- Repeat the action of count ONE.

(d) FOUR-- Recover to the starting position.

c. Exercise 3: Squat Thrust

(1) Starting Position. Position of attention. (See fig. 3-6, C.)

(2) Cadence Moderate.

(3) Movement. A four-count exercise: at the count of—

(a) ONE-- Assume the squatting position; heels together, placing the hands flat on the ground, shoulder width apart, elbows locked and inside the knees.

(b) TWO-- Thrust the legs to the rear, assuming the front leaning rest position, body in line from head to toe, heels and toes together.

(c) THREE-- Return to the squatting position as in ONE.

(d) FOUR-- Return to position of attention.

d. Exercise 4: Side Bender

(1) Starting Position. Feet are spread more than shoulder width apart, arms are raised sideward and overhead, thumbs interlocked palms to front, fingers extended and joined, elbows locked. (See fig. 3-6, D.)

(2) Cadence. Slow.

(3) Movement. An eight-count exercise: at the count of—

(a) ONE-- Bend to left as far as possible, then recover slightly.

(b) TWO-- Again bend to the left as far possible, then recover slightly.

(c) THREE-Repeat the action of count TWO.

(d) FOUR-- Recover sharply to the starting position.

(e) FIVE-- Bend to the right as far as possible, then recover slightly.

(f) SIX-- Again bend to the right as far as possible, then recover slightly.

(g) SEVEN-- Repeat the action of count SIX.

(h) EIGHT-- Recover sharply to the starting position.

NOTE:Keep the elbows and knees locked throughout the exercise. The bend should occur to the side and not the front.

e. Exercise 5: Knee Bender

(1) Starting Position. Feet are spread less than shoulder-width apart, hands on hips, thumbs in small of back, elbows back. (See fig 3-6, E.)

(2) Cadence. Moderate.

(3) Movement. A four-count exercise: on the count of—

(a) ONE-Do a knee bend, lean trunk forward at the waist, thrust arms between legs until the extended fingers touch the ground palms to the ground, hands 6 inches apart.

(b) TWO~Recover sharply to the starting position.

(c) THREE-Repeat the action of count ONE.

(d) FOUR-Repeat the action of count TWO.

f. Exercise 6: Bottoms Up

(1) Starting Position. Front leaning rest position. A silent one-two count is used as in the pushups. (See fig. 3-6, F.)

(2) Cadence. Moderate.

(3) Movement A four-count exercise: at the count of-

(a) ONE-- With the weight on the hands, and knees locked, jump forward bringing the feet as close to the hands as possible; look to the rear.

(b) TWO--Keeping the knees locked, thrust the legs backward assuming the front leaning rest position.

(c) THREE-- Repeat the action of count ONE.

(d) FOUR-- Repeat the action of count TWO.

g. Exercise 7: Stationary Run

(1) Starting Position. Position of attention. (Sec fig. 3-6, G.)

(2) Cadence. Fast.

(3) Movement

(a) At the command of execution, start running in place at double time, lifting the left foot first. Follow the instructor as he counts two repetitions of cadence; for example: 1, 2, 3, 4--1, 2,3, 4. The instructor then gives informal commands such as FOLLOW ME, running on the toes and balls of the feet, keeping the back straight, speeding up the cadence to a sprint, raising the knees high, leaning forward at the waist, and pumping the arms vigorously.

(b) To halt the exercise, the instructor will count two repetitions of cadence as the left foot strikes the ground: 1, 2, 3, 4 - 1, 2, 3, HALT.

NOTE: When counting cadence, the instructor counts only as the left foot strikes the ground. The duration of the exercise is approximately 1 1/2 minutes.

Section III. Rifle and Log Drills

3301. RIFLE DRILL

a. Description and Objective. Rifle exercises are conditioning exercises performed with a rifle. Each drill contains six exercises (fig. 3-7) and they are numbered in a set pattern. The drill takes 15 minutes to complete. The objective of rifle drills is to exercise the arms, shoulders, and back muscles in order to develop strength and endurance, particularly in the upper body. In units without rifles, log drills may be substituted.

b. Area and Equipment. Any level area is satisfactory for conducting this drill. Each Marine completes these exercises from a standing position and no ground contact is required. Each Marine will need a rifle and, if the group exceeds a platoon in size, then the instructor will need an instructor's stand.

c. Formation. The extended rectangular formation is used in this drill. (See app. A.)
d. Starting Positions. Starting positions vary with the exercises and are explained in each exercise. As in all set conditioning drills, the command used to start the exercise is STARTING POSITION, MOVE. The following directions apply to rifle drill.
(1) In those exercises which start from the rifle downward position, on the command MOVE, execute port arms as prescribed in NAVMC 2691, Drill and Ceremonies Manual, and then assume the starting position. The command to return to the men to the position of attention at the conclusion of the exercise is POSITION OF ATTENTION, MOVE.
(2) In exercises which terminate in the rifle downward position, on the command MOVE, execute the position followed by order arms as prescribed in NAVMC 2691.
(3) In the exercises which terminate in a position other than the rifle downward position, Marines first assume the rifle downward position before executing port arms and order arms.
(4) These movements are executed without command. This procedure promotes uniformity, but precision is not expected. To be effective, rifle exercises must be strenuous enough to tire the arms, but not to the point where the arms cannot move with precision.

e. Leadership. A principal instructor demonstrates and leads the drill. He must be familiar with leadership techniques for conditioning exercises and the peculiar techniques for rifle drill.

3302. EXERCISES PERFORMED WITH RIFLES

The exercises of rifle drill are outlined in the following paragraphs,
a. Exercise 1: Foreup, Behind Back

(1) Starting Position. Rifle downward, feet together.(See fig. 3-7, A.)

Figure 3-7 is available in the illustration section at the end of the book.

(3) Movement. A four-count exercise: at the count of—
(a) ONE-- Swing the arms forward and upward to the overhead position. Inhale.

(b) **TWO**-- Lower the rifle to the back of the shoulders. Exhale.

(c) **THREE**-- Recover to position ONE and inhale.

(d) **FOUR**-- Recover to the starting position and exhale.

b. Exercise 2: Lunge Side, Turn and Bend

(1) Starting Position. Rifle downward, feet together.(See fig. 3-7, B.)

(2) Cadence. Moderate.

(3) Movement. An eight-count exercise:at the count of—

(a) **ONE**-- Lunge sidewards to the left, swing the rifle forward and upward to the overhead position.

(b) **TWO**-- Turn the trunk to the left and bend forward over the left hip. At the same time, swing the rifle to a **low** horizontal in front of the left ankle.

(c) **THREE**-- Recover to position ONE.

(d) **FOUR**-- Recover to the starting position.

(e) **FIVE, SIX, SEVEN, and EIGHT**-- Repeat on the right side.

c. Exercise 3: Foreup, Back Bend

(1) Starting Position. Rifle downward, feet together.(See fig. 3-7, C.)

(2) Cadence. Moderate.

(3) Movement. A four-count exercise:at the count of—

(a) **ONE**-- Swing the arms forward and upward to the overhead position.

(b) **TWO**-- Bend backward, emphasizing the bend in the upper back. The face is up.Keep the knees straight.

(c) **THREE**-- Recover to position ONE.

(d) **FOUR**-- Recover to the starting position.

d. Exercise 4:Up and Forward

(1) Starting PositionRifle downward, feet together.(See fig. 3-7, D.)

(2) Cadence. Fast.

(3) Movement. A four-count exercise:at the count of—

(a) **ONE**-Swing the arms forward and upward to the overhead position.

(b) **TWO**- Swing the arms forward to shoulder level.

(c) **THREE**-Recover to position ONE.

(d) **FOUR**-Recover to the starting position.

e. Exercise 5: Foreup, Full Squat

(1) Starting Position. Rifle downward, feet in narrow stance. (See **fig.** 3-7, E.)

(2) Cadence. Moderate.

(3) Movement. A four-count exercise: at the count of—

(a) **ONE**-- Swing the arms forward and upward to the overhead position.

(b) **TWO**-- Swing the arms down to shoulder level and assume the squatting position.

(c) **THREE**-- Recover to position ONE.

(d) **FOUR**-- Recover to the starting position.

f. Exercise 6: Arms Forward, Side Bend

(1) **Starting Position.** Sidestraddle, regular stance, rifle forward.(See fig. 3-7, F.)
(2) **Cadence.** Moderate.
(3) **Movement.** A four-count exercise:at the count of-

(a) **ONE**-- Bend the trunk to the left. Keep the knees straight.
(b) **TWO**-- Recover to the starting position.
(c) **THREE**-- Bend the trunk to the right. Keep the knees straight.
(d) **FOUR**-- Recover to the starting position.

NOTE:Keep the rifle on the same level as the shoulders throughout the exercise.

3303. LOG DRILL

a. Description and Objective. Log exercises are conditioning exercises performed with a log. Each drill contains six exercises and they are numbered in a set pattern. The drill takes 15 minutes to complete. The objective of log drills is to develop strength and muscular endurance and, in this instance, under maximum loads. Log exercises also develop teamwork. Log may be used in lieu of conditioning drills after the Marines have become somewhat conditioned.

b. Area and Equipment. Any level area is satisfactory for conducting this drill. Each Marine completes these exercises from a standing position and no ground contact is required. If the group exceeds a platoon in size, then the instructor will need instructor's stand. Each six-person group or eight-person group will need a log. The logs should be from 6 to 8 inches in diameter. They may vary in length from 14 feet (for 6 people) to 18 feet (for 8 people).The logs should be skinned, smoother, and dried. The 14-foot logs should weigh approximately 300 pounds and the 18-foot logs, approximately 400 pounds. Rings should be painted on the logs to indicate each person's position. When not in use, the logs should be stored on a rack to keep them off the ground.

c. Formation. All the Marines assigned to the same log team should be about the same height at the shoulders. The recommended method of dividing the platoon is to have the Marines form a single file or column with short people to the front and tall people to the rear. Have the Marine assume their positions in the column according to shoulder height, not head height. When they are in position, they are given the command COUNT OFF BY SIXES (OR EIGHTS), COUNT OFF, to divide them into six- or eight-person log teams. Each team in turn, can then proceed to the log rack, shoulder a log, and carry it to the designated exercise area. The log teams form in columns in front of the instructor. With the Marine holding the log in the chest position, have them face the instructor and ground the log at least 10 yards between columns and 10 yards between log teams within the columns.

d. Starting Positions. The Marines fall in, facing the log, with their toes about 4 inches from it. The basic starting positions and commands are as follows:(See fig. 3-8.)

(1) **RIGHT-HAND STARTING POSITION, MOVE.** It the command MOVE, move the left foot 12 inches to the left, and lower the body into a flat foot squat. Keep the back straight, head up, and arms between the legs. Encircle the far side of the log with the left hand. Place the right hand underneath the log.(See fig. 3-8, A.)

(2) LEFT-HANDED STARTING POSITION, MOVE. These commands are executed in the same manner as in paragraph a. except that the left hand is underneath the log and the right hand encircles its far side. (See fig. 3-8, B.)

Figure 3-8 is available in the illustration section at the end of the book.

(3) RIGHT SHOULDER POSITION, MOVE. At the command MOVE, pull the log upward in one continuous motion to the right shoulder. At the same time, move the left foot to the rear and stand up, facing left. Balance the log on the right shoulder with both hands.(See fig. 3-8, C.)This movement cannot be performed from the left-hand starting position because of the position of the hands.

(4) LEFT SHOULDER POSITION, MOVE. These commands should be given from the left hand starting position. At the command MOVE, pull the log upward in one continuous motion, to the left shoulder. At the same time, move the right foot to the rear and stand facing right. Balance the log on the left shoulder with both hands.(See fig. 3-8, D.)This movement cannot be performed from the right-hand starting position.

(5) WAIST POSITION, MOVE. From the right hand position pull the log waist high. Keep the arms straight and fingers laced underneath the log. The body is inclined slightly to the rear and the chest is lifted and arched. (See fig. 3-8, E.)

(6) CHEST POSITION, MOVE. This command should be given after the waist position has been assumed .On the command MOVE, shift the log to a position high on the chest, bring the left arm under the log and hold the log in the bend of the arms.(See fig. 3-8, F.)Keep the upper arms parallel to the ground.

(7) To move the log from the right shoulder to the left shoulder, the command is: LEFT SHOULDER POSITION, MOVE. On the command MOVE, push the log overhead and lower it to the opposite shoulder.

(8) To return the log to the ground from any of the above positions, the command is: STARTING POSITION, MOVE. At the command MOVE, slowly lower the log to the ground. The hands and fingers must be kept from under the log.

e. Leadership. A principal instructor demonstrates and leads the drill. He must be familiar with the leadership techniques for conditioning exercises and the peculiar techniques for log drill.

3304. EXERCISE PERFORMED WITH LOGS

The exercises of log drill are outlined in the following paragraphs. Figure 3-9 graphically explains log drill.

a. Exercise l:Two-Arm Pushup

(1) Starting Position. Right or left shoulder position. Regular stance. (See fig. 3-9, A.)

(2) Cadence. Moderate.

(3) Movement. A four-count exercise:at the count of—

- **(a) ONE--** Push the log overhead until the elbows lock
- **(b) TWO--** Lower the log to the opposite shoulder.
- **(c) THREE--** Repeat the action of count ONE.
- **(d) FOUR--** Recover to the starting position.

b. Exercise 2: Forward Bender

(1) Starting Position. Chest position. Regular stance.(See fig. 3-9, B.)

(2) Cadence. Moderate.

(3) Movement. A four-count exercise:at the count of-

(a) **ONE**-- Bend forward at the waist, keeping the back and legs straight.

(b) **TWO**-- Recover to the starting position.

(c) **THREE**-- Repeat the action of count ONE.

(d) **FOUR**-- Recover to the starting position.

c. Exercise 3: Straddle Jump

(1) Starting Position. Right or left shoulder position, feet together, fingers interlaced on top of the log.(See fig. 3-9, **C.)**

(2) Cadence. Moderate.

(3) Movement. A four-count exercise:at the count of—

(a) **ONE**-- Jump to a side-straddle. Pull down on the log with both hands to keep it from bouncing on the shoulder.

(b) **TWO**-- Recover to the starting position.

(c) **THREE**-- Repeat the action of count ONE.

(d) **FOUR**-- Recover to the starting position.

Figure 3-9 is available in the illustration section at the end of the book.

d. Exercise 4: Side Bender

(1) Starting Position. Right shoulder position, feet regular stance.(See fig. 3-9, D.)

(2) Cadence. Moderate.

(3) Movement. A four-count exercise:at the count of—

(a) **ONE**-- Bend sideward to the left as far as possible, bending the left knee.

(b) **TWO**-- Recover to the starting position.

(c) **THREE**-- Repeat the action of count ONE.

(d) **FOUR**-- Recover to the starting position.

(4) After completing the required number of repetitions, change shoulders and execute an equal number of repetitions to the other side.

e. Exercise 5: Knee Bend

(1) Starting Position. Right or left shoulder position. Narrow stance. Fingers interlocked on top of the log.(See fig. 3-9, E.)

(2) Cadence. Slow.

(3) Movement. A four-count exercise:at the count of—

(a) **ONE**-- Flex the knees to a quarter-squat position.

(b) **TWO**-- Flex the knees to a half-squat position.

(c) THREE-- Lower the body to a three-quarter squat position.(Lean slightly forward.)

(d) **FOUR**-- Recover to the starting position.

NOTE:Pull forward and downward on the log throughout the exercise.

f. Exercise 6: Overhead Toss

(1) Starting Position. Right or left shoulder position, regular stance. The knees are bent to a quarter-squat.(See fig. 3-9, F.)

(2) Cadence. Moderate.

(3) Movement. A four-count exercisc:at the count of-

(a) ONE-Straighten the knees and toss the log into the air approximately 12 inches overhead. Catch the log with both hands and lower it toward the opposite shoulder. As the log is caught, lower the body into a quarter-squat.

(b) TWO-- Again toss the log into the air and when caught return it to the original shoulder.

(c) THREE-- Repeat the action of count ONE.

Section IV. Grass Drills

3401. GENERAL

a. Description and Objective. Grass drills are extremely strenuous exercises and are performed at top speed for only short periods of time. No cadence is counted but the Marines continue to execute the multiple repetitions of the command until the next command is given. The grass drill consists of two drills:Drill One and Drill Two. Each drill contains six exercises. The objective of the drills is to decrease reaction time, to develop cardiovascular endurance, and to provide a vigorous workout for all major muscles.

b. Area and Equipment. Any level area suitable for ground contact and of a size to accommodate the group is adequate. No equipment is needed.

c. Formation. All movements are executed in place. The extended, rectangular formation is recommended for a platoon- or company-size unit. The circle formation is suitable for groups of squad or section size. At the beginning of an exercise program, 2 to 3 minutes of grass drills will insure a good workout.

d. Progression. Progression is gained by gradually increasing the length of time devoted to the drills. As the physical condition of the Marines improves, the periods should be gradually lengthened to 5 minutes. As the second drill is more difficult than the first, some progression can be attained by initially executing Drill One; then as the program and the Marines progress, introduce Drill Two. To extend the duration of the drill, it may be necessary to repeat the drill.

e. Starting Position

(1) The drills are started from the GO position. Other basic positions are FRONT, BACK, and STOP. (See fig. 3-10, A.)

(a) GO. Running in place (top speed): on the toes and balls of feet, knees raised high, arms pumping, body bent forward at waist.

(b) FRONT. Prone position: elbows bent (along body), palms flat on ground directly under the shoulders, legs together and straight.

(c) BACK. Supine position (flat on back): arms extended near side on ground with palms down, legs together and straight, feet toward the stand or instructor.

(d) STOP. Football lineman stance: feet spread and staggered, left arm across left thigh, right arm straight, knuckles on ground, head up, back parallel with ground.

(2) To assume the FRONT or BACK position from the STANDING, GO, or STOP position, vigorously get into the prescribed position as quickly as possible. (See fig. 3-10, B.)

(3) To change from the FRONT to the BACK position, quickly do a pushup, move the feet several short steps to the right or left, lift the arm on the side toward which the feet move, and thrust the legs vigorously to the front. (See fig. 3-10, C.)

Figure 3-10 is available in the illustration section at the end of the book.

(4) To move from the BACK to the FRONT position, sit up quickly, place both hands on the ground to the right or the left of the legs. Move the feet several short steps to the rear on the side opposite the hands. When the feet are opposite the hands, thrust the legs vigorously to the rear and lower the body to the ground.(See fig. 3-10, D.)

f. Leadership. A warm up activity of lesser intensity should proceed grass drill. During the instructional phase and conduct of these drills, the following points should be applied.

(1) The instructor executes only GO and STOP with the Marines. This allows the instructor to supervise the drill.

(2) The commands for grass drills are given in rapid succession without the usual preparatory command.

(3) To prevent confusion, the instructor should give the commands sharply to distinguish them from comments or encouragement.

(4) As soon as the Marines know the drill, they should respond to the instructor's commands and perform all exercises vigorously and as rapidly as possible. All exercises are executed continuously until the next command is given. Insist on top speed performance; anything less is not effective.

(5) The commands peculiar to each exercise are identical to the name of the exercise.

(6) Marines are not to be required to assume the position of attention once the drills are started. To halt the drill for instructions or for rest, the command UP is used. At this command, the Marines assume a relaxed standing position. Do not demand formality. At the conclusion of a fast and vigorous 5-minute grassdrill, it is physically impossible for people to stand at attention.

(7) The sequence of commands for the execution of grass drills should occur as follows:Drill One. GO, FRONT, Bouncing Ball; GO, BACK, Bicycle; GO, Full Squatter; GO, BACK, Situps; GO, FRONT, Mountain Climber; GO, FRONT, Roll Left; GO, STOP, UP.

g. Use With Other Programs. Since grass drills can be executed in a short period of time, they may be executed where only a few minutes are available for exercise or in conjunction with another type of activity. Grass drills are an excellent substitute for running when time is a factor.

3402. GRASS DRILL ONE AND TWO

a.Grass Drill One

(1) Bouncing Ball. From the FRONT position, push up, supporting the body on the hands (shoulder-width apart) and feet. (See fig. 3-11.)Keep the back and legs in line and the knees straight. Bounce up and down by a series of short, upward springs from the hands, hips, and feet simultaneously,

(2) Bicycle. From the BACK position, raise the legs and hips. Keep the elbows on the ground and support the hips with the hands. Move the legs vigorously as if pedaling a bicycle.

(3) Full Squatter. From the STOP position, assume a full knee bend, the feet on line, hands on hips. Bounce up and down in place by short, bouncing jumps.

Figure 3-11 is available in the illustration section at the end of the book.

(4) Situps. From the BACK position and with arms stretched overhead, sit up, reach forward, and touch toes. Return to the supine position.

(5) Mountain Climber. From the STOP position, place both hands on the ground directly under the shoulders. Thrust the right leg to the rear, knee straight. The left foot should be close to the left hand, the left knee outside the left arm. Shift the weight to the hands, thrust off with the rear (right) foot and bring that foot up close to the right hand, the right knee outside the right arm. At the same time, thrust the left leg vigorously to the rear, knee straight. Continue at a fast cadence, alternating the legs.

(6) Roll Left. From the BACK or FRONT position, make one complete roll in the direction commanded. On completing the roll, return to the FRONT or BACK position.

b. Grass Drill Two

(1) Legs Over. .From the BACK position and with arms stretched overhead, palms up, raise the legs upward and then swing them backward over the head until the toes touch the ground behind the head. Return legs to the starting position. (See fig. 3-12.)

(2) V-Up and Touch Toes. .From the BACK position, raise the legs with the knees straight, sit up until the trunk and legs form a V, and touch the toes with the hands. Return to the BACK position.

(3) Rocker. .In the FRONT position, clasp the hands behind the back, arch the body, holding the head back. Start rocking, using the front part of the trunk as a rocker.

(4) Bounce and **Clap Hands**. The procedure is the same as for bouncing ball, but while in the air, clap the hands. This requires a more vigorous bounce or spring. (See fig. 3-12.)

(5) Leg Spreader. .From the BACK position, raise the legs so that the heels are 10 to 12 inches from the ground, spread them apart as far as possible, then close them together. Open and close legs as rapidly as possible.

(6) Forward Roll. .For forward roll from the STOP position, place both hands on the ground, tuck the head, and do one complete forward roll, keeping the legs tucked as you roll, and come back to the STOP position.

Figure 3-12 is available in the illustration section at the end of the book.

Section V.Guerrilla Exercise

3501.GENERAL

a. Description and Objective. Guerrilla exercises are individual exercises of various basic skills that are performed rapidly while moving forward in a circle formation. There are two tables of guerrilla exercises. Each table takes 15 minutes to complete. The objective of this exercise is to increase strength and endurance, aid flexibility, and develop coordination.

b. Area and Equipment. Any level area is suitable for conducting guerrilla exercises. No ground contact is required other than the hands. There is no equipment requirement.

c. Formation

(1) The circle formation (app. A) is used for guerrilla exercises. Each platoon forms its own circle and engages in guerrilla exercises under a platoon instructor. If the platoon exceeds 30 Marines, double or concentric circles may be used.

(2) When the circle is formed, the instructor steps into the center of the circle and moves clockwise in a small circle. He commands:QUICK TIME, MARCH, 1-2-3-4. (Rapid cadence of approximately 130 counts per minute. Cadence and step are maintained between exercises.)

(3) To reform the platoon after completing guerrilla exercises, the instructor halts the Marines and places the base man or platoon guide where he wishes and commands:

(a) BASE MAN (or platoon guide),POST.

(b) FALL OUT AND FALL IN ON THE BASE MAN (or platoon guide).

d. Progression. Progression may be attained by moving from table I to table II or by shortening the quick time marching periods between exercises and performing all exercises a second time.

e. Leadership

(1) To execute the exercises, the Marines continue at quick time while the instructor simultaneously explains and demonstrates the exercise to be performed, and then commands the Marines accordingly. In each instance, the preparatory command will be the name of the exercise and, in all instances, the command of execution will ***MARCH***.. *To* terminate each exercise, the command *is* ***QUICK TIME, MARCH.*** *The* Marines immediately pick up the step as the instructor counts cadence.

(2) Unless specified differently, each exercise should be continued for 20 to 40 seconds depending upon the vigor of the exercise. The leader can determine the duration of each exercise by observing its effect upon the Marines.

(3) To form for double guerrillas, the commands for pairing the Marines (who are in circle formation) are-

(a) PLATOON, HALT.

(b) FROM (designate an individual), BY TWO'S, COUNT OFF(,Example1-2; 1-2; 1-2; etc.)

(c) EVEN NUMBERS MOVE UP BEHIND ODD NUMBERS(At this time, adjust pairs according to height and weight.)

(d) YOU ARE NOW PAIRED UP FOR DOUBLE GUERRILLAS. To change the Marines position, merely command CHANGE.)
(e) FORWARD, MARCH.

f. Place in the Program. Many Marines have not had the opportunity to perform the simple skills involved in guerrilla exercises. The conduct of these exercises is a simple matter since they can be performed easily and quickly in almost any situation. The tables of exercise are applicable to all personnel. The tables can constitute a station within a 1-hour period or be completed within a separate 15-minute period.

3502.GUERRILLA TABLES

a. Table I
(1) Double Time.(See fig. 3-13, A.)Hold arms in the thrust position. Execute a double-time run, maintaining the circle formation and the prescribed distance between your fellow Marines. Duration-1 minute.
(2) All Fours.(See fig. 3-13, B.) Face downward. Support the body with the hands and feet. Walk forward hands first.
(3) Crab Walk.(See fig. 3-13, C.) Get in the sitting position, face upward, and lift the hips. Support the body with the hands and feet. Walk forward feet first.
(4) Squat Walk.(See fig. 3-13, D.) Assume a full knee bend position. Grasp the ankles (left ankle with the left hand, right ankle with the right hand). Walk forward.
(5)Broad Jump.(See fig. 3-13, E.) Jump forward on both feet in a series of broad jumps. Swing the arms vigorously to assist the jumps.
(6)Toe-Touch Walk.(See fig. 3-13. F.) Walk forward, bending at the waist and touching one hand to the toe of the opposite foot while it is on the ground. Raise the trunk to the vertical position between steps. Keep the knees straight.
(7)Bottoms-Up Walk.(See fig. 3-13, G.) Assume the front leaning rest position and move the feet toward the hands in short steps, keeping the knees locked. When the feet are as close to the hands as possible, walk forward on the hands to the front leaning rest position.
(8) Straddle Run.(See fig. 3-13, H.) Run forward, leaping to the right from the left foot and to the left from the right foot.
(9) Fireman's Carry .(See fig. 3-13,1.)
(lO)Single-Shoulder Carry(See fig. 3-13, J.) Two men execute the carries as indicated by the diagram. No. 1 man executes one type; No. 2 man executes the other.

Figure 3-13 is available in the illustration section at the end of the book.

b. Table II
(1) Double Time.(See fig. 3-14, A.)Hold arms at the thrust position. Execute a double-time run, maintaining the circle formation and the prescribed distance between your fellow Marine.Duration-1 minute.
(2) Toe-Grasp Walk.(See fig. 3-14, B.) Bend forward and grasp toes. With knees slightly bent, walk forward.

(3) Hand-Kick Walk..(See fig. 3-14, C.) Walk forward, kicking the moving foot upward on every step. At the same time, lean forward and touch the elevated toe with the hand of the opposite arm.

(4) Pike Juinping.(See fig. 3-14, D.) Jump forward and upward from both feet, keeping the knees straight, and at the same time, swing the legs forward and touch the toes with the hands at the top of each jump.

(5) Squat Jump.(See fig. 3-14, E.) Leap forward from the squatting position, with the hands on the ground and the arms between the legs. Land on the ground with hands and legs extended. Bring up the legs to the squatting position.

(6) Steam Engine..(See fig. 3-14, F.) Lace the fingers behind the neck and walk forward in the following manner: as the left leg moves forward, raise the knee high, bend the trunk forward, and touch the outside of the right elbow to the outside of the knee. Then lower the left leg and step forward on the left foot and raise the right leg. Repeat with the right leg and left elbow.

(7) Knee-Touch Walk.(Scc fig. 3-14, G.) Walk forward, bending the knees and touching the ground on each step. The knees are bent and straightened on each step.

(8)Hobble Hopping.(See fig. 3-14, H.) Hold foot behind back with opposite hand and hop forward. On the command CHARGE, grasp the opposite foot with opposite hand and hop **forward.**

(9) Cross Carry.(See fig. 3-14,1.)(10)Saddle Back Carry. (See fig. 3-14, J.) Two Marines execute the carries as indicated in the diagram. No. 1 Marine executes onetype;No. 2 Marine executes the other.

Figure 3-14 is available in the illustration section at the end of the book.

Section VI. Running and Orienteering

3601. RUNNING

a. General

(l)Description. The general form and technique for all types of running are fairly constant. (See fig. 3-15.)The head is erect, body slightly forward without bending at the waist, and the arms are at a loose thrust position alternating from front to rear in straight planes. A cross-body arm movement wastes energy .The movement of the legs and feet will be discussed in subsequent paragraphs dealing with the different types of running. Of primary importance is the fact that in all types of running, the toes should be pointed straight ahead. Toeing out is a common error in both running and walking and should be an item of individual correction.

(2) Objective. The objective of running is to develop cardiovascular endurance. Despite the fact that Marines have developed their muscle structure and the strength of their muscle tissue, unless they have developed cardiovascular endurance to a satisfactory degree, they are not entirely physically fit or combat ready. Cardiovascular endurance (wind) depends on the efficiency of the lungs and heart. The efficiency of the lungs and heart depend on the amount of oxygen the lungs can absorb with each breath inhaled and the amount of carbon dioxide the lungs can expel. The process of absorbing oxygen and expelling carbon dioxide (cardiopulmonary process) is performed by the blood that circulates through the lungs. The condition of this process will determine the amount of effort a person can exert over a period of time. Running is one of the best activities to develop this vitally important endurance.

b. Running Skills. In the development of running skills, individuals may require instruction to improve their proficiency. Some important skills to consider are-

Figure 3-15 is available in the illustration section at the end of the book.

(1) Action of the Arms. Arm action is important. Check to see that arms are held loosely and that the action is relaxed. The faster the run, the more rapid the arm action.

(2) Breathing. Allow the individual to breathe through the mouth as the body demands a large supply of oxygen. Oxygen can be inhaled in greater quantities through the mouth.

3602. TYPES OF RUNNING

a. Double Time

(1) Description and Objective. Double timing is marching at the rate of 180 steps per minute, each step being 36 inches in length. It takes practice to double time with precision in formation. The Marines should keep in step, placing their feet flat on the ground. This, however, should not be a stamping motion, but should be done with as slight a jolt as possible. Double timing is like a jog, the difference being that in a jog the feet are lifted well off the ground and the running motion is bouncy. In double timing, the feet skim the ground and there is no bounce to the ran. Double timing is a vehicle for teaching proper running form and for the development of the cardiovascular system.

(2) Area and Equipment. This type of running can be completed over a variety of surfaces. Usually a Marine uses a field or road. There is no equipment requirement.

(3)Progression. There is no set standard for alternating quick time and double time in the early conditioning of Marines. A general rule is to begin with enough quick-time marching to ensure a thorough warming up, then double time about 100 paces. Change again to quick time until the individuals have made a reasonable recovery from the running, then double time another 100 paces. The amount of double time can be increased and the quick time decreased from week to week, until the individuals are double timing about 1800 yards. This type of training should be given at least twice a week, but by no means is it adequate as the sole means of conditioning.

(4) Leadership

(a) The instructor should be to one side of the column or group and toward the rear so there is a full view of all Marines. Inexperienced instructors have a tendency to supervise from a position too far forward.

(b) Select an individual who can maintain the proper cadence to act as the guide during double-time running.

(c) There are several ways for the instructor and group to count cadence while double timing. If not contrary to local policy, learn several methods and use them for variety.

b. Wind Sprints

(1) Description and Objective. This type of running involves a series of 30- or 40-yard dashes, usually conducted in successive waves of squads. Each squad is in line and the squad leader on the right flank. Wind sprints assist indeveloping speed and cardiovascular endurance. Any flat and level area may be used which will permit the squad to form a line and run the required distance.

(2) Progression. One or two 30-yard sprints will be adequate at the beginning. As time passes, sprints can be lengthened and up to six or seven sprints may be used.

(3) Leadership

(a) At the command READY (given by squad leader), each runner assumes the sprinter's starting position. At the command GO, the squad sprints approximately 30 yards, takes 10 yards to stop, and lines up immediately with the squad leader who repeats, READY, GO, and again the squad sprints. At the conclusion of the third sprint, the squad waits until all the squads of the platoon have made three sprints. Then they all line up and the squad leaders conduct three more wind sprints in the opposite direction.

(b) Valuable time is gained by having each squad ready to go when the preceding squad has moved off its second sprint mark.

c. Cross-Country Running

(1) Description and Objective. Cross-country running is a distance run conducted on a course laid out along roads, across fields, over hills, through woods, and on any irregular ground. The cross-country ran may be utilized as a conditioner or as a competitive event; the objective is to cover the distance in the shortest possible time. The course should be 2 to 2.5 miles in length and be laid out to avoid heavy vehicular traffic. The course should be marked by directional arrows until the runners know the course. These runs build leg muscles, increase lung capacity, and develop endurance. Any local area of varied terrain is suitable.

(2) Progression. In the mass training of a large group, leaders should be stationed at the head and the rear of the column and should make every effort to keep the individuals together .After determining the abilities of the individuals in cross-country running, it is advisable to divide the unit into three groups. The poorest conditioned group starts first, and the best conditioned group, last. The starting time of the groups should be staggered so that all of them finish about the same time. In preliminary training, the running is similar to ordinary road work in that it begins with rather slow jogging, alternating with walking. The speed and distance of the run is gradually increased. As the condition of the individual improves, occasional sprints may be introduced. At first the distance run is from 1/2 to 1 mile. It is gradually increased to 2 or 2.5 miles. Well-conditioned personnel can run 2 to 2.5 miles within a 15-minute period.

(3) Leadership. Marines should not be required to take part in distance running until they have been through a progressively scheduled training program which requires a considerable amount of running. Cross-country runs should be scheduled occasionally to provide variety in the program. Cross-country running has the advantage of allowing mass participation. Interest can be stimulated by putting the runs on a competitive basis. As a single activity, short cross-country runs can be scheduled once a week, gradually increasing the distance as the physical conditioning improves.

(4)Use With Other Programs. Cross-country running can be combined with other activities such as conditioning exercises.

d. Fartlek Training

(l)Description and Function. Fartlek training is various running exercises conducted along an intense combat-like course. The running is conducted to overwork the lungs, allowing only partial recovery which is followed by another intense period of overload. This sequence is repeated for the duration of the workout. Fartlek training is a useful combat training method and general cardiovascular conditioner. The entire unit must be in uniformly good to excellent condition to qualify for this training.

(2) Progression. The following is an example of one fartlek session:

(a) Warm up by stretching 3-5 minutes.

(b) Warm up running easily 5-10 minutes.

(c) Run at a fast, steady pace for 3/4 to 1 3/4 miles (dependent on terrain).

(d) Walk/jog at a moderate pace for 5 minutes (recovery).

(e) Run easily sprinting 15 to 20 meters occasionally.

(f) Run full speed uphill for 175 to 200 meters.

(g) Warm down by running easily for 1/2 to 1 mile.

(h) Warm down by stretching 3 to 5 minutes.

(3)Leadership. To add variety, try incorporating combat movements, terrain association, a series of exercises or carrying table of equipment weapons (hit and roll). This will take some initiative in setting the course, but it allows for variety. In this manner, the fartlek course can be used as a training session or part of one. It works extremely well with highly motivated, competitive Marines.

3603.ORIENTEERING

Orienteering is land navigation over a prescribed course as a timed, competitive event. It is an excellent way to integrate land navigation training with physical conditioning and requires Marines to be skilled in both areas. Commanders adjust the training as needed to change the degree of difficulty or to emphasize either navigation or conditioning. Taken to the extreme, an orienteering exercise is conducted in full combat gear over rugged terrain against aggressors. Orienteering requires the following physical skills: endurance, speed, strength, and all-around conditioning.

Section VII. Circuit Training

3701. GENERAL

a. Description and Objective. A strength circuit consists of a series of stations where individuals in small groups exercise vigorously for a short period of time and then move (on signal) to the next station where a different form of exercise is conducted. This rotation continues until all groups move through all stations. Strength circuits contain no set or specific types of exercise stations within the circuit. The objective of circuit training is to develop strength. There are three general types of circuits.

(1) Fixed Circuit. This is a circuit in which apparatus of an immovable type (fixed into the ground) is used.

(2) Movable Circuit. This circuit consists of individual exercise apparatus which is portable and can be moved to and from the training area.

(3) Simplified Circuit. This circuit requires no equipment or apparatus.

b. Formation. The exercises are done at will, but rapid, steady, and continuous work is required of all. Each Marine's nervous and muscular system reacts differently to timed vigorous exercises. His performance should be measured on how many movements per exercise he can complete as an individual. For example, one Marine may be able to complete 5 movements, while another may be able to complete 20, and yet each is receiving the maximum benefit. All three circuits contained in this chapter are designed for platoon-sized groups. Expansion beyond this capacity requires a large amount of equipment, as each Marine in the fixed and movable types of circuits must have an item of equipment available for exercise at each station. A group larger than a platoon could be exercised through use of the simplified type of circuit; however, the group would be unwieldy and control could be a problem.

c. Place in the Program. All circuits illustrated can be completed in a 15-minute period. This feature allows the exercise of a platoonor smaller group on the circuit for a single 15-minute period, or the scheduling of the circuit as a 15-minute period within a longer period. A circuit can thus be utilized within a rotating activity system of scheduling. Choice of a circuit by the unit depends upon area, facilities, and other local factors; however, there is a circuit for every need.

3702. FIXED STRENGTH CIRCUIT

a. Description and Objective. The strength circuit is an arrangement of various types of exercise apparatus which are fixed in position. (See fig. 3-16.)Seven basic exercises are used and each exercise requires an apparatus. All apparatus of one type are positioned together to constitute a station. Each station will accommodate 10 Marines. The circuit is designed to be accomplished in 15 minutes when conducted on a time-rotation basis, as normally executed for unit training. For individual use, a Marine may complete the circuit by executing a specific number of repetitions for each exercise. In this case, the time required for completion of the circuit would vary slightly depending on the number of repetitions the Marine completed. The objective of this circuit is to provide a series of exercises which will improve and maintain the strength of the body's major muscle groups.

Figure 3-16 is available in the illustration section at the end of the book.

b. Starting Level. Marines must be thoroughly warmed up prior to participating in the circuit system. If Marines have not engaged in vigorous exercise immediately prior to starting the circuit, then an instructor should execute the following warm up exercises. These exercises should be conducted in the normal formation for set drills. Seven repetitions of each exercise will normally provide sufficient warm up. The exercises are-

(1) High jumper, Conditioning Drill 1.
(2) Bend and reach, Conditioning Drill 1.
(3) Squat bender, Conditioning Drill 1.

c. Progression. The instructor can adjust the circuit system through several methods to accommodate participants of varying physical ability. One method of adjusting an exercise is to change the method of executing the leg lift or to select a heavier weight for the barbell curls. When it becomes apparent during a unit program that the overall fitness of the group has Improved, then the exercise may be made more strenuous in two additional ways. First, the instructor can have one or more of the supplementary stations added. Second, the time spent exercising at each station can be increased in 5-second increments to a maximum of 60 seconds. Also, as Marines learn the circuit, the 45-second period for movement and instruction between stations can be eliminated, thus allowing only enough time to change stations.

d. Starting Positions. The Marine starts on any station, exercises steadily for a certain period (45 seconds initially), then moves on command to the next type of apparatus where he again exercises steadily for an equal period. The Marine continues until the required exercise is completed at each station. If it is desired to expand the number of stations in the circuit to accommodate more participants at one time, the instructor can provide four additional supplementary exercises, requiring no equipment. This will increase the amount of time required to complete the circuit.

e. Leadership. Close control of all Marines is necessary to ensure that a minimum amount of time is spent in moving them to their initial stations and in moving them between stations. One instructor can control the activity on the strength circuit. A stop watch or wrist watch with a second hand is required. When Marines arrive at the strength circuit, the instructor will form them for exercise and conduct the warm up drill. The group is then reassembled and formed into a number of files equal to the number of stations being used in the circuit. Each file is then directed to a station. As soon as all participants have reached an exercise position at a station, the instructor gives the command READY, GO. After 45 seconds of exercise, the instructor gives the command STOP, CHANGE OVER. The instructor allows Marines 45 seconds for moving to the next station and for preparing for the next exercise before the command EXERCISE is again given. In lieu of verbal commands, a whistle may be used to stop and start the exercises. For large groups, a megaphone or loudspeaker is useful.

f. Individual Conditioning Program. For the Marine working alone on the strength circuit, it would be impractical to time the exercise periods. For individual exercise, the Marine should select a number of repetitions of each exercise to complete, then rotate to the next stations after completing these repetitions. The number of repetitions selected should be at or near the maximum that the Marine is capable of doing without halting for rest.

g. Fixed Circuit Stations

(1) Primary Stations (See fig. 3-16.)

(a) Pullups/Chinups. A horizontal bar placed 8 feet above the ground is necessary for each Marine at this station. The Marine will also need a space on the bar that is 45 inches wide. On the command**EXERCISE**,the Marine grasps the bar with both palms facing either forward or to the rear, arms fully extended, and feet free of the ground and executes the exercise as described in MCO 6100.3_.The exercise is repeated as many times as possible until the command **STOP** is given. Then the Marine moves to the next exercise. If a Marine has done his maximum number of pullups/chinups prior to the command **STOP**, he will remain in the "dead" handing position until the command **STOP** and move to the next station.

(b) Twist Grip. The apparatus is a horizontal bar, free to turn, held between uprights placed 30 inches apart. The bar is 52 inches above the ground. A weight of 20 pounds is attached to the center of the bar by a light rope long enough to permit the weight to rest on the ground. The Marine stands at arm's length from the bar and grasps it with his hands on either side of the rope, palms down, thumbs under the bar. On the command **EXERCISE,** the hands are rotated so that the backs of the hands are rotated away from the body, thus winding the rope on the bar. The elbows are kept straight to ensure that the exercise is performed by the hand and forearm. When the weight is drawn up to the bar, the bar is then rotated in the opposite direction to lower the weight to the ground. This exercise is continued until the command **STOP** is given. The Marine then moves to the next station.

(c) Pulley Weights. The apparatus is a T frame with a system of pulleys that suspends a weight of about 90 pounds. The weight is attached to a light steel cable which has a drawbar attached to the other end. The Marine grasps the drawbar and sits down directly under the bar, legs extended to the front and arms extended overhead. The exercise is executed by pulling the drawbar down behind the head, then extending the arms slowly again until they are fully extended overhead. The exercise is repeated as many times as possible until the command STOP is given. The Marine then will move to the next exercise. Upon completion of the exercise, the weight is lowered slowly to the ground.

(d) Barbell Curls. A barbell is necessary for each Marine at this station. The barbell is constructed of 1 1/4-inch pipe 5 feet long, and two concrete-filled No. 10 cans. Each barbell should weigh about 40 pounds. Variance in the weight of the barbells, up to about55 pounds, will allow appropriate overload to be applied to Marines who are above average in strength or weight. The Marine grasps the bar with the palms forward and assumes a standing position with the barbell held in front of the hips, hands approximately shoulder's width apart. On the command **EXERCISE,** the Marine flexes the elbows and draws the barbell up until it touches the upper chest. The elbows remain at the sides. Breath is inhaled with the upward movement and exhaled as the barbell is lowered to the starting position. The exercise is repeated as many times as possible until the command **STOP** is given. The Marine moves to the next apparatus.

(e) Step-Up. The apparatus is a platform or ledge 18 inches high and of such size. At accommodate 10 Marines. The Marine faces the platform and on the command **EXERCISE,** steps up onto the platform, bringing his trailing foot up beside the leading foot. He then steps back down to the original position, stepping down first with the same foot he initially used in stepping up. After 10 repetitions of the exercise, he changes the order of moving the feet to use the opposite leg for stepping up. Repeat this exercise until the command **STOP** is given. The Marine then moves to the next station.

(f) Leg Lift. The apparatus is a horizontal bar constructed as described in (a). To prevent the body from swaying, a horizontal back support is added 40 inches below the horizontal bar. The arms are kept fully extended. On the command **EXERCISE,** the Marine jumps up, grasps the bar with the palms forward and the back support behind him. The exercise is executed by raising the legs to a horizontal position then slowly lowering them to the vertical position. The Marine does not flex his knees. He does not swing the legs to the rear of a vertical position to gain momentum for raising them in the next repetition of the exercise. The movement is repeated until the command **STOP** is given. The Marine then moves to the next exercise. If unable to raise his legs to a horizontal position without flexing his knees, the Marine flexes his knees and draws them up to the chest, then lowers his legs to the vertical position.

(g) Rope Climb. The rope climb is 20 to 30 feet high with five ropes suspended from a horizontal bar which forms the uppermost part of the framework. To prevent the horizontal bar from sagging and to provide safety, only five ropes are attached to it. There are two frameworks per station. The ropes are 6 feet apart. Any method may be used to climb the rope, and the Marines climb as high as possible. Marines who are proficient should climb therope several times during the time allotted. Inexperienced Marines should be cautioned to take care during descent to avoid rope burns on their hands.

(2) Supplementary Stations. The following exercises are designed to expand the basic circuit by being inserted in specific places within the system. For each supplementary station used, there should be adequate room for 10 Marines to exercise.

(a) Bent Knee Situp or Bottoms-Up. These calisthenics are designed to strengthen the abdominal muscles. These exercises will be inserted between the pullup/chinup and twist grip stations. The primary stomach exercise is the situp. In case of inclement weather or other conditions that make ground contact undesirable, the bottoms up exercise is used. In situps on the command **EXERCISE,** the Marine lies on his back (supine position) with knees flexed and both feet flat on the ground and executes the exercise (minus an assistant) as described in MCO 6100.3_.The Marine then returns to the starting position, repeating the exercise until the command **STOP** is given. The Marine then moves to the next station. In bottoms up on the commandEXERCISE,the Marine assumes the front leaning rest and executes the bottoms-up exercise as described in Conditioning Drill 3. He continues this exercise at a moderate cadence until the command **STOP** is given. The Marine then moves to the next station.

(b) Pushup. This exercise is designed to strengthen the arm- and shoulder-girdle muscles. It should be included between the twist grip and pulley weight stations. Upon the command **EXERCISE,** the Marine executes the pushup as described in Conditioning Drill 1. He continues this exercise at a moderate cadence until the command **STOP** is given. The Marine then moves to the next station.

(c) Knee Bender. This exercise is designed to build leg muscles and is included between the pulley weight and barbell curl stations. On the command **EXERCISE,** the Marine executes the knee bender as described in Conditioning Drill 3.He continues this exercise at a moderate cadence until the command **STOP** is given. The Marine then moves to the next station.

(d) Trunk Twister. This exercise strengthens the major muscles of the trunk and is included between the step-up and pullup/chinup stations. On the command EXERCISE, the Marine executes the trunk twister as described in Conditioning Drill 1. He continues this exercise at a moderate cadence until the command **STOP** is given. The Marine then moves to the next station.

3703. MOVEABLE STRENGTH CIRCUIT

a. Description and Objective. The exercises in this circuit are progressive and the course is planned to gain and hold the interest of the participating groups. The circuit consists of a series of stations, with each station designed to develop a particular group of muscles. Along with muscular development, correct posture and deep rhythmic breathing should be stressed on this circuit at all times.(See figs. 3-17 and 3-18.)

Figure 3-17 is available in the illustration section at the end of the book.

b. Equipment. The equipment is set up in files. Six files of 8 stations will accommodate a platoon

c. Formation. The platoon marches to the area where the equipment is positioned and forms a file within each lane of stations, covering down on a piece of equipment. Movements are made on the double, the important factor being that no time is wasted in getting to work.

d. Progression. Initially 40 to 45 seconds per station is adequate. As individuals become stronger the time should be increased in 5-second increments until a minute to a minute and a half is reached.

e. Leadership

(1) The leader stands in front of the barbell station and controls the rotation from this position. The leader supervises the entire group, with the assistance of several instructors who move about in the platoon correcting and encouraging the Marines.

(2) The leader starts each group but does not count cadence nor lead them through the exercises. Each individual exercises rapidly but individually.

(3) As each Marine finishes his repetitions with the barbell, he places the barbells on the ground. The leader calls, READY, followed by the command, FALL OUT ONE. All will double-time to the station directly in front of them, while the individuals on the barbell stations will do an about face to the rear station in their lane.

f. Movable Circuit Stations. The Marine can obtain best results on the movable circuit if the exercises on the various stations are given in the following manner:

(1) Station 1—The Barbell. (See fig. 3-17, 1.)The exercises at this station stress proper posture; deep, rhythmic breathing; and development of the muscles of the arms, shoulders, and upper body. An instructor teaches the proper methods of lifting before the exercises begin. The methods are to lift with the legs, to keep the back straight, and to merely grip with the hands. Two recommended exercises are given below. The instructor will use only one exercise per period .He may use either exercise. At successive periods, the other exercise is used.

(a) Exercise 1--Two-Hands Military Press. (See fig. 3-18, 1.)Grasp the barbell with both hands, knuckles up at shoulder width, and lift to the chest. Steadily press to arm's length overhead; lower to the chest resisting weight all the way. Inhale as the weight is pressed up and exhale as the weight is brought down.

(b) Exercise 2--Two-Hands Regular Curl. (See fig. 3-18, 1.) Lift the weight to the waist, with the palms of the hands out, heels together, stomach in, chest lifted and arched, shoulders back, elbows in close to the sides; inhale deeply and curl the weight to the shoulders, using the arms only, at the same time keeping the elbows close to the sides; exhale Emphasize posture and the use of the arms only. This is a very valuable exercise for the development of the biceps and the grip and should be repeated from 8 to 16 times, depending on the ability of the Marine.

(2) Station 2--The Jump Rope. (See fig. 3-18, 2.) This exercise develops strength and agility in the legs and stamina of the whole body. It makes the Marine agile on his feet and increases his footwork efficiency and timing. Each Marine should progress until able to jump rope at least 3 minutes at top speed.

(3) Station 3--The Twist Grip. (See fig. 3-18, 3.) The twist grip is an excellent exercise for the hands and forearms, and adds greatly to the Marine's ability in hand-to-hand combat. The handle is gripped and twisted, winding the rope until the weighted can is level with the height of the hands, which are held horizontal. The weight is lowered in the same manner; the Marine resists the weight all the way, occasionally stopping the twisting motion and alternately removing first one hand, then the other, from the handle. A variation of the above exercise is to wind the handle with the palms up and the arms bent and with the elbows held close in to the sides. Each Marine should maintain a good posture and keep the stomach muscles taut throughout this exercise.

(4) Station 4--The Incline Plane. (See fig. 3-18, 4,)The use of the incline plane is a very strenuous exercise and well designed for the development of the abdomen. Although 6 to 10 repetitions are sufficient for the beginners, more can be added as ability increases. Marines with hernias or recent operations will be excused from participation at this station.

(5) Station 5--The War Club. (See fig. 3-18, 5.) The war club is a simple and effective means of exercising the principal muscle groups of the body, especially those of the trunk, back, and shoulders. To gain the maximum benefit from this exercise, the Marine must keep both feet flat on the ground at all times. Throughout the exercise period, the weight is swung from arm's length as follows:

(a) As in chopping wood, first on one side, then on the other.

(b) As a batter warming up with a number of bats.

(c) In large circles, first with one hand and then with the other.

(6) Station 6--The Bicycle Ride. (See fig. 3-18, 6.) The bicycle ride is well suited to exercising many of the muscle groups of the body, particularly those of the abdomen. Vary the speed of the exercise, but keep everyone "riding" the entire period. A variation exercise may be performed by placing the legs together, raising them slowly to a height about 2 feet from the ground, and then lowering them slowly to the ground.

(7) Station 7--The Step-up. (See fig. 3-18, 7.) The step-up exercises the legs. The step-up is performed by initially stepping up with the left foot, followed by the right, then stepping down with the left foot followed by the right. Continue for 20 seconds, then change to the right foot as the lead foot for 20 seconds.

(8) Station 8--The Isometric Pull. (See fig. 3-18, 8.) Two trainees work at this station with a cable pull and perform the following exercises:

(a) Initially start with one Marine in the supine position and one Marine sitting. The sitting Marine lowers the upper body to the ground and pulls the partner up to the sitting position. The partner then performs this same action and this is continued for 20 seconds at a rapid rate.

(b) During the last 20 seconds, the same action takes place but, in this case, the Marine in the supine position resists the pull of the partner for approximately 5 seconds before allowing to be pulled up into the sitting position.

3704. CIRCUIT-INTERVAL TABLE

a. Description and Objective. The circuit-interval table is designed to develop strength and endurance within a short period of time, with no equipment requirement, through a rapid and vigorous routine of exercise. Fifteen minutes is an adequate period to execute all exercises and to secure a vigorous workout with the circuit-interval principle.

Figure 3-18 is available in the illustration section at the end of the book.

b. Formation and Starting Positions

(1) A leader forms platoons or smaller groups in an oval or circular formation with 3- to 5-yard intervals between Marines. The Marines face to the right and move forward at quick time and then double time. (See fig. 3-19, A.)After running several platoon circle laps, the leader calls out the name of an arm and shoulder exercise from the list below, orders quick time and commands, for example, PUSHUPS. On this command, all Marines immediately hit the ground and individually and rapidly begin doing pushups. No cadence is counted. (See fig. 3-19, B.) After 30 seconds of exercise, the leader commands. **QN YOUR FEET, FORWARD, MARCH** The platoon resumes the quick time cadence and the leader, when ready, gives the necessary commands for double time. The double time is continued for one or more laps and the leader calls out the name of the next exercise and the process is repeated. This continues, with running between each exercise, until every body part has been exercised.

(2) The instructor controls the running and quick time to observe the effects of the exercise upon the Marines. Cadence, step, and precision are not important to the objective and the instructor should not use them. What is important is speed and the instructor should stress this. After the exercise period is started, the Marines do not stop. This circuit method emphasizes stress and recovery, the recovery occurring during the quick time periods.

c. Activities. The leader can use the following exercises. These exercises can be repeated if necessary, during a second round. The leader can use other calisthenic exercises in the circuit-interval table.

(1) Arms and shoulders-push-ups.
(2) Stomach situps.
(3) Back-squat thrusts.
(4) Legs-bicycle (on back).

d. Progression. Ihe progress is controlled by the leader, who must pace the running, quick-time movement, and exercise in such a way that Marines will receive a vigorous workout yet be able to participate throughout the 15-minute period. Marines who are in the initial stages of physical condition will not be able to double time or exercise as long as those who are better conditioned. The idea is to set a pace which can be increased during each workout, thus progressing gradually to a higher level of physical fitness.

e. Leadership. The platoon leader, platoon sergeant, or section leader can lead the group. The leader must execute the exercise with the unit to feel the effects and thereby adjust the pace.

f. Place in the Program. This activity may be scheduled whenever a short period of time is available. The only requirement is that enough space, indoors or out, be available to form the circle.

Figure 3-19 is available in the illustration section at the end of the book.

Section VIII. Basic Physical Skills and Obstacle Courses

3801. GENERAL

The purpose of this section is to list the basic military skills and the methods for their development. Many of these skills are best developed by obstacle courses but other drills are also discussed. The section explains types of obstacle courses, details of construction, and methods of negotiating the various obstacles.

3802. BASIC PHYSICAL SKILLS

a. Objective. The objective is to develop proficiency in the various military physical skills which are essential to personal safety and effective combat operations. In travel by foot over rugged terrain and in the execution of combat duties, Marines must be trained to perform basic skills, such as running, jumping, climbing, and carrying. During training, Marines will develop agility and coordination in these skills .Fast and skillful execution of these skills may mean the difference between success and failure on the battlefield.

b. Place in the Program. These skills are practiced throughout the entire program in many different activities. Many of these skills are best practiced on obstacle courses.(See par. 3803.)

c. Description. These basic skills are the minimum skills required by all Marines. The basic skills are as follows:
(1) Running. Running is used to strengthen the legs and develop cardiovascular endurance. Marines should be exposed to running in various situations: on roads, over rough ground, up and down hills, cross-country, and running over low obstacles.
(2) Jumping. In broad jumping, the take-off foot is planted firmly and the spring comes from the extension of this leg as the other leg reaches for the far side of a ditch or similar obstacle. (See fig. 3-20.) The arms are forcibly raised forward and upward to assist in propelling the body up and forward. Landing may be on one or both feet depending upon the length of the jump. In vertical jumping downward from a height, the jumper should aim his feet at the desired landing spot and he should also jump with the knees slightly bent and feet together, with the trunk inclined slightly forward. As the feet touch the ground, the shock is absorbed by bending the knees into a full squatting position. If the height is too great or the ground too hard to absorb the shock, then the jumper should forward roll or side roll thus eliminating some of the momentum.

Figure 3-20 is available in the illustration section at the end of the book.

(3) Dodging In combat situations, it is often times necessary to change directions quickly. To execute this movement while running, a lead foot is firmly planted, left foot if the direction is to the right and right foot if the direction is to the left. The opposite foot is moved toward the new direction. The knees are slightly flexed during the movement and the center of gravity is low and balanced. (See fig. 3-21.) At the time of the change of direction, the head and trunk are turned quickly in the new direction.

Figure 3-21 is available in the illustration section at the end of the book.

(4) Climbing and Surmounting. All Marines should know how to climb and surmount various types of obstacles. (See fig. 3-22.)

Figure 3-22 is available in the illustration section at the end of the book.

(a) Vertical Climbing, as in Climbing a Rope or Pole. The hands grasp the rope or pole overhead with the palms toward the face. Gripping the object, the body is pulled upward with the arms and shoulders, assisted by the feet which grip the object and assist by pushing downward. If shoulder girdle strength and body coordination are not adequate to permit alternating the hands, the arms act together in pulling upward.

(b) Climbing as in Surmounting a Wall. Ingoing over a wall, the body should be kept as close to the top as possible, since in combat operations it is important to offer as small a target as possible to the enemy. If an individual climbs a wall while carrying a rifle, both hands should be freed by slinging the rifle over the back. There are two methods commonly used for surmounting a wall of moderate height, but only one for dropping from it.The methods are as follows:

1 Run, Jump, and Vault. Approach the wall at a run, jump forward and upward at the wall and place one foot against it as high up as possible. Use the foot in contact with the wall to help push the body upward while grasping the top of the wall with the hands. Pull the body up with the arms, assisted by pressure of the foot against the wall and swing the legs over, propelling the body weight over the wall.

2 Hook and Swing. Approach the wall at a run and jump forward and upward. Hook one elbow over the wall, locking the arm in place by pulling up until the top of the wall is underneath the armpit. Grasp the top of the wall with the other hand. Draw the leg which is closer to the wall up as far toward the top as possible. Then swing the other leg over the top of the wall.The body is then carried over with a rolling motion. A variation of this leg action can be used by Marines who are unable to draw up the leg as described. While hanging with both legs fully extended, start a swinging motion with the legs together. When the legs have enough momentum, swing the outside leg over the top of the wall with a vigorous kick, then follow with the body.

3 **Dropping.** All drops from a height are executed in the same manner regardless of the method used to gain the top. One hand is placed against the far side of the wall while the other hand grasps the top. From this position, the body is rolled over the wall and vaulted away from it with the legs swinging clear. As the body passes over the wall and drops, it should at all times face the wall. This will keep the rifle and other equipment clear. Break the fall by retaining a grasp on the top of the wall as long as possible.

(c) Climbing Ladders and Cargo Nets. Rope ladders, stationary vertical ladders, and cargo nets employ the same general technique. The important element is to grasp the side supports firmly in the hands about shoulder height and place the feet on a rung which would cause the body to be fully extended. In movement upward, one hand is moved upward and a new grasp is secured and, at the same time, the opposite leg moves up a rung. As the knee straightens, the body is elevated. This process is repeated using the opposite arm and leg. Alternation continues in this manner until the climber reaches the objective.

(5) Traversing Horizontal Objects. The traversing of horizontal objects puts heavy stress on the arms and shoulder girdle area as the feet are usually suspended in the air with all of the body weight on the arms and shoulders.(See fig. 3-23.)

(a) Traversing Horizontal Ropes or Pipes. The hands grasp the horizontal support overhead with the palms facing. To propel the body forward, one hand is released and moved forward to secure a new grasp. At the same time, the opposite side of the body is swung forward (some people are able to "walk" in the air, keeping the body to the front and moving the legs in time with the arms as in walking on the ground). The other hand is then released and moved forward; this alternation is continued until the objective is reached.

Figure 3-23 is available in the illustration section at the end of the book.

(b) Traversing Horizontal Ladders. In this situation, the movement is the same as used in traversing a rope or pipe. The hands, however, are placed on the rung with the palms away from the face. Other than this difference the technique is the same.

(6) Crawling. Crawling in combat situations is an often used skill. Crawling may be high or low.(See fig. 3-24.)

Figure 3-24 is available in the illustration section at the end of the book.

(a) High Crawl.

In the high crawl, the Marine moves on hands and knees, moving one hand and the opposite knee and then continuing to move the hands in alternation with the opposite knee following the companion hand.

(b) Low Crawl. The Marine is in the prone position usually with the forearms and palms of the hand on the ground. He propels forward by bending the knee of one leg and pushing with the inside edge of the shoe. At the same time, the opposite arm moves forward and pulls to the rear. The body remains low and movement is continued by bending the opposite knee and pushing, and at the same time sliding the opposite arm forward and pulling. Alternation of hands and legs continues until the objective is reached.

(7) Throwing. Throwing may be executed from the kneeling or standing positions. The object to be thrown is held in the hand, and the throwing arm is bent at the elbow; the hand is then moved to the rear until the hand is behind the ear. The body is turned so that the lead foot and balance arm on the side toward the target are pointing at the target. (See fig. 3-25.) The balance arm is used to sight over and align the throwing hand and the target. When properly aligned, the elbow is moved rapidly forward until it is at a pointjust in front of the body where the arm is straightened and the wrist snapped. This whip motion propels the object to the target. Underhand throws secure momentum by the thrower bending his knees and swinging the throwing arm to the rear. As the knees are straightened, the arm is forcefully swung forward from the shoulder and the object released.

Figure 3-25 is available in the illustration section at the end of the book.

(8) Vaulting. Vaulting is employed to overcome low barriers or fences. (See fig. 3-26.)The object to be surmounted is approached at an angle. The hand on the side next to the obstacle is placed on the top of the obstacle and, with a straight arm, the body weight is pushed upward. At the same time, the leg on the side next to the obstacle is thrown upward and over the top followed by the other leg (side approach). In landing, the weight comes down on the leading leg first followed by regaining the balance on both legs. The free arm serves as a balance. A direct (front) approach can be used at which time both legs go over the object together.

(9) Man Carrying. There are three basic individual means of carrying personnel in combat situations and one of these methods may be used in carrying objects.

Figure 3-26 is available in the illustration section at the end of the book.

(a) Fireman's Carry."A" stands sideways in front of "B", "A" bends his knees and leans forward, placing one arm through "B's" crotch, grasps the wrist of "B's" arm, which is hanging over the shoulder, and then "A" runs forward.

(b) Saddle-Back Carry. With his back toward "B", "A" stands in front of "B". "B" mounts "A's" hips and clasps his arms in front of "A's" chest. "A" grasps "B's" thighs.

(c) Single-Shoulder Carry. "A" stands facing "B"."A" assumes a semi-squatting position. "B" leans forward until "B" lies across "A's" right shoulder. "A" clasps his arms around "B's" legs and straightens up, lifting "B" from the ground. "A" then runs forward. This method may also be used to carry heavy objects.

(10) Balancing. Balancing the body while walking or running on a narrow object when crossing obstacles is a skill which requires practice and confidence. Balance is required in negotiating a log placed across a stream, in crossing a narrow beam or rail, and in similar situations.(See fig. 3-27.) To perform this skill, place the feet on the object to be crossed, hold the arms to the sides at shoulder level, and fix the eyes on the object approximately 5 yards in front of the feet. Generally, it is not a good practice to look down at the feet. Walk the beam by placing first one foot and then the other in the center of the beam, thereby moving forward, using the arms to aid in maintaining balance.

Figure 3-27 is available in the illustration section at the end of the book.

(11) Falling. Injury can be avoided if Marines are taught to fall properly. They should know how to use the body momentum to their advantage during a fall rather than to try resisting that force. (See fig. 3-28.) If enough force is present, such as occurs during a fall while running or in jumping downward from a height, individuals can extend their hands to catch the weight. At the same time, duck the head and roll forward onto the feet. The key to falling without injury from the standing position is relaxation and rolling on the outside of the leg, hip, and buttocks to take the brunt of the fall.

Figure 3-28 is available in the illustration section at the end of the book.

3803. OBSTACLE COURSES

a. General. Obstacle courses are a valuable part of physical readiness training. The challenge presented by the obstacles assists in developing and testing the basic physical skills. In many combat situations, success will depend upon the Marine's ability to perform one or more of these skills, often while carrying field equipment and when fatigued. In this section, two types of obstacle courses will be discussed-Conditioning Obstacle Course and Confidence Obstacle Course.

b. Course Safety Precautions. Commanders and course instructors should take certain precautions to prevent injury on obstacle courses. A few of the precautions are:
(1) Inspect the course for faulty construction of obstacles, protruding nails, rotten logs, condition of the landing pits, and other hazards to safety.
(2) Conduct warm up exercises before the unit runs the course.
(3) Explain and demonstrate the correct techniques for negotiating all the obstacles before allowing the Marines to try them.
(4) Give Marines at least two weeks of conditioning exercises before scheduling the obstacle and/or confidence courses.
(5) Ensure that negotiation of the higher and more difficult obstacles are under the supervision of an instructor.
(6) Do not permit individuals who have neither practiced the basic skills nor run the conditioning obstacle course to participate in the confidence obstacle course.
(7) Weather conditions may cause footing or handhold surfaces to be slippery. If such is the case, postpone training on the course.

3804.CONDITIONING OBSTACLE COURSES

a. Description and Objective. The Conditioning Obstacle Course is commonly known as the Obstacle Course. This course consists of fairly low obstacles which are designed to be negotiated quickly. The obstacles serve to test various basic skills, and running the course is a test of the Marine's physical condition. After receiving instruction and an opportunity to practice the skills, Marines run the course against time.

b. Area and Equipment
(1) Complete standardization of obstacle courses should not be attempted since topographical conditions always vary. Commanders should use ingenuity in constructing a course, making good use of streams, hills, trees, rocks, and other natural obstacles. Since the course is eventually run at high speed, it should not be dangerous.

(2) The course should be wide enough for six or eight men to run simultaneously, encouraging competition. The lanes for the first several obstacles should be wider and the obstacles themselves easier than those that follow. This avoids congestion until the contestants scatter out over the course. The last two or three obstacles should not be too difficult and should not involve high climbing. This prevents injuries and falls resulting from fatigue.
(3) The total distance of the course should range from 300 to 450 yards and include from 15 to 25 obstacles. Normally the obstacles should be 20 to 30 yards apart and arranged so that those which exercise the same groups of muscles are separated.
(4) The obstacles should be substantially built. Peeled logs, 6 to 8 inches in diameter, are ideal for many of the obstacles. Sharp points and corners should be eliminated. Landing pits for jumps or vaults should be filled with sand or sawdust to prevent injuries.
(5) The course should be constructed and marked so that it is not possible to sidestep or detour obstacles. However, it is desirable to provide alternate obstacles of varying degrees of difficulty.
(6) The course should be in the shape of a horseshoe or figure eight so that the finish is close to the start and signs should be placed to indicate the course route.

c. Leadership. Before Marines run an obstacle course, they should be instructed in the proper technique of negotiating each obstacle. In each case this technique should be explained and demonstrated in detail, with emphasis on avoiding injury. Every individual should be given an opportunity to practice on each obstacle until he becomes reasonably proficient at negotiating it. Before the course is run against time, it is advisable for individuals to make several runs at a slower pace. During such practice or trial runs, the instructor should observe the performances and make appropriate corrections. Marines should never be permitted to run the course for time until they have practiced on all obstacles. The best method of timing the runners is to have the timer stand at the finish and call out the minutes and seconds as each individual finishes. If several watches are available, each wave may be timed separately. If only one watch is available, the different waves should be started at regular intervals, such as every 30 seconds. If an individual fails to negotiate an obstacle, a previously determined penalty should be exacted.

d.Types of Obstacles

(1) Jumping-Type Obstacles. These obstacles may be ditches which are cleared with one leap, trenches which the individuals can jump into, heights which require jumping downward, or hurdles which an individual can jump over. (See fig. 3-29.)

Figure 3-29 is available in the illustration section at the end of the book.

(2) Dodging-Type Obstacles. Obstacles of this type are usually mazes consisting of posts set in the ground at irregular intervals. (See fig. 3-30.)The intervals between posts should be rather narrow so that the Marines must pick their way carefully through and around them. Lane guides may be constructed to guide the Marines to dodge and change direction. Obstacles may be put into a maze pattern to cause the Marines to change direction.

Figure 3-30 is available in the illustration section at the end of the book.

(3) Vertical Climbing and Surmounting Type Obstacles. These obstacles may be climbing ropes, either plain or knotted and 1-5 inches in diameter; cargo nets or walls 7 or 8 feet high; or vertical poles 6 to 8 inches in diameter and 38 feet high.(Seefig. 3-31.)

Figure 3-31 is available in the illustration section at the end of the book.

(4) Horizontal Traversing-Type Obstacles. Horizontal obstacles may be pipes, beams, ladders, or ropes. (See fig. 3-32.)

(5) Crawling Type Obstacles. Obstacles which require crawling may be constructed of large pipe sections, low rails, and wire. (See fig. 3-33.)

Figure 3-32 is available in the illustration section at the end of the book.

(6) Vaulting-Type Obstacles. Obstacles of 3 to 3.5 feet in height such as low walls or fences may be used as a vaulting obstacle.(See fig. 3-34.)

(7)Balancing-Type Obstacles. Beams, logs, and planks may be used as balancing-type obstacles.(See fig. 3-35.)These items may be used to span water obstacles and dry ditches, or raised off the ground somewhat to simulate natural depressions.

Figure 3-33 is available in the illustration section at the end of the book.

3805.CONFIDENCE OBSTACLE COURSES

a. Description and Objective. The Confidence Obstacle Course is commonly known as the Confidence Course. This course is composed of higher and more difficult obstacles than those used in the conditioning course. The confidence obstacle course is designed to give Marines confidence in their mental and physical capacities and to cultivate their spirit of daring. They are encouraged but not compelled to negotiate this course and the course is not run against time. The negotiation of a confidence course, however, is strenuous enough to be an excellent physical conditioner. Marines should NEVER attempt to take the obstacles at high speed and should not compete for speed. The obstacles vary from fairly easy to extremely difficult. Some are of considerable height to accustom Marines to climbing such heights without fear. Considerable emphasis is placed on obstacles that train and test an individual's balance.

Figure 3-34 is available in the illustration section at the end of the book.

b. Area and Equipment

(1) The confidence course accommodates four platoons, one platoon at each group of six obstacles. The course should be made up of about 24 obstacles, numbered and marked as follows:! to 6, white numbers on red background; 7 to 12, black numbers on a white background;

13 to 18, white numbers on a blue background; and 19 to 24, white numbers on a black background.

(2)A few simple pieces of equipment should be provided for individuals who do not have the strength, courage, or ability to negotiate the obstacles.

Figure 3-35 is available in the illustration section at the end of the book.

c. Formation. The obstacles should be divided into groups of six, and each group is designated by a different color. Each platoon starts at a different color. Individuals are separated into groups of 8 to 12 at each obstacle. At the starting signal from the company commander, they proceed numerically through. Anyone may skip an obstacle who is afraid to try. Individuals proceed from obstacle to obstacle until time is called, then assemble as ordered.

d. Leadership. If the Marines are new to the confidence course, an instructor will demonstrate or will give a brief orientation at each obstacle, including an explanation and demonstration of a method of negotiating it. Marines are encouraged to try the various obstacles, but they are not compelled to do so. No compulsion is to be used. The manner of negotiating any obstacle is left to the discretion of the Marine. However, the instructor assists anyone who experiences difficulty. Instructors must supervise closely at all times to prevent injuries, as some of the obstacles are quite high. Also, some of the obstacles should not be used when slippery or wet. The example of instructors and especially selected demonstrators will serve to inspire the individuals to greater effort.

e. Negotiating the Obstacles. Although personnel need not conform to any one method of negotiating the obstacles, there should be some uniformity in the approach to them. A general method of negotiating the obstacles is indicated below.

(1) Red Group This group contains the first six obstacles. (See fig. 3-36.)

(a) The Belly Buster Individuals may vault, jump, or climb over. Warn them that the log is not stationary.

(b) Reverse Climb. Climb the reverse incline and go down the other side to the ground.

(c) The Weaver. Move from one end of the obstacle to the other by weaving the body under one bar and over the next.

(d) Hip-Hip. Step over each bar, either alternating legs or using same lead leg each time.

(e) Balancing Logs. Step up on log, retaining the balance, walk or run along log.

(f) Island Hopper. Jump from one log to another until the obstacle is negotiated.

(2)White Group. This group is composed of the second six obstacles. (See fig. 3-37.)

(a) Tough Nut. Step over each X in the lane.

(b) Slide for Life. Climb the tower, grasp the rope firmly, and swing the legs upward. Hold the rope with the legs to distribute the weight between them and the arms. Braking the slide with the feet and legs, proceed down the rope. Warn Marines that there is danger of getting rope burns on their hands. When the rope is slippery or wet, this can be a dangerous obstacle.

(c) **Low Belly Over.** .Mount the low log and jump onto the high log, both arms grasping over the top of the log, the stomach area in contact with it. Swing the legs over the log and lower the body to the ground.

(d) **Belly Crawl** Move forward under the wire, belly down, to the end of the obstacle.

(e) **The Dirty Name** Mount the low log and jump to or reach the higher logs in succession, then jump or drop to the ground. Warn the Marines about the height of the final log.

(f) **The Tarzan** Mount the lower log and walk the length of it and each successive, higher log until reaching the horizontal ladder. Grasp two rungs of the ladder and swing the body into the air. Negotiate the length of the ladder by releasing one hand at a time and swing forward, grasping a more distant rung.

(3) **Blue Group.** This group is formed by the third group of six obstacles. (See fig. 3-38.)

(a) **High Step over**. Step over each log, alternating the lead foot or using the same lead foot.

(b) **Swinger.** Climb onto the swinging log and over to the ground on the opposite side.

(c) **Low Wire.** Move under the wire on the back, using the hands to raise the wire to clear the body.

(d) **Swing,** Stop and Jump. Gain momentum with a short run, grasp the rope, and swing the body forward to the top of the wall. Release the rope while standing on the wall and jump to the ground.

(e) **Six Vaults.** Vault over the logs, using one or both hands.

(f) **Easy Balancer.** Walk up one inclined log and down the one on the other side to the ground.

(4) **Black Group.** The last group is formed by the final six obstacles. (See fig. 3- 39.)

(a) **Incline Wall.** Approach the underside of the wall, jump up and grasp the top and pull the body up and over. Slide or jump down the incline to the ground.

(b) **Skyscraper**. Jump or climb to the first floor, climb up the corner posts or assist each other to any desired floor. Descend to the ground in any desired manner.

(c) **Jump and Land**. Climb up the ladder to the platform and jump to the ground.

(d) **Confidence Climb**. Climb the inclined ladder to the vertical ladder. Go to the top of the vertical ladder, then down the other side to the ground.

(e) **Belly Robber.** Step on the lower log and assume the prone position on the horizontal logs. Crawl over the logs to the opposite end of the obstacle.

(f) **The Tough One** Climb the rope or pole on the higher end of the obstacle, then go down the ladder and across the log platform. Climb over or between the logs at the end and go down the rope or pole to the ground. Vault over the final log.

Figure 3-36 is available in the illustration section at the end of the book.

Figure 3-37 is available in the illustration section at the end of the book.

Figure 3-38 is available in the illustration section at the end of the book.

Figure 3-39 is available in the illustration section at the end of the book.

Section IX Individual Exercise Programs

3901.GENERAL

Marines at times will be stationed on independent duty and consequently will be responsible for their own physical fitness program. This section will assist them in understanding the need for exercise and will aid in the planning and execution of an individual exercise program. Exercise activities included are the Bench Conditioner, 6-12 Plan, Weight Training, and Isometric Contraction. Keeping physically fit is a problem that faces every Marine. Even though we are frequently engaged in training that requires some physical effort, in many cases, it is not enough to prepare us to meet the intense physical demands of combat. Attaining a satisfactory level of physical readiness is not an insurmountable objective. Available time appears to be the most difficult obstacle to the development of physical readiness. In most cases, regular physical training programs are centralized, requiring the individual to temporarily leave the work area. The problems involved in setting an hour aside two or three times each are numerous. However, most of us can devote a few minutes each day to physical fitness with little, if any, impact on our daily work schedule, especially if it does not require us to leave our work area.

a. Type of Program. There are many good physical fitness programs available to the individual or group. Regardless of the type or duration, to be effective, the program must contain exercises that are strenuous and are challenging to the individual. Space will not permit the inclusion of all available means of individual exercise. The programs selected for this chapter have met the requirement of minimum space and minimum time.

b. Need to Augment Program. These programs are quite strenuous and will develop a satisfactory level of physical readiness. However, if the Marine desires additional development of endurance, it is recommended that he supplement these programs with a 15- to 30-minute period of wind sprints and double timing on an alternating daily basis.

c. Progressive Training. If Marines are performing duties which require little or no physical activity, they must plan a physical conditioning program that assures a moderate beginning, moderate but steady progression, and sufficient warmup before starting the vigorous exercise. To avoid organic of bodily harm, a Marine activity without adequate warmup. He should conduct conditioning programs on a daily basis over an extended period of time, never on an unduly accelerated or crash basis.

3902.THE BENCH CONDITIONER

a. Description and Objective. The bench conditioning program uses a modified bench to employ both isotonic (moving) and isometric (stationary) exercises are as the nucleus of the program. The exercises are designed to develop strength and endurance in all the major muscle groups of the body. The principles of progression, overload, and balance are employed when the exercises are performed properly.

b. Area and Equipment. The conditioning apparatus can be constructed in any unit motor pool with welding equipment found in most salvage yards.(See figs. 3-40, 3-41.)Additionally, there are available any number of commercially produced apparatus available. The important thing is that in utilizing this equipment the following exercise routine be adhered to.

c. Starting Level and Progression.. The program consists of two tables, each with 10 exercises. The Marine controls progression by required repetitions or, in some cases, by application of maximum effort. Each table can be completed within 15 minutes.

d. Starting Position. To start the program, the Marine begins with the Table 1 and executes each exercise for the required number of repetitions as indicated. The Marine controls the starting level and progression. When the Marine executes the maximum repetitions for Table I within a 15-minute exercise period, he progresses to Table II. To maintain this level of development, the Marine should also execute the maximum repetitions for Table II within a 15-minute exercise period. The Marine should keep substitution of exercises to a minimum. However, it he completes a full 15 minutes of strenuous exercise and exercises all muscles, then there should be no appreciable difference in the overall development.

e. Bench Conditioning, Table I

(1) Exercise 1: Side-Straddle Hop. This is a two-count warmup exercise done at moderate cadence. The starting position is the position of attention. On count ONE jump slightly into the air, swinging the arms out to the sides and up to a vertical position, hands touching.(See fig. 3-40, A.)At the same time, spread the feet wider than shoulder-width apart. On count TWO, using a slight flexing of the knees and ankles, jump slightly into the air and return to the starting position by swinging the arms back down to the sides. Twenty repetitions of this exercise is the standard dosage throughout the program.

(2) Exercise 2: Hand Walk. Remove the lower horizontal bar. Adjust the upper horizontal bar so that it is high enough to permit a "dead hanging" position, with the feet off the ground.(See fig. 3-40, B.j From the "dead hanging" position, release one hand and drop the arm to the side of the body. Then raise that arm and regrasp the horizontal bar. Release the bar with the other hand and drop that arm to the side. Repeat this as many times as possible.

(3) Exercise 3: Situps. Lie down with the fingers interlocked and placed behind the head. Hook the toes under the foot braces. Raise the trunk and upper body to an upright sitting position, twisting it to the left and then forward and downward until the right elbow touches the left knee.(See fig. 3-40, C.)Lower the body to the starting position .Sit up again but twist the body to the opposite direction as before, touching the left elbow to the right knee. Again lower the body to the starting position. The starting dosage is 20 situps. The Marine should continue the progression until he has attained 40 situps.

Figure 3-40 is available in the illustration section at the end of the book.

(4) Exercise 4: Double Step-upStarting at one end of the bench, step up onto the bench, and walk across it. Step down from the other end; turn around and repeat the process to return to the starting point.(See fig. 3-40, D.)Each return to the starting point constitutes a repetition. The starting dosage is 20 repetitions. Maximum dosage is 35 repetitions. This exercise should be done at a rapid cadence.

(5) Exercise 5: Isometric Bar Lift. Adjust the lower bar so that it is slightly higher than the beltline. Placing the feet on the footplates at the base of the frame, grasp the lower bar so that the hands are spread shoulder-width apart. Assume a crouched position and lift with maximum effort using the arms, back and legs.(See fig. 3-40, E.)Starting dosage is 4 repetitions of a stress time of 5 seconds followed by a 5-second rest prior to the next repetition. The Marine obtains progression by lengthening stress periods to 6 and later 7 seconds. Do not increase the number of repetitions.

(6) Exercise 6: Knee Lift. Adjust the upper bar to the same height used in Exercise 2.Adjust the lower bar so that is stops rearward movement of the hips when the "dead hanging" position is assumed.(See fig. 3-40, F.) Keeping the arms extended, flex the legs and raise the knees as high as possible. Hold this position for 5 seconds, then return to the starting position. After 2 seconds in the starting position, raise the knees again. Each return to the starting position constitutes one repetition. The dosage is five repetitions. Progression is the same as in Exercise 5.

(7) Exercise 7: Isometric Pull. Adjust the lower horizontal bar to a position where it is slightly higher than the beltline. Grasp the handles and pull outward.(See fig. 3-40, G.)Apply maximum effort and hold for approximately 5 seconds. Relax for 5 seconds between repetitions; perform four repetitions. Moving the body closer to or farther away from the bar will change the stress from the upper arms to the forearms. Progression is the same as in Exercise 5.

(8) Exercise 8: Isometric Compression. Maintain the position as in exercise 7.(See fig. 3-40, H.)Grasping the handles in the same manner, press in **with** maximum effort and hold for approximately 5 seconds. Relax for 5 seconds between repetitions. Progression is the same as in Exercise 5.

(9) Exercise 9: Isometric Press. Remove the lower horizontal bar. Adjust the upper horizontal bar until it is about 6 inches lower than the extended arms can reach. Stepping on the footplates at the bottom of the frame, grasp the bar with both hands and push up.(See fig. 3-40,1.)Keep both the legs and arms slightly flexed and the back straight. Apply maximum effort for 5 seconds then relax for 5 seconds. Complete four repetitions. Progression is the same as in Exercise 5.

(10) Exercise 10: Pushups. Grasping the foot braces with both hands, assume the front leaning rest position.(See fig. 3-40, J.)Keeping the back and legs straight, lower the body until the chest is lower than the **hands,** then return to the starting position. The Marine should complete the maximum possible number of repetitions.

f. Bench Conditioning, Table II. There is no limit on the maximum number of repetitions attainable in Exercises 3, 4, and 6 of Table II. The only limit imposed is that the entire program of 10 exercises outlinted in either table should not exceed 15 minutes.

(1) Exercise 1: Side-Straddle Hop. This is a two-count warm up exercise done at a moderate cadence. The starting position is the position of attention. On count ONE, jump slightly into the air, swinging the arms out to the sides and up to a vertical position, hands touching.(See fig. 3-41, A.)At the same time, spread the feet wider than shoulder-width apart. On count TWO, using a slight flexing of the knees and ankles, jump slightly into the air and return to the starting position by swinging the arms back down to the sides. Twenty repetitions of this exercise is the standard dosage throughout the program.

(2) Exercise 2: Pull-up. Adjust the horizontal bar so that is is high enough to permit a "dead hanging" position with the feet off the ground. Grasp the bar with both hands, palms facing forward. By flexing the arms, raise the body to a position where the chin is higher than the bar.(See fig. 3-41, B.)Then lower the body to the "dead hanging" position. Repeat as many times as possible.

(3) Exercise 3: Bench situps. Sit on the bench and hook the feet under the foot braces. With the fingers interlocked behind the head, lean back until the head touches the floor.(See fig. 3-41, C.)Return to the starting position. The starting dosage is 15 situps.

(4) Exercise 4: Step-Up. Face the bench and step up on it with one foot, bringing the trailing foot up next to the leading foot. Step back down again, leading with the same foot used first in stepping up.(See fig. 3-41, D.)Perlorm half of the total repetitions, then change the sequence of moving the feet to use the other leg in stepping up, and repeat the same amount of exercise. The starting level is a total of 40 step-up. **This exercise should be done at a rapid cadence.**

(5) Exercise 5: Isometric Bar Lift\djust the lower bar so that it is slightly higher than the beltline. Placing the feet on the footplates at the base of the frame, grasp the lower bar so that the hands are spread shoulder-width apart. Assume a crouched position and lift with maximum effort using the arms, back, and legs.(See fig. 3-41, E.)Starting dosage is four repetitions of a stress time of 8 seconds followed by a 5-second rest prior to the next repetition. Progression is obtained by lengthening stress periods to 10 seconds. Do not increase the number of repetitions.

Figure 3-41 is available in the illustration section at the end of the book.

(6)Exercise 6: Leg Lift. Adjust the bars and assume the starting position as shown in figure 3-41, F. Keeping arms and legs extended, raise the legs to a horizontal position and hold in that position for 2 seconds. Then lower the legs slowly to the starting position. Five repetitions is the starting level.

(7) Exercise 7: Isometric Pull. Adjust the lower horizontal bar so that it is slightly higher than the beltline. Grasp the handles and pull outward.(See fig. 3-41, G.)Apply maximum effort and hold for approximately 8 seconds. Relax for 5 seconds between repetitions; perform four repetitions. Moving the body closer to or farther away from the bar will change the stress from the upper arms to the forearms. The Marine obtains progression by lengthening the stress period to 10 seconds.

(8) Exercise 8: Isometric Compression. (See fig. 3-41, H.)Maintain the position as in exercise 7.Grasping the handles in the same manner, press in with maximum effort and hold for approximately 8 seconds. Relax for 5 seconds between repetitions; perform four repetitions. The Marine obtains progression by lengthening the stress period to 10 seconds.

(9) Exercise 9: Isometric Pres. Remove the lower horizontal bar. Adjust the upper horizontal bar until it is about 6 inches lower than the extended arms can reach. Stepping on the footplates at the bottom of the frame, grasp the bar with both hands and push up.(See fig. 3-41,1.)Keep both the legs and arms slightly flexed and the back straight. Apply maximum effort for 8 seconds, then relax for 5 seconds. Complete four repetitions. The Marine obtains progression by lengthening the stress period to 10 seconds.

(10) Exercise 10: Inclined Pushup. Assume the front leaning rest position with the feet on the bench.(See fig. 3-41, J.)Keeping the back and legs straight, lower the body until the nose touches the ground. By extending the arms, raise the body to the starting position. Repeat as many times as possible.

3903. THE 6-12 PLAN

a. Description and objective. The 6-12 plan of physical fitness has been developed to assist in regulating quantity and progression and to provide a convenient set of exercises. This is a basic program and will take 18 weeks to complete if you follow the moderate progression as to the time prescribed for each level of achievement. The time can be shortened as explained below. This plan consists of six basic exercises a day which can be completed in 12 minutes. There are six tables of six exercises each, thus allowing you to progress from table to table. The plan is progressive, fits any age group, contains balance and variety, and applies the principle of overload in a safe and gradual manner. Begin at Table I, Progression Guide, with the number of repetitions as indicated by age.

b. Formation. If just starting an exercise program, do not rush through the first table. Remember, individuals should remain at each level for about a week before moving upward. The time allotment stated for each exercise at the bottom of the tables is a guide; some people may take more and some less time on the individual exercises. At the end of a one-week period, when the individual can comfortable perform the six exercises in 12 minutes, move on to the next level. To a certain degree, the individual must be the judge of his ability to progress form level to level and table to table. If starting with a certain degree of fitness, some of the beginning tables may present little challenge.

c. Starting Level **and Progression.** There are three levels of achievement for each age group, indicated as A, B, and C. Start at the C level for the appropriate age group. At the end of a one-week period, or when the individual can do all exercises at that level within 12 minutes, progress to the B level. At the end of the second week, or when the individual can accomplish that level within 12 minutes, progress to the A level. At the conclusion of the third week or when the individual can achieve the A level within the time limitation, move on to table II.(See Tables I through VI, Progression Guide.)

d. Maintenance Level. Attempt to work through all six tables. If this proves to be too difficult, then maintain exercise at the—

(1) A-level on Table IV, Progression Guide it in the 45 to 49, 50 to 59, or over 60 age group.
(2) A-level on Table V, Progression Guide if in the 17 to 29, 30 to 39, or 40 to 44 age group.

e. Precautions. To achieve the maximum benefit, perform each exercise exactly as specified. Read the descriptions and study the illustrations. Do not slight the movements. Use a sensible approach and follow these four points as they apply before starting or during your exercise program.

(1) If you have the slightest doubt about your ability to participate in this exercise program, consult a physician.

(2) Stop immediately if you notice unusual breathlessness or chest pain while taking part in these exercises. If these conditions persist, consult a physician.

The following Tables of Progression Guides and Exercises are the progression of the 6-12 Plan program.

Table I. Progression Guide is available in the illustration section at the end of the book.

Table I. Exercises.is available in the illustration section at the end of the book.

Table II. Progression Guide.is available in the illustration section at the end of the book.

Table II. Exercises.is available in the illustration section at the end of the book.

Table III. Progression Guide.is available in the illustration section at the end of the book.

Table III. Exercises.is available in the illustration section at the end of the book.

Table IV. Progression Guide.is available in the illustration section at the end of the book.

Table IV. Exercises.is available in the illustration section at the end of the book.

Table V. Progression Guide.is available in the illustration section at the end of the book.

Table V. Exercises.is available in the illustration section at the end of the book.

Table VI. Progression Guide.is available in the illustration section at the end of the book.

Table VI. Exercises.is available in the illustration section at the end of the book.

(3) Unless you have exercised regularly and know yourself to be in good physical condition, start at table I with the C level appropriate to your age.

(4) If you are out of shape, admit that fact to yourself. Hide your pride; after all, you are in the privacy of your own quarters. Set your goal for the longer, steadier pull toward fitness. Resist the urge to pass over the lower numbered tables to find a table that will test your fitness. You are not trying to test, but rather to develop.

3904. WEIGHT TRAINING

Weight or barbell training should not be confused with the more common types of weight lifting used as a competitive sport. Weight lifting is designed to develop specific muscle groups so that the individual is capable of lifting a large amount during a single lift. In contrast, weight training is the systematic development of all the major muscle groups by the use of calisthenics reinforced with weight to provide resistance.

3905. WEIGHT LIFTING

a. Description and Objective. The weight lifting program is progressive and applies the principle of overload in a safe, gradual manner. The exercises of the table can be completed in 15 minutes. The objective of the exercises is to develop strength and muscular endurance, and muscle tone of the five major muscle groups:legs, arms, back, trunk, and shoulder girdle.(See fig. 3-42.)

b. Warmup. A warm up exercise is important to prepare the body for the more vigorous exercises that are to follow. Ten repetitions of the high jumper exercise are excellent for a warm up period. Also, an exercise which will require fast body movement is needed.To provide such exercise, 3 to 5 minutes of rope skipping is recommended to increase the individual's development potential.

c. Formation. Marines should take care in the completion of these weight lifting exercises. They should ensure that their back is straight during the lifting phase of all exercises. When exercises require assuming the standing position with the weight, they should always grasp the weight while in a squatting position and then rise to a standing position.

d. Progression. Each exercise has a starting number of repetitions and specified pounds of weight. After each 4th or 5th day of exercise, the Marine should increase the repetitions by one until he has reached the maximum of 10 repetitions. At this time, the Marine should increase the weight by 5 pounds. Then he should repeat the process again with the initial number or repetitions.

e. Weight Training

(1) Exercise 1:Squat(See fig. 3-42, A.)Starting level—six repetitions, 50 pounds (commonly called the flatfoot deep knee bend). Place the bar upon the shoulders. Stand with feet about 18 inches apart. Keeping the feet flat, lower the body into the low squat position. Stand erect and repeat. Exhale while lowering into the squat position and inhale while standing. This constitutes one repetition.

(2) Exercise 2:Waist Bender(.See fig. 3-42, B.)Starting level-6 repetitions, 40 pounds. Assume the standing position with the bar across the shoulders and the feet a shoulder-width apart .Bend forward at the waist until the upper body is parallel to the ground; return to the starting position. Each return to the upright position constitutes one repetition.

(3) Exercise 3:Curl (See fig. 3-42, C.)Starting level—6 repetitions, 40 pounds. Grasp the barbell with the palms facing to the rear and assume the standing position with the feet a should-width apart. With the barbell held in front of the hips, flex the elbows and lift the weight until the bar touches the upper chest. Lower the barbell back to the hip level position. Inhale deeply with the upward movement and exhale on the downward movement. Each time that the bar touches the chest will constitute one repetition.

(4) Exercise 4:Side Bender(.See fig. 3-42, D.)Starting level—6 repetitions per side, 40 pounds. Assume the standing position, with the bar across the shoulders, with feet a shoulder-width apart. Bend to the left as far as possible and return to the starting position. Repeat six times and then execute the same procedure to the right for six repetitions.

(5) Exercise 5:Standing Press(See fig. 3-42, E.)Starting level—6 repetitions, 45 pounds. Grasp the bar with the palms facing forward and assume the starting position. Curl the weight to the upper chest position. Inhale deeply and press the bar upward to an overhead position. Exhale while lowering the bar to the chest position. Each time that the bar is pressed upward constitutes one repetition.

(6) Exercise 6:Upward Row. .see fig. 3-42, F.)Starting level—6 repetitions, 40 pounds. Grasp the bar, hands close together, palms to the rear, and assume the standing position. Starting with the bar held in front of the hips, flexing the elbows and the shoulder girdle muscles, lift the bar straight up to an overhead position. Inhale deeply while lifting the bar. Exhale while lowering the bar to the hip position. Each time that the bar returns to the hips constitutes one repetition.

(7) Exercise 7:Shoulder Curl. See fig. 3-42, G.)Starting level—6 repetitions, 25 pounds. Grasp the bar, palms down, and assume the standing position. Keeping the elbows locked, curl the bar, pivoting the arms at the shoulders until the bar is in an overhead position and as far to the rear as possible. Return the bar in the same manner to the hip position. Each time that the bar returns to the hip position constitutes one repetition.

Figure 3-42 is available in the illustration section at the end of the book.

3906.ISOMETRIC EXERCISING

a. Description and Objective. Isometric exercising is the application of maximum effort during an exercise period. The isometric principle is to apply force gradually over a 5- to 10- second period until the maximum effort is applied. Relaxation follows for approximately 5 seconds and then force is applied again. The Marine continues this process at the prescribed level for each exercise for a period of 15 minutes or less. The objective of isometric exercises is to create muscle growth. Isometric exercises are the fastest means of creating muscle growth. They are founded on the fact that a muscle will grow only so fast regardless of the type or duration of the activity. The principle of overload—that the muscles develop commensurate with demand—reinforces that fact. However, isometrics will not develop cardiovascular or muscular endurance. Consequently, Marines who choose to use isometrics in their exercise routine must also include running or some other type of aerobic activity.

b. Area and Equipment isometric exercises may be designed to be performed with or without equipment.

c. Use With Other Programs. The isometric exercises presented in this section will provide a variety from which to choose. In addition to the exercises contained in this section, the application of isometric force is used in some of the exercises in section II.

d.Door Frame Exercises. The following exercises are designed for use with a standard door frame found in all offices or barracks.(See fig. 3-43.)

(1) Exercise l:Arm Press. Stand in the doorway with the legs straight, knees locked. Using the arm muscles, press hard upward against the top of the door frame. Repeat for three repetitions applying gradual effort to maximum contraction.

(2) Exercise 2:Leg Press. Stand in the doorway with the hands on the top of the door frame, elbows locked .With knees bent, press hard with the leg muscles .Repeat for three repetitions beginning with a gradual effort and increasing to maximum contraction. A low platform may be necessary to reach the top of the door frame and still maintain a bent knee position.

(3) Exercise 3:Side Press. Extend both arms to the side of the doorway. Palms are shoulder high, facing outward. With both arms, press hard against the sides of the door frame. Repeat for three repetitions. Begin gradually and increase to maximum contraction.

(4) Exercise 4:Lateral Raise. Extend both arms to the sides of the doorway, arms down, palms facing inward. With the back of the hands, press hard against the sides of the door frame .Repeat for three repetitions. Begin with a gradual effort and increase to maximum contraction.

Figure 3-43 is available in the illustration section at the end of the book.

(5) Exercise 5:Neck Press?lace the forehead against the door frame, hands clasped behind the back Using the neck muscles, press hard against the door frame. Repeat for three repetitions, then reverse position so that the back of the head is resting on the door frame. Again do three repetitions .Begin gradually with both exercises and increase to maximum contraction.

(6) Exercise 6:Door Pull Stand facing the edge of the open door and grasp the doorknobs. Pull outward with both arms (if doorknobs are not available grasp the edge of the door).Applying outward pressure, move the body toward and away from the door .Repeat the three repetitions. Begin with gradual effort and increase to maximum contraction.

Chapter 4
COMBAT WATER SURVIVAL
Section I. Marine Corps Water Survival Program

4101.MILITARY SWIMMING

Military swimming emphasizes strokes that result in staying power rather than speed. Marines need to swim easily, quietly, and with adequate vision. Strokes should be used that allow the Marine to carry basic combat equipment and to tow or push a wounded buddy while keeping the face out of the water to allow breathing. The best strokes to use are the side stroke and the breast stroke. Marines must be proficient in floating, drown-proofing, and in using their basic combat equipment to make flotation devices. They must be drilled on using the Kapok and Mae West buoyancy compensators and on emergency egress from helicopters, landing craft, and amphibious assault vehicles.

4102.PROGRAM DEVELOPMENT

Commanders of the Landing Force Training Commands (Atlantic/Pacific) are tasked with developing and maintaining the individual water survival and swimming training program. They are responsible for training all Marine Corps water safety and survival instructors and integrating this training in to their combat readiness training program. All Marines are deployable and as soldiers of the sea must constantly practice the skills necessary to achieve mission accomplishment. Commanders must be imaginative in program development, integrating basic skills development which enhances the total mission accomplishment. An example is teaching waterproofing of the Marine's backpack so that it can be used as a flotation device in streams or rivers. Once learned, this skill would be practiced first in a pool and then in a stream or a river. Routine inspections during deployments would ensure that backpacks could be used as flotation devices.

4103.QUAL STDS AND TRAINING GUIDELINES

The inherent nature of Marine Corps operations and training requires that Marines achieve an ability to survive in water. Water survival and swimming training is designed to reduce fear of water, instill self-confidence, and develop a Marine's ability to survive in water. As a minimum, every Marine should be qualified as swimmer, third-class (S3).This qualification should be met during recruit training or Officer Candidate School. Once qualified, a Marine need not be requalified unless requalification at a specific level (S3 or higher) is required by unit mission, military occupational specialty or duty assignment, or for other reasons determined by the commander. The following classifications and standards apply to Marine Corps water survival and swimming training. Abbreviations will be used for service, medical, and training record entries citing MCO 1510.29A, Individual Water Survival and Swimming Training, as authority.

a. Qualification standards and test procedures are described in paragraph 4104.

b. Swimmer classification and abbreviations are:

(1) Unqualified - UQ.

(2) Swimmer, third-class (S3).Minimum aquatic skill level for all Marines.

(3) Swimmer, second-class (S2)-

(a) Minimum standard for all naval aviators, naval flight officers, and aircrewmen with written waiver from the commanding officer.

(b) Requirements for military occupational specialty 1803 (assault amphibian vehicle officer), 1833 (assault amphibian vehicle crewman), 1302 (engineer officer), 1371 (combat engineer), 1381 (shore party specialist), and 1379 (engineer operations chief).

(4) Swimmer, first-class (S1)-

(a) Naval aviator, naval flight officer, and aircrewman qualification requirement.

(b) Requirements for military occupational specialty 0321 (reconnaissance man), 8652 (reconnaissance man, parachute jump qualified), 8653 (reconnaissance man, self-contained underwater breathing apparatus qualified), 8654 (reconnaissance man, parachute/self-contained underwater breathing apparatus qualified), 9952 (self-contained underwater breathing apparatus Marine (officer/enlisted)), and 9953 (parachutist/self-contained underwater breathing apparatus Marine (officer/enlisted)).

(5) Water Survival Qualified (WSQ).Ultimate water survival/swimming goal for Marines.

c.Instructor and assistant instructor classifications are-

(1) **American Red Cross Advanced Lifesaver.** Individuals who are currently certified as American Red Cross advanced lifesavers may assist qualified instructors during Marine Corps water survival/swimming training and testing. Certification is valid for 3 years.

(2) American Red Cross Water Safety Instruction. Individuals who are currently certified as American Red Cross water safety instructors may conduct swimming training and qualify individuals as 3d, 2d, and 1st class swimmers. Certification is valid for 2 years.

(3) United States Marine Corps Water Safety and Survival Instruction. Those personnel certified as water safety and survival instructors by one of the landing force training commands may conduct Marine Corps water survival and swimming training and American Red Cross swimming and water safety training. These instructors may also participate as assistant instructors during the conduct of landing force training command-sponsored water safety and survival instructor courses. Certification is valid for2 years.

(4) American Red Cross Water Safety Instructor Traine. Individuals qualified who have been certified as instructor trainers by the American Red Cross may train American Red Cross water safety and basic swimming instructors. Certification is valid for 2 years.

(5) United States Marine Corps Water Safety and Survival Instructor Traine. Individuals qualified who are certified as instructor trainers through special training courses conducted jointly by the American Red Cross and the Marine Corps, under the direction of the Commandant of the Marine Corps, may train water safety and survival instructors and American Red Cross water safety and basic swimming instructors. Certification is valid for 2 years.

d. Training Guidelines Instructors engaged in training Marines as swimmers will use the current edition of the American Red Cross swimming and water safety courses instructors' manuals as the authoritative textbooks for swimming procedures, strokes, breaks, and holds.

4104.WATER SURV/SWIM QUAL STD AND TEST PROCS

a. The following qualification standards and test procedures are applicable to Marine Corps water survival and swimming training. The utility uniform without cover, boots/oxfords or socks will be worn for all swimming tests through swimmer, first class. The utility uniform with boots/oxfords will be worn as prescribed for the water survival qualification.

(1) Swimmer, Third-Class (S3) To qualify as a swimmer, third-class, a Marine must enter the water feet first from a minimum height of 5 feet and remain afloat for 5 minutes. During this time the Marine must swim 50 yards using any stroke or combination of strokes. This test should be taken by all Marines as early as possible in their initial training. Those unable to pass this test will be classified as UQ and should be given instruction in fundamental swimming skills. Those who barely meet requirements are swimmers who need help and should receive additional instruction in fundamental swimming skills, whenever possible.

(2) Swimmer, Second-Class (S2.) To qualify, a Marine must enter the water feet first from a minimum height of 10 feet and remain afloat for 10 minutes. The Marine must swim 100 yards and use the three basic survival swimming strokes (side stroke, backstroke, and breast stroke) for a minimum distance of 25 yards each.

(3) Swimmer, First-Class (S1)A prerequisite to qualification as swimmer, first-class, is successful completion of the test for swimmer, second-class. To qualify as a swimmer, first-class, a Marine must be able to do each of the following:

(a) Approach a person of approximately the same size while in the water, demonstrate a release, get that person in a carry position, and tow the "distressed swimmer" 25 yards.

(b) Enter water feet first and immediately swim underwater for 25 yards. Swimmer is to break the surface twice for breathing during this distance at intervals of approximately 25 feet.

(c) Remove trousers in water, inflate for support, and remain motionless for a minimum of 1 minute.

(d) Swim 220 yards using any survival swimming stroke or any combination of survival strokes.

(4) Water Survival Qualification (WSQ)A prerequisite to water survival qualification is successful completion of the test for swimmer, first-class. To be water survival-qualified, a Mannemust:

(a) Enter water by jumping from a height of 10 feet.

(b) Stay afloat in full utility uniform (less cover) for 1 hour. Boots/oxfords will be removed after 5 minutes in the water but will be retained by the swimmer.

(c) Traverse 75 yards in deep water in full utility uniform (less cover) with boots/oxfords.

b. All water survival and swimming training will include instruction in the following:

(1) Employment of safety lines or other comparable expedients to assist in swift stream/river crossings.

(2) Adverse physiological effects caused by cold water (hypothermia) and the precautionary measures to be taken prior to exposure to such an environment.

(3) Employment of a standard and expedient flotation devices in a water survival situation.

(4) Removal of individual field equipment after unscheduled water entry. Entry will be made from a minimum height of 10 feet with a complete set of field equipment, properly packed as an expedient flotation device and waterproofed. The ALICE pack will then be removed, the rifle laid across the top of the pack, the helmet remaining or put back on the individual's head, and the individual will traverse 25 yards to the side of the pool or water's edge.

c. The additional water survival qualification (flight physiology training) and requalification required of personnel assigned to flight status will be conducted in accordance with OPNAVINST 3710.7_,Promulgation of NATOPS General Flight and Operating Instructions. Swimming qualification (first-class, SI) is a prerequisite to water survival (flight physiology) training and is a one-time requirement.

Section II. Water Sports and Competitive Activities

4201.GENERAL

The commander, when developing the combat readiness program, needs to look carefully at the mission. If the mission requires Marines who must perform mission-essential water-related activities of long duration, he may want to look to water sports/competitive activities as a method of developing endurance. Endurance and confidence can only be developed by extended time in the water. While defined training can accomplish this goal, the goal can be enhanced through training which is integrated with the physical conditioning program. This approach will provide variety and enthusiasm. Two activities which will accomplish this goal are water relays and water polo. The important consideration is that the unit achieve total participation by all members of the unit.

4202.WATER RELAYS

Development of water relays will generally conform to the guidelines in chapter 5, section IV. When contemplating relays, the commander should review chapter 5 and adapt the principles to the size of the unit and the facility to be used.

4203.WATER POLO

The official rules provided in this paragraph will need to be adapted to the unit's unique situation, its size, and the size of the facility. Commanders should make every effort to have all members of the unit participate.

a. NCAA Rules This style of play is based on two popular sports: swimming and basketball. Playing is done on the surface of the water by teams of seven players each. The size of the playing area is 25 yards by 14 yards, which is the size of the typical high school or college indoor pool. The deeper the water, the better. If the entire playing area is deep-6 feet or more—this is ideal, although most indoor pools have a shallow end. At each end of the pool is a goal. In deep water, the goal is 10 feet across and 3 feet high. In shallow water, where the goalie can stand on and jump from the bottom, the goal is 10 feet across and 5 feet high. One of the seven players on each team is the goalie. He is the only player on the team who can stand on or jump from the bottom (if it is shallow enough to permit this), catch and pass the ball with both hands at the same time, or hit the ball with a clenched fist. However, the goalie may not swim across or pass the ball across the middle of the pool.

b.Equipment. The goalie on the visiting team must wear a white cap with the number 1 clearly marked on it; his teammates, who are designated as guards and forwards, must wear white caps numbered 2 through 7, with the substitutes wearing white caps numbered 8 through 21.The goalie on the home team must wear a dark-colored cap with the number 22 clearly marked on it; his teammates designated as guards and forwards, must wear similar dark-colored caps numbered 29 through 42.The guards and forwards on each team may swim freely up and down the pool as they see fit, interchanging positions as often as they wish. They may not stand on or jump from the bottom, touch the ball with more than one hand at a time when catching or passing or shooting it, or enter inside the opposing team's 2-yard line unless preceded by the ball. The ball is similar to a soccer ball, except that it is yellow and covered with a waterproofing substance that makes it easier to handle with one hand.

c. Playing Time. In intercollegiate competition the game consists of four 7-minule quarters, the teams changing ends after every period of play. In high school competition the game consists of four-,five-, or six-minute quarters, depending on the ages of the participants. Here again the teams change ends after every period. There is a very brief interval between each quarter. At the start of each period the teams line up at their respective ends, and with ablast of the whistle, the head referee throws the ball into the middle of the pool. The fastest swimmers on each team then race to gain possession.

d. Officials. The head referee walks along one side of the pool watching for infractions. He is aided by the assistant referee, who patrols the opposite side of the pool. The referees are both equipped with whistles and two official's flags. These flags should be 12 inches square and should correspond in color to the caps worn by the respective teams. Whenever a referee sees an infraction, he blows his whistle and signals with his flag indicating that the player nearest to the ball with the appropriate colored flag will be given possession of the ball. Whenever the whistle is blown, all players should react by looking immediately to see which color flag is being waived which will indicate which team should be moving into the defense. Other officials are a timer and a scorer. As in all sports, it is important that the officials be skilled and competent.

e. Technical Fouls. The following are some of the common technical fouls: starting before the referee blows his whistle to open play; holding onto or pushing off from the side of the pool during play; taking hold of the ball underwater when tackled by an opposing player; swimming inside an opponent's 2-yard line unless preceded by the ball; touching the ball with both hands at the same time (goalie excepted); standing, walking on or jumping from the bottom when taking an active part in the game. When a technical foul occurs, thereferee blows the whistle and with his flag awards possession. The player on the team awarded possession nearest the point of infraction then has 5 seconds to put the ball into play by passing to a teammate or by dropping the ball into the water and swimming after it.

f. Personal Fouls. The following are some of the more common personal fouls: committing any of the technical fouls mentioned above for the purpose of scoring or preventing a goal; holding, ducking, pulling, pushing off from, swimming over, or impeding the arm or leg movement of any opponent who is not touching the ball; and splashing water in the face of an opponent. When a personal foul occurs, the referee will follow the same procedure as with a technical foul. While doing this, the referee will also call out clearly and loudly the number of the player who was guilty of the infraction and a personal foul will be marked against that player by the official scorer. When a player accumulates five personal fouls, he fouls out of the game and must be replaced by a substitute.

g. Penalty Shot. A penalty shot can be awarded by either of the officials or the scorer when:

o An offensive player inside the opponents' 4-yard line not touching the ball is held, ducked, pulled-back, kicked, or struck.

o A team has accumulated a total of ten personal fouls. When the former occurs, the head or assistant referee should immediately blow the whistle and by holding a two-flagged stick in a vertical position above his head, signal that a penalty shot has been awarded. When the latter occurs, the scorer should use a buzzer located at the scorer's tabic to signal that one team has accumulated ten personal fouls, thereby entitling the other team to a penalty shot. A penalty shot is taken from the 4-yard line in front of the goal. All players except the defending goalie must leave the 4-yard line until the shot is taken. No player can be within 1 yard of the shooter. After ascertaining that the shooting player is on the 4-yard line and the goalie is onThegoal line, the referee will ask the shooter to lift up the ball. When he docs, the referee will give a sharp blast from the whistle. At the whistle, the shooter must shoot without delay and without any faking at the goal. The goalie may try to block the shot. If he does, or the shot is otherwise missed, it is immediately in play and action continues.

h. When a Goal is Scored. When a goal is scored, either from a shot taken by a player out in the field or by penalty shot, the defending goalie must then pass the ball to the nearest referee. The referee will then pass the ball back to the goalie, who puts it into play by a pass to one of his teammates as soon as possible.

Section III. The Battle Swimming Test

4301.GENERAL

The battle fitness test is the evaluation contained in the Guidance for Basic Warrior Training (BWT) as part of the Basic Warrior Training Concept Plan.

4302.THE BATTLE SWIMMING TEST

The battle swim test consist of the following:
a. Be able to climb a three meter high board, while wearing a field uniform, carrying a M-16, web gear, and four canteens. Jump off blindfolded, recover and swim 100 meters with the M-16.
b. Repeat the same process, without the blindfold, and upon entering the water a second time, recover and retrieve all submerged gear.
c. Remain afloat and tread water for five minutes.
d. Demonstrate a knowledge of how to cross a stream using field expedient techniques and a single strand rope bridge.
e. Successfully complete helicopter egress training.

Chapter 5
COMPETITIVE CONDITIONING ACTIVITIES
Section I. Organization of Competitive Activities

5101.LEADERSHIP OF COMPETITIVE ACTIVITIES

a. Place in the Program. Competitive conditioning activities consist of dual combatives, relays, team contests, and team sports in which individuals or teams compete against an opponent to win. Competition is one of the best ways of maintaining interest in the physical training program during the sustaining stage. Organized competition provides enjoyable, vigorous physical activity that has proved to be one of the best supplements to conditioning drill activities. The benefits of competition are the development of aggressiveness, unit pride and identity, teamwork, and the will to win. These activities help develop cardiovascular and muscular endurance, strength, and coordination. Scheduling competitive activities in an orderly and progressive manner is desirable. The progression should be from relays to dual combatives, to team contests, and finally to team athletics. These activities should be conducted as part of the program after a basic level of conditioning has been developed, usually during the slow-improvement stage of conditioning. Muscles and joints should be strengthened by preconditioning to withstand the strain placed upon them by sudden stops and turns, body contact, bearing of weight, and falls. Competitive activities, however, should not be allowed to dominate the physical training program.

b. Area and Equipment. Some of the competitive activities included in this chapter require specific types of areas and equipment; others do not. The area requirement can usually be satisfied on available training fields. When items of equipment are required or specific courts or field layouts are to be marked off, such information is included.

c. Leadership. The principal factor for success as a leader of competitive activity is an energetic, dynamic, enthusiastic approach. The leader's attitude is reflected by the group, so the commander must carry on the activity in a snappy and vigorous manner. Confidence on the part of the commander will create an impression of decisiveness and certainty. Confidence grows out of experience and a thorough knowledge of the activity. Mastery of subject is the first step in developing confidence, assurance, and poise.

(1) The following suggestions are offered for leaders of competitive combatives, contests, and sports:

(a) Get the activity underway quickly by selecting and teaching only the essentials.

(b) Use rules to add to the enjoyment of the activity and do not allow them to interfere with the spirit of competition.

(c) Stop the activity before interest begins to lag.

(d) In team contests, clearly distinguish sides by attempting to maintain unit identity.

(e) Always insist on fair play, enforcing the rules impartially.

(2) The following procedure is recommended for presenting a competitive activity.

(a) Name the activity.

(b) Briefly explain the objective of the activity and give only the pertinent rules.
(c) Have a demonstration at slow speed and answer questions.
(d) Organize groups into teams and appoint captains.
(e) Arrange teams in the proper starting positions.
(f) Conduct the activity.

d. Competitive Units. Units for competition should be the same organizations in which Marines train: battalions, companies, platoons. In most situations, the unit is the squad.

e. Provisions for Instruction. One of the most effective methods of maintaining interest and participation in competitive activities is to provide instruction in activities with which most Marines are unacquainted. Such instruction can be conducted during regular physical training periods. Careful planning is required to keep all individuals continuously engaged in vigorous activity. The materials in the following sections should be used as a guide for instructional purposes.

f. Officiating and Control. Every effort should be made to provide good officials for all competitive activities. Poor officials quickly cause dissatisfaction among participants in team activities and create a situation which can turn an organized activity into a brawl. Each company should have or develop several competent and qualified officials available for games on company and platoon levels.

5102.MILITARY FIELD MEETS

When units reach the latter part of the slow-improvement and sustaining stage of training, interest in the program may lag. A change in course content can arouse the desire to participate. An event which does not require a high degree of skill yet demands strenuous activity is ideal for the military field meet. This is a series of team contests conducted on a station-to-station basis during a given period of time. Team contests carried on simultaneously provide essential training, vigorous exercise, and stimulating competition in an atmosphere that is enjoyable for all. Because of the healthy rivalry that a field meet arouses, it is an excellent form of inter-unit competition.

a. Objective and Advantages. The objective of the military field meet is to provide activity for everyone in the participating units. Activities are chosen that will develop aggressiveness, teamwork, a will to win, and competitive spirit, and that will stimulate interest and build esprit de corps. The military field meet can **be** included as part of the physical training program or as part of the off-duty recreational program. It is a form of contest that can be conducted in nearly all circumstances because it can be easily modified, requires a minimum amount of equipment, and can be readily organized.

b. Level of Competition. The military field meet is flexible. It can be adjusted to large or small groups. A company-size unit is the most desirable, but it may also be administered within a larger unit. If it is conducted within a company, the participating units will be squads. If it is held within a larger unit, platoons compose the teams. A larger-size unit requires more extensive organization and administration.

c. Selection of Events. In organizing a military field meet, select events that are simple and easy to administer. All rules and regulations should be clearly understood by everyone. No event should require previous practice. In selecting the events, consider the interest and capabilities of the Marines and available equipment and facilities. Select events which will require that all members of the units participate. Events should not be dominated by the athletes; rather, MOS-related events which have significance in producing skills and combat readiness should be used.

d.Equipment and Facilities. The site must be large enough to permit events to be grouped about a central control point. Use available facilities such as parade decks, Softball fields, or tracks. A public address system is desirable at the control point for the initial orientation of teams and for subsequent announcements of time lapses, cumulative scores, and final standings. A tally board is necessary so that all units can constantly monitor how the events are progressing.

e.Administrators. Efficient administration of the military field meet depends on the referees, judges, and scorers. For these positions, choose individuals within the units who have had athletic or officiating experience. Prior to the day of the event, all administrators should be briefed and assigned a specific task in order that they may become familiar with the rules and organization of the contests they will conduct. The following officials should be available:

(1) A primary instructor or supervisor who is in charge of the control center and who is responsible for the successful operation of the athletic carnival. The supervisor must have an assistant to act as a timer and scorer.

(2) One assistant instructor in charge of each event. This person should be-

(a) Familiar with the rules of the event.

(b) Effective as a leader to ensure proper supervision and control over the participating teams.

(c) Enthusiastic to provide proper motivation.

(d) Self-confident of the ability to judge infractions of the rules. The assistant instructor must be fair in judgment and penalize without hesitation when infractions occur.

(3) Runners between event stations and the control point. These individuals collect and deliver scores.

f. Team Organization for Competition. The size of the teams is determined by the level on which the field meet is organized. Maintaining the integrity of the unit promotes esprit de corps, but this does not preclude grouping two squads into one team.

g.Conduct of Events

(1) The assistant instructor at each station takes charge of the group and gives a brief explanation of the major rules of the event. The assistant instructor speaks clearly and distinctly from a position to be seen and heard by all. A short demonstration is desirable if it will help clarify the event.

(2) The assistant instructor should make certain that teams can be clearly distinguished by the use of, for example, T-shirts and fatigue jackets, or caps and no cap, or colored jerseys. The assistant instructor should teach a whistle response (teams stop play immediately upon hearing whistle), get the event started as quickly as possible, and make any necessary corrections as the contest progresses. The rale of good officiating is to use a minimum of calls, yet maintain control of the contest. Penalize when necessary, but refrain from disqualifying contestants or teams.

(3) Keep the activity moving as rapidly as possible. When the central control point sounds the whistle to stop the play, all competition ends immediately. The assistant instructor then assembles the group, forwards the team scores to the central point, and, upon a signal from the central control point, rotates the teams to their next station. It is essential that rotation and orientation of teams be quick and orderly to reduce time spent between contests.

(4) Upon completion of the final event, assistant instructors move their teams to the central control point for the announcement of winning teams, presentation of awards (if any), and final critique.

h. Scoring System. The system for determining the winner of the military field meet should be simple and efficient. At the completion of competition, the scorer totals the points that each team has scored in all contests. The scorer then subtracts the number of points scored against a team (penalty points for nonparticipation,etc.) from the total. The resulting scores are placed in a column with the highest score at the top and the lowest at the bottom. The team with the highest total is the winner. This type of scoring system encourages a team to ensure total unit participation.

Section II.Combatives

5201.DESCRIPTTON AND OBJECTIVE

Combatives are strenuous, short competitive contests in which two individuals attempt to overcome each other in a bout of skill and strength. These contests help to develop Marines' resourcefulness, confidence, strength, agility, coordination, and will to win. Any level ground area can be used. However, extremely hard ground should be avoided as some of the combatives require ground contact. A whistle is needed to control the bouts during the competition since voice commands may go unheeded. The extended rectangular formation is used for dual combatives. Combatives are conducted on an informal basis. Marines are allowed to remain at ease between activities and are allowed to brush themselves off after being on the ground. To get the most out of combatives, individuals must be urged to overcome their opponents as quickly as possible as would be required in combat.

a. Benefits for Marine Training. The possibility of close contact with an enemy in combat faces Marines at all times. They must be trained to react aggressively and violently. Combatives may be used as an introduction to such hand-to-hand contact and should be followed by hand-to-hand combat training. Marines enjoy competition, and this type of activity is a welcome change from formal conditioning activities.

b. Instructor Responsibility

(1) The instructor tells the individuals that all combatives begin and end on the whistle signal. The instructor demonstrates each activity before having it performed, explaining it in simple terms.

(2) After stopping one activity, the instructor gets everyone in place for the next.

(3) The instructor must closely supervise combatives to insure that contestants do not use unfair or unsportsmanlike tactics. To avoid unnecessary injury, instructors must see that the bouts are closely controlled and opponents equally paired. Adjustments should be made in apparent cases of mismatched abilities.

5202.COMBATIVES TABLES

There are three tables of combatives, each of which can be completed in 15 minutes. These tables become progressively more difficult from lower to higher numbers.

a.Combatives Table I

(1) Open-Hand Slap Boxing(See fig. 5-1, A. individuals assume a boxer's stance, palms open, fingers extended and joined. Each contestant tries to slap the opponent about the head and upper body with the open hand. This is a good warm- up activity.

(2) Wrist Tug-of-War(See fig. 5-1, B.)Two individuals sit on the ground with the soles of their feet in contact. Each grasps the opponent's wrists so that the hands are directly over the feet. At the whistle, each individual tries to pull the opponent from sitting to standing position.

(3) Arm-Lock Wrestling.(See fig. 5-1, C.)Two individuals sit back-to-back with legs spread and arms locked at the elbows. Each right arm is inside the opponent's left. Each tries to force the opponent's left arm or shoulder to the ground. The individual who first wins three bouts is the winner.

(4) Bulling(See fig. 5-1, D.)Two individuals assume the Westmoreland wrestling hold, each grasping the opponent's neck with the right hand and the opponent's right elbow with the left hand. Each tries to force the other to move one foot by pushing, pulling, or otherwise manipulating. The individual who first wins two boutsis the winner.

(5) Indian Wrestling(See fig. 5-1, E.)Two individuals lie on the ground, side by side, with their heads in opposite directions. They link right elbows. On the instructor's signal or by mutual agreement, each individual raises the right leg approximately straight and far enough to engage the opponent's heel. To start the contest, each person usually raises the leg three times rhythmically and on the third time engages the opponent's heel. Each tries to roll the other over backward. The right leg is used for three bouts, then the left leg for three bouts.

Figure 5-1 is available in the illustration section at the end of the book.

b. Combatives Table II

(1) Wrist Bending.(See fig. 5-2, A.) Opponents pair off, face each other, raise their arms forward, and, with palms forward, interlock their fingers. At the starting signal, each individual attempts to bend the opponent's wrist. The arms are kept up and forward and are not allowed to swing around and down to the sides. The individual who first wins two bouts is the winner.

(2) Back-to-Back Push.(See fig. 5-2, B.)Two individuals stand back-to-back with elbows locked. Each right arm is inside the opponent's left arm. At the starting signal, each pushes backward, trying to move the opponent. Opponents are not allowed to lift and carry each other; only pushing is permitted. The one who pushes the opponent the farthest wins the bout. The individual who first wins two bouts is the winner.

(3) Hop and Pull Hand.(See fig. 5-2, C.) The individuals are matched in pairs. Each grasps the opponent's right hand and, hopping on the right foot, attempts to pull the opponent off balance. Contestants automatically lose if they touch their free hand or their lifted foot to the ground. For successive bouts, they alternate hands and feet.

(4) Westmoreland Wrestling. (See fig. 5-2, D.)Each contestant grasps the back of the opponent's neck with the right hand and the opponent's right elbow with the left hand. In this position, each attempts to pull, push, or force the opponent to touch the ground with any part of the body other than the feet. The individual who first wins two bouts is the winner.

(5) Crab Fight.(See fig. 5-2, E.)Two individuals sit on the ground facing in opposite directions with their hands on the ground behind them. At the whistle, they raise their hips and push with their shoulders and bodies, each trying to make the other's hips touch the ground. The individual who first wins two bouts is the winner.

Figure 5-2 is available in the illustration section at the end of the book.

c. Combatives Table III

(1) Hand Wrestling.(See fig. 5-3, A.) Two individuals stand facing each other. Their right feet are forward and braced side by side. They grasp right hands for the first bout, left hands for the second bout. Each pulls, pushes, makes sideward movements, and otherwise maneuvers to force the opponent to move one or both feet from the original position. The contestant who first wins two bouts is the winner.

(2) Back-to-Back Tug.(See fig, 5-3, B.) Two individuals stand back-to-back with both arms linked at the elbows. Each has the right arm inside the opponent's left arm. At the starting signal, each attempts to pull the opponent. Lifting and carrying are permitted. The contestants must maintain their original direction and keep their arms linked. After a predetermined time, the player pulled or carried the farthest is the loser.

(3) Wrestling to Lift off Feet.(See fig. 5-3, C.) Contestants face each other. Each places the right arm under the left arm of the opponent and around the body. The left arm is over the opponent's right shoulder. Each tries to lift the other off the ground. The individual who first wins two bouts is the winner.

(4) Arm Pull Between Legs.(See fig. 5-3, D.) Two individuals are paired off, back-to-back. Each bends forward and, extending the right arm between the legs, grasps the opponent's right wrist. At the starting signal, each person attempts to pull the opponent After a predetermined time, the player who has pulled the opponent the farthest is the winner of the bout. The one who first wins two bouts is the winner. Repeat with the left hand and then both hands.

(5) Rooster Fight.(See fig.5-3, E.) Each contestant grasps the left foot with the right hand from behind, and the right arm with left hand. Each hops on the right foot, and, by shoulder-butting the opponent or by feinting and sudden evasions, forces the opponent to let go of the foot or arm. The contestant who first wins two bouts is winner.

Figure 5-3 is available in the illustration section at the end of the book.

Section III .Relays

5301. DESCRIPTION AND OBJECTIVE

Relays are races in which each member of a team runs one leg of the race. The team effort decides the winner. Relays provide stimulating competition and contribute to the conditioning of personnel. They also develop aggressiveness, team spirit, and the will to win. Relays should be dispersed throughout the program for short periods of time to provide a change of activity.

a. Team Organization. Relays are conducted most efficiently in platoon-size groups. Teams of equal size must be organized. Competitive spirit is encouraged and teams are organized better by basing teams on units such as squads, crews, or sections. Team captains should be designated. Extra Marines may be used as officials. The number of individuals on a relay team should be limited to squad size. If larger teams are used, the runners will spend too much time awaiting their turns and too little time actually participating. Two to six teams are ideal for relay competition. It is difficult to keep track of winners when more teams compete.

b. Administration of Relays

(1) The time spent on any one relay should be relatively short. If one team achieves a substantial lead in a long relay, the competitive spirit and enthusiastic participation of the other teams may decrease. Several short relays are generally better than one long relay.

(2) To maintain competitive spirit throughout a number of relays, determine the teams that win, place, and show in each relay and their total points for all relays. This can be done by awarding points to all teams on the basis of position at the finish of each relay. The team with the greatest number of points is the winner of the entire set of relays.

(3) Difficulties commonly encountered in conducting relays may be avoided by the following procedures:

(a) The last player in a relay race should be conspicuously identified. For example, the last runner can tie a handkerchief around the head or arm, take off the shirt, put on a hat or take it off, or use some other means.

(b) Another way to keep track of the progress of the race is to have each player sit or squat as soon as each is finished.

(c) Judges at the starting line can keep the runners from starting too soon.

(d) To prevent contestants from turning before they run the full distance, they should be required to run around a peg, pole, or assistant instructor.

(e) Batons, handkerchiefs, tent pegs, or other objects should be passed from one runner to the next when relays are run on a circular track.

(4) Before a unit's first participation in a relay, inform participants of the rules and scoring system. Violation of the rules should not result in disqualification. Instead, impose point penalties. A point penalty is imposed by subtracting a point from the team total at the conclusion of the relay.

(5) Careful administration will prevent most violations. For efficient conduct of relays, follow this procedure:

(a) Announce the name of the relay.

(b) Form the individuals in relay position.

(c) Briefly explain the relay and the rules for running it.
(d) Demonstrate.
(e) Have a definite finish line, and ensure that all know where it is.
(f) Answer questions.
(g) Conduct the relay.
(h) Determine winner and award points.

5302. RELAY TABLES

a. Events. The following relays are grouped into a table of activities. Each relay table can be completed in 15 minutes. Thus, relays can be used as a sole activity or as a part of a longer period. Each table is planned for a platoon-size group (30-60Marines). Adequate warmup for participants is provided by conducting several repetitions of exercises 1 and 2 of a conditioning drill. The recommended relay tables require an area 40 by 60 meters in size. Each table provides a variety of activity. The tables are progressive in the overload applied and should be scheduled in numerical order although not necessarily on successive days.

b. Relay Table I
(1) **60-Meter Lane Relay** (See fig. 5-4, A.) Each team is assembled in single file behind the starting line. On signal, the first individual of each team runs to the turn-around line 30 meters away, then runs back and touches the next teammate waiting at the starting line. The winning team is the first team to get its last member across the finish line. If an individual starts before being touched by the preceding runner, the team may be penalized.
(2) Wheelbarrow Race.(See fig. 5-4, B.) The players of each team pair off and line up in single file behind the starting line. The first individual walks on the hands while the partner grasps the ankles. They advance to the distance line (25 meters) behind which they exchange positions and return to the starting line. The rear individual must always hold the partner's ankles. After the first pair returns across the starting line, the next pair starts.

Figure 5-4 is available in the illustration section at the end of the book.

(3) Squad Front Relay .(See fig. 5-4, C.) The teams form in a line along the starting line with a 10-foot interval between teams. The members of each team lock elbows so that they are linked together. At the starting signal, the teams run to the distance line (20meters) where the left flank individual acts as a pivot. The team swings around on the pivot and returns to the base line. If a team breaks its links, it must reform before continuing. The first team to completely cross the base line intact is the winner.
(4) Crab-Walk Race.(See fig. 5-4, D.) The players of each team line up in single file. The first person of each team assumes the crab-walk position with feet forward on the starting line, hands on the ground behind it. At the starting signal, contestants move forward to the distance line (10 meters). They touch the line with their feet and then return to the starting line in the reverse position with the head and hands leading. The second person may not start until the first individual touches the finish line.

c.Relay Table II

(1) 100-Meter Lane Relay.(See fig. 5-5, A.)This relay is conducted exactly as the first relay of table 1 except that the start and turn-around line are 50 meters apart. This relay provides progression in sprinting.

(2) Frog-Jump Relay. (See fig. 5-5, B.) Each team lines up in single file. The first individual assumes a squatting position on the starting line. At the starting signal, the contestant progresses to the distance line (15 meters) and back by leaping forward, catching the weight on the hands, and bringing up the legs to the squat position for the next leap forward.

(3) Simple Relay. (See fig. 5-5, C.)Each team lines up in single file. Place a marker on the distance line (20 meters) in front of each team. Each team member, behind the first one, grasps the belt of the individual ahead. At the starting signal, each team runs as a unit to the marker, circles it, and returns to the starting line. The first team to completely cross the starting line intact is thewinner.

(4) Fireman's Carry Relay.(See fig. 5-5, D.)The players of each team line up in pairs, one behind the other. One individual in each pair carries the partner to the distance line (30 meters), using the fireman's carry. At the distance line, individuals exchange places and return to the starting line. As a variation, the person to be carried lies on the ground. The one carrying lifts the partner to the proper position. This relay may be performed with the other carries described in guerrilla exercises in chapter 3.

d.Relay Table III

(1) 200-Meter Circle Relay .(See fig. 5-6, A.)A course is laid out in a circular, rectangular,or oval pattern that is 200 meters around. Each team provides one runner on the starting line. On signal, the runner races around the 200-meter track and touches the next teammate, waiting at the starting line, who runs the same course. Each team member runs one lap of the course.

Figure 5-5 is available in the illustration section at the end of the book.

(2) Bear-and-Crab Race.(See fig. 5-6, B.) Each team lines up in single file. At the signal to start, the first contestant in each column assumes the bear-walk position (hands on ground in front), walks to the distance line (15 meters), and then runs back to the starting line. The returning contestant touches off the second individual and goes to the rear of the line. The second contestant assumes the crab-walk position (hands on ground behind) and crab-walks with feet leading to the distance line. This contestant also runs back to the starting line and touches off the next individual, who walks bear fashion. The rest of the members of each team alternate in this manner. The relay ends when the first person is back at the head of the line.

(3) Pilot Relay.(See fig. 5-6, C.) The players are grouped in threes, arms interlocked at the elbows, and outside players facing backwards. The middle person runs forward; the two outside individuals run backward. They run to the turning point (15meters), where they start back, this time with the middle person running backward and the two outside individuals running forward. The next set of three players starts when the first set crosses the starting line.

(4) Saddle-Back Relay.(See fig. 5-6, D.) Mark two parallel lines 15 meters apart. Each team selects a rider. The remaining members of each team count off. The even-numbered players from each team form in single files behind one line and the odd-numbered players from each team form in single files behind the other line directly across from their teammates. At the starting signal, the rider mounts the back of the first player of the team who carries the rider across the other line. There the rider changes mounts to the first player in the second line without touching the ground. This person carries the rider to the next player waiting in the first line. The relay continues until all of the mounts have carried the rider. A rider who falls off must mount again at the point of the fall. A rider who falls in changing mounts must get back on the original mount before making the change.

e. Relay Table IV

(1) 100-Meter Circle Relay (See fig. 5-7, A.)A course is laid out in either a circular, rectangular, or oval pattern that is 200 meters around. Each team is divided in half with each half positioned at starting lines on opposite sides of the track. Each runner races halfway around the track and touches a teammate who completes the lap. Each runner then waits in file at the halfway point until touched by the runner and then completes the second half of the lap. The first team to return all runners to their original starting line is the winner.

Figure 5-6 is available in the illustration section at the end of the book.

(2) In-and-Out Relay.(See fig. 5-7, B.) Each team lines up in a file with players 2 meters apart. At the starting signal, the first player runs back through the column in a zigzag fashion. He alternates, going to the right of one teammate and to the left of the next. Upon completing the run, the player lines up 2 meters behind the last one in the column. As soon as the first runner has passed the second person, the latter starts to run. This continues until all the players have realigned their original order. The team that finishes first is the winner. It may be desirable to have this relay continue until all individuals have run through their entire team two or three times in succession.

(3) Circle Race.(See fig. 5-7, C.) Each team forms a circle and holds hands with all individuals facing out except one who faces in and is the "driver."At the starting signal, the teams race to the distance line (20 meters) and back, keeping the circle intact. All the individuals in the circle must completely cross the distance line. The "driver" gives directions and orders. When the circle breaks, it must be reformed before it can continue. The first team completely over the starting line is the winner.

(4) Horse-and-Rider Relay.(See fig. 5-7, D.)Each team lines up in a single file. At the signal to start, the second player in each column leaps upon the back of the first who carries the rider across the distance line (30 meters). At the distance line, the rider dismounts and runs back to the starting line. There the rider picks up the third individual in the column, and carries this player to the distance line where the first player has remained. This continues until the last person is carried across the distance line.

Figure 5-7 is available in the illustration section at the end of the book.

Section IV. Team Contests and Athletics

5401. DESCRIPTION AND OBJECTIVE

Team contests are competitive activities in which Marines as a team compete with another team to win. They are guided by simple rules and organization. The function of team contests is to provide competition and an opportunity for body contact and to contribute to the development of physical readiness. In competing and working together as a team, individuals develop aggressiveness, the will to win, and teamwork. Team athletics deserve a prominent place in the physical training program because they contribute to increased combat efficiency. Because of the competitive nature of athletics and their natural appeal, individuals take part with enthusiasm. Athletic teams formed at intramural and higher levels are a strong unifying influence and provide one of the best means of developing esprit de corps.

a. Area and Equipment A level training field is sometimes the only area required. Many contests need no equipment. In contests requiring equipment, the need is for standard items such as logs, balls, nets, goals, and similar types of equipment. Specific requirements for area and equipment are listed with each contest description.

b. Progression. Team contests of a strenuous nature should be introduced after a basic period of conditioning has been completed and individuals are in the slow improvement stage of conditioning. Progression can take place from the less active to the more vigorous contests, and then from the non-contact to the contact or combative-type contests.

c. Necessity for Preconditioning Individuals must undergo conditioning prior to participation in athletics. Muscles, organs, joints, and ligaments not accustomed to stress and strain from sudden stops and starts, falls, body contact, rapid turns, prolonged running, and other rigors of athletic competition are subject to injury. Although athletics should not be introduced until players are physically prepared, there is still opportunity to engage in competition through lead-up contests. Individuals learn many of the skills required for athletics while participating in team contests.

d. Benefits. Athletics are beneficial primarily to sustain interest in the program and to maintain an achieved level of physical fitness. Athletics are a supplement, not a substitute for other types of conditioning activities which should continue. All of the desirable traits of physical fitness cannot be developed through athletics, yet their contribution is significant. For athletics to make a proper contribution to physical conditioning, the selected sports must bevigorous. The team contests recommended here are carefully selected for their simplicity, aggressiveness and applicability to squad- teaching teamwork. They are generally strenuous and involve the whole team throughout the entire duration of the contest.

5402. TEAM CONTESTS

a. Pushball

(1) **Players.** 10 to 50 players on a side. (See fig. 5-8.)

(2) Equipment A large pushball 5 to 6 feet in diameter.
(3) Area. A field 240 to 300 feet long, 120 to 150 feet wide. Mark a center line in the middle of the field parallel to the end lines. Mark a line 45 feet on either side of this center line and parallel to it, extending it across the width of the field. Mark another parallel line 15 feet from each end line, extending it across the width of the field.
(4) Game. Four 10-minute quarters are played. Give 2-minute rests between quarters and 5-minute rests between halves. The object of the game is to propel the ball over the opponent's goal line by pushing, rolling, passing, carrying, or any other way except kicking.

(a) The ball is placed on the center line with the opposing captains 3 feet from the ball. The rest of the players are 45 feet from the ball, on their half of the field. On the referee's starting whistle, the captains immediately play the ball with their respective teams coming to their assistance.

(b) At quarter time, the ball remains dead for 2 minutes at the spot where it was when the quarter ended. At half time, the teams exchange goals. The play is then started as it was in the beginning.

Figure 5-8 is available in the illustration section at the end of the book.

(c) Players may use any means of interfering with an opponent's progress except striking and clipping. Clipping is throwing the body across the back of an opponent's leg, while the opponent is running or standing. Legal use of force may be applied to all opponents whether or not they are playing the ball. For striking an opponent, the offender is removed from the game. The team penalty is half the distance to their goal. The penalty for clipping is the same.

(d) When any part of the ball goes out of bounds, it is a dead ball. The teams line up at right angles to the side lines and 3 feet apart at the point where the ball went out. The referee then tosses the ball between the teams.

(e) When, for any reason, the ball remains in one spot for more than 10 seconds, the referee declares the ball dead. The ball is then put into play as it is for an out-of-bounds situation.
(5) Scoring. A goal is scored when the ball, or any part of it, is propelled across the opponent's end line. A goal counts 5 points. The team scoring a goal has the privilege of trying for a point after the goal. To try for this extra point, the ball is placed on the opponent's 5-yard (or 15-foot) line. The teams line up on either side of this line separated by the width of the ball. Before the whistle blows, one player on each team may place hands on the ball. On the referee's signal, the ball is put into play for 1 minute. If any part of the ball is driven across the goal line in this 1-minute period, the offensive team scores 1 point. The defending team may not score during the opponent's try for the extra point.

b. Line Rush
(1) Players. Any number up to 50 on each side. (See fig. 5-9.)
(2) Equipment. None.
(3) Area. A field, 75 by 100 feet.
(4) Game. One team lines up behind one goal line and the other in midfield. On the starting signal, the team standing behind the goal line seeks to cross the goal line at the opposite end of

the field within 30 seconds. The team in the center seeks to prevent it by catching and holding the runners. At the end of 30 seconds, the teams change.

(5) Scoring. Count the number of individuals who have crossed the far goal at the end of 30 seconds. After each team has had from three to five tries, the scores are added and the winner declared. A player scores 1 point when any part of the body is across the goal line.

Figure 5-9 is available in the illustration section at the end of the book.

c. Human Tug-of-War

(1) Players. 10 to 20 on a team.(See fig. 5-10.)

(2) Equipment. None.

(3) Area.40 to 60 feet.

(4) Game. Draw a line in the center of the area. Divide the players into two equal teams. Place them in single file on opposite sides of the center line facing each other. Each individual places the arms around the waist of the teammate in front. The two leaders of the opposing teams grasp each other around the waist. On signal, the teams try to pull each other over the center line within 30 seconds.

(5) Scoring. The team pulled across the center line loses. If neither team is pulled over the center line, but one team breaks its file, that team loses the match.

(6) Variation. Use a 3/4- or 1-inch rope. Space the leading players on each team 10 feet apart, each holding one end of the rope. The team pulled across the center line loses.

Figure 5-10 is available in the illustration section at the end of the book.

d. Master of the Ring

(1) Players. Any number. (See fig. 5-11.)

(2) Equipment. None.

(3) Area. A clearly marked circle large enough to contain all the players.

(4) Game. All the players stand inside the circle. At the signal, all players attempt to throw each other out of the circle. All tactics are fair except unnecessary roughness. When any part of the body touches across the line, the player is out and leaves the circle at once. Several officials are needed to spot the players who cross the line.

(5) Scoring The player who remains in the circle when all the others are out is the master of the ring.

(6) Variation 1. The players are divided into two equal teams. Each team is clearly marked. On signal, each team tries to throw the opponents out of the circle. The winning team is the team that eliminates all opponents from the circle.

(7) Variation 2. The players are divided into two equal teams. Each team sends only one individual into the circle. When one person has been forced out of the circle, the losing side sends in only one player. The team which eliminates all opponents is the winner.

(8) Variation 3.A pit, approximately 4-feet deep, is used rather than a circle on level ground. The contest may then be an individual or team activity.

Figure 5-11 is available in the illustration section at the end of the book.

e. Log Pivot Circle. Space teams far enough apart so each team can pivot in a circle without colliding with any other team. Each team holds a log in the bend of the arms in front of the chest. (See fig.5-12.) At the command **CIRCLE RIGHT, MOVE**,the left-flank individual holds the pivot while the log is carried around 360 degrees in a counter- clockwise motion, back to the original position. This movement may also be performed to the left by facing in the other direction and pivoting clockwise (**CIRCLE RIGHT, MOVE**).Other commands may be used such as: **CIRCLE HALF-RIGHT, CIRCLE HALF- LEFT** and so on. The first team to complete the prescribed movement is the winner.

Figure 5-12 is available in the illustration section at the end of the book.

f. Log-Rolling Race. Each team tries to roll its log a measured distance by pushing it with the hands and driving the body forward with the legs. (See fig. 5-13.)The first team to get the entire length of the log across the finish line wins.

Figure 5-13 is available in the illustration section at the end of the book.

g. Prone Pushing Contest.. Two teams lie prone, facing each other with a log between them. (See fig. 5-14.)Both teams place their hands against the log, keeping their arms straight. Then, by driving with the legs, each team attempts to push the other a measured distance to the rear.

Figure 5-14 is available in the illustration section at the end of the book.

h. Shuttle Relay Race. Each team in this race divides into two parts, A and B. Members of part A run 50 yards with a log held under their right arms. (See fig 5-15.)At the distance line, they give the log to their teammates in part B who bring it back to the starting line. The team pair finishing first is the winner.

Figure 5-15 is available in the illustration section at the end of the book.

5403.CROSS-COUNTRY AND DISTANCE RUNNING

a. Value of Running. Long-distance running gives some benefits that cannot be obtained in the same degree from any other sport. It builds powerful leg muscles, increases lung capacity, and develops endurance. For these reasons, cross-country and distance running should be included in the physical training program. These sports require only a few miles of open space which is generally available. Short cross-country runs and middle-distance runs can be used to supplement other activities, particularly team sports or the sports that develop precision or agility rather than endurance. Short cross-country runs can be scheduled once a week, gradually increasing the distance as the physical condition of the runners improves.

b. Cross-Country Runs. A cross-country run is a distance run held on a course laid out along roads, across fields, over hills, through woods, and on any irregular ground. A flat cinder or dirt track is not a suitable surface for cross-country running. Some runs are as long as 6 miles while others are as short as 3 or 4 miles. If cross-country running is being used to supplement other activities, the 3-mile course is long enough for most. Cross-country runs should be used only after individuals reach the sustaining stage of conditioning. This type of running should then be scheduled occasionally to provide variety. Cross-country running has the advantage of allowing mass participation. Interest can be stimulated by putting the runs on a competitive basis.

c. Practice Methods. Conditioning is essential to distance and cross-country running. Championship distance running depends on stamina, which can be developed only through constant training. An individual of only average ability can become an outstanding distance runner by steady and careful training. Hiking is an excellent method for getting into condition before the season opens. Long walks build up leg muscles. During the first month of the season, training should be gradual, starting with short distances, and increasing day by day. At first the legs will become stiff, but the stiffness gradually disappears if running is practiced for a while every day. To prevent strain, it is essential to limber up thoroughly each day before running.

(1) In training a large group, leaders should be stationed at the head and the rear of the column. They should make every effort to keep the runners together. After determining the abilities of the unit in cross-country running, it is advisable to divide them into three groups. The poorest-conditioned group is started first, the best-conditioned group, last. The starting time of the groups should be staggered so that all of them come in at about the same time.

(2) In preliminary training, running is similar to ordinary road work in that it begins with rather slow jogging, alternating with walking. The speed and distance of the run is gradually increased. As conditioning improves, occasional sprints may be introduced. Atfirst, the distance run is from one-half to 1 mile. It gradually increases to 2 or 3 miles. On completing runs, require runners to walk for 3 or 4 minutes before stopping to permit a gradual cooling off.

d. Facilities and Equipment. There should be at least one timer with a stopwatch (preferably three) for timing the runners. A course 3 or 5 miles long should be measured and marked by one of the three following methods:

(1) Fasten directional arrows to the top of tall posts and place them at every point where the course turns. Such signs should also be placed at every other point where there may be doubt as to the direction of travel.

(2) Place a lime line on the ground over the entire course.

(3) Place flags, clearly visible to the runners.

- **(a)** A red flag indicates a left turn.
- **(b)** A white flag indicates a right turn.
- **(c)** A blue flag indicates the course is straight ahead.

e. Rules/Scoring. A cross-country team consists of seven individuals, unless otherwise agreed. In dual meets, a maximum of 12 on one team may be entered, but 7 or fewer are scored. First place scores 1 point, second place 2, third place 3, and so on. All who finish the course are ranked and tallied in this manner. The team score is then determined by totaling the points scored by the first five of each team to finish. The team scoring the least number of points is the winner. If fewer than five (or the number determined prior to the race) finish, the places of all members of that team are disregarded. If two or more teams score the same number of points, the event is called a tie.

5404.TEAM ATHLETICS

Team athletics add variety to the combat readiness program. At no time should they be allowed to replace intramural activities or to provide practice time for intramural or base teams. If team athletics are integrated into the program, the goal should be total participation and competition. Therefore, the games selected can be played by the skilled or the unskilled. They require limited equipment, yet will teach competitiveness, aggressiveness, and teamwork while they improve conditioning.

a.Soccer. Soccer is one of the best athletic activities for developing endurance, agility, leg strength, and great skill fulness in using the legs. The most popular sport in the world, soccer is the national game of many European, Asian, and Central and South American countries. In recent years, it has become popular in the United States. A soccer ball is the only equipment needed for the game, and Marines can learn to play it easily. Players do not need much skill to participate, but the amount of skill they can develop is almost unlimited.

(1) Place in the Program. Soccer should be introduced during the latter part of the slow improvement stage of physical training. It can be used as a competitive activity in the sustaining stage. It is primarily a spring or fall sport. Any level field is suitable for competition. The boundaries for the soccer field are similar to the dimensions for a football field. (See fig. 5-16.) Goal posts are essential, but they are easily constructed and usually temporary so that they may be removed when not in use.

Figure 5-16 is available in the illustration section at the end of the book.

(2) Basic Skills

(a) Passing. Passing with the feet is the primary means of moving the ball. Short passes are easier to control. Emphasis should be continually placed on the skill of passing.

(b) Dribbling. The ball is dribbled by a series of kicks with the inside or outside of the foot. Do not kick with the toe. Keep the head over the ball when kicking and propel it only a short distance at a time. Keep it close to the feet. If a player lets a ball move far from the feet, an opposing player can easily take it away.

(c) Instep Kicking. The instep kick is the basic soccer kick. The toe does not come in contact with the ball. The toe is pointed downward, and the instep (the area under the shoe laces) is applied to the ball with a vigorous snap from the knee.

(d) Inside-of-the-Foot Kicking. The ball is kicked with the inside of the foot, and the leg is swung from the hip. The toe is turned outward, and the sole of the foot is parallel with the ground as the foot strikes the ball. This kick is used for short passes and for dribbling.

(e) Foot Trapping. The foot trap is the method of stopping the ball by trapping it between the ground and the foot. Place the sole of the foot on top of the ball at the instant it touches the ground, keeping the foot relaxed. This is an effective way to stop a fast-moving ball.

(f) Shin Trapping. .The shin trap is a method of stopping the ball with the shins. Stand just forward of the spot where the ball should strike the ground and allow it to strike the shins in flight or on the bounce. Use either one or both legs from the knee down so as to absorb the bounce.

(g) Body Trapping. .Intercept the ball with any part of the upper body except the arms and hands. Keep the body relaxed and inclined toward the ball. To keep the ball from bouncing, move backwards from it as it strikes the body. This drops the ball at the feet in position for dribbling or passing.

(h) Heading. Heading is a technique for changing the direction of the flight of a ball by butting it with the head. Tense the neck muscles and jump up to meet the ball. Butt the ball with the forehead at about the hairline to reverse its direction. Use the side of the head to deflect it to the side.

(3) Offensive and Defensive Positions. Forwards usually play on the offensive half of the field. Fullbacks usually play on the defensive half of the field. Halfbacks are the backbone of the team. They move forward on the offense and back on defense. The goalkeeper almost always remains within a few feet of the goal.

(4) Abridged Rules

(a) A soccer team is composed of 11 players. (See fig. 5-16.)

(b) The player propels the ball by kicking it with the feet or any part of the legs, by butting it with the head, and by hitting it with any portion of the body except the arms or hands.

(c) The goalkeeper is the only individual allowed to place hands on the ball, but only in the goalkeeper's area. The term "hands" includes the whole arm from the point of the shoulder down.

(d) A goal is made by causing the ball to cross completely the section of the goal line lying between the uprights and under the cross bar.

(e) Each goal scores 1 point for the team scoring the goal.

(f) The penalty for a foul committed anywhere on the playing field (except by the defensive team in its penalty area) is a free kick awarded to the opposing team.

(g) All opponents must be at least 10 yards from the ball when a free kick is taken.

(h) The penalty for a foul committed by the defensive team in its penalty area is a penalty kick.

(i) A penalty kick is a free kick at the goal from the spot 12 yards directly in front of the goal. The only players allowed within the penalty area at the time of the kick are the kicker and the defending goalkeeper.

(j) An official game consists of two 30-minute halves,

(k) Teams change goals at the end of every quarter.

(l) In the event of a tie, an extra quarter may played. This may be followed by a sudden death period. If the tie still exists, the game can be decided by penalty kick.

(m) After a team has propelled the ball across a side line and out of play, the ball is put back into play by a throw-in from the side line by a member of the opposing team. The ball is thrown in from the point where it crossed the side line as it went out of bounds. The ball must be thrown in with both hands using an over-the-head motion, keeping both feet on the ground.

(n) When the offensive team propels the ball across the defensive team's goal line but not in the goal, the defensive team is awarded a goal kick-a free kick taken from within its own goal area. The ball must come out of the penalty area to be in play.

(o) When the defensive team causes the ball to go behind its own goal line, excluding the portion between the goal posts, the offensive team is awarded a corner kick—a free kick taken by a member of the offensive team at the quarter circle, at the cornerflag post nearest to where the ball went behind the goal line. The flag post must not be removed.

(p) The game is started and, after a goal has been scored, is resumed by placing the ball in the center of the midfield line. Players must be on their side of the line until the ball is kicked. The ball must be kicked forward and must move at least 2 feet to be legal. The first kicker may not touch the ball twice in succession at the kick off. The opposing team must be 10 yards from the ball until it moves.

b.Speedball. Speedball offers vigorous and varied action with plenty of scoring opportunities. It is easy to learn and provides spontaneous fun. Little equipment is needed-a ball is all that is absolutely necessary. Speedball combines the kicking, trapping, and intercepting elements of soccer; the passing game of basketball; and the punting, drop kicking, and scoring pass of football.

(1) Place in the Program.Speedball, like soccer, should be introduced into the physical training program during the latter part of the toughening stage and used as a competitive activity in the sustaining stage. Speedball may be played any time weather permits but is primarily a spring or fall activity.

(2) Method of Play.The game is played as follows:

(a) Two teams of 11 each play the game under official rules, but any number of players may successfully constitute a team.An inflated leather ball, usually a soccer ball, is used. The playing field is a football field with a football goal post at each end. (See fig. 5-17.)

(b) The game starts with a soccer-type kickoff. The kicking team tries to retain possession of the ball and advance it toward the opposite goal by passing or kicking it. Running with the ball is not allowed, with the result that there is no tackling or interference. After the ball touches the ground, it cannot be picked up with the hands or caught on the bounce but must be played as in soccer until it is raised into the air directly from a kick. Then the hands may be used.

(c) When the ball goes out of bounds over the sidelines, it is given to a player of the opposing team and is put into play with a basketball throw- in. When it goes over the end line without a score, it is given to a player of the opposing team who may either pass or kick it onto the field.

(d) When two opposing players are contesting possession of a held ball, the official tosses the ball up between them as in basketball.

(e) Points are scored by kicking the ball under the crossbar of the goal post, drop kicking the ball over the crossbar, or completing a forward pass into the end zone for a touchdown.

Figure 5-17 is available in the illustration section at the end of the book.

(3) Basic Skills

(a) Shared Skills. Skills include kicking, passing, heading, and trapping (from soccer); punting, dropkicking and forward passing (from football); and passing, receiving, and pivoting (from basketball).

(b) Kickups. The kickup is a play in which a player lifts the ball into the air with the feet so as to legally play the ball with the hands. The kickup generally makes the transition from ground play to aerial play. The technique of making the play depends upon whether the ball is rolling or stationary. If the ball is rolling or bouncing toward the player, the foot is held on the ground with the toe down until the ball rolls onto the foot. Then the foot is raised, projecting the ball upward. If the ball is stationary, the player rolls it backward with one foot. Then the player places a foot where the ball will roll onto it. The player then raises the foot, projecting the ball as before. If a ball is rolling away, the player should stop it with a foot and play it as a stationary ball. The player can also raise the ball by standing over it with a foot on either side. The player presses the feet against the ball and jumps into the air, propelling the ball into the hands.

(4) Offensive and Defensive Play. The positions of players in speedball are much the same as in soccer. However, some of the positions have different names. Each team has 11 players. The forward line is composed of five players, the right end, right forward, center, left forward, and left end. The second line consists of right halfback, fullback, and left halfback. In the next line is the right guard and left guard. The player who defends the goal is the goal guard. The strategy employed in speedball during offensive play is very similar to that of soccer. There are two types of defensive formations in speedball: man-for-man and position defense. Man-for-man defense is recommended for beginning players.

(5) Abridged Rules

(a) Field. A speedball field is 100 yards long with additional 10-yard end zones on each end and 55 yards wide, the same as a regulation football field.

(b) Players. Eleven players are on a team. The goal guard has no special privileges.

(c) Time. Ten-minute quarters. Two minutes between quarters. Ten minutes between halves. Five minutes for extra overtime periods.(Begin the first overtime by a jump ball at center, same goals; change goals in the event of a second overtime period.)

(d) Winner of Toss. The winner of the toss has the choice of kicking, receiving, or defending a specific goal.

(e) Kickoff. A kick off starts the first and second half of a game and starts play after a score. The kick off is made from any place on the 50-yard line. The kicking team must be behind the ball when it is kicked. The receiving team must stay back of its restraining line (10 yards distance) until the ball is kicked. The ball must go toward the opposing team's goal before it can be played. Kick off out of bounds goes to the opponent at that spot. A kick off touched by the receiving team before going out of bounds, with no impetus added, still belongs to the receiving team. A kick off, in possession and control of the receiving team and then fumbled out of bounds, belongs to the kicking team at that spot.

(f) Starting Second and Fourth Quarters. The ball is given to the team that had possession at the end of the previous quarter, from out of bounds, as in basketball.

(g) Second Half. The first-half receiving team kicks off at the beginning of the second half.

(h) Fly and Ground Ball. The most characteristic playing rule of speedball is the difference between a fly ball (or aerial ball) and a ground ball. A player is not permitted to touch a ground ball with the hands and must play it as in soccer. A fly ball is one that has risen into the air directly from the foot of a player (for example, a punt, drop kick, place kick, or kickup).Such a ball may be caught with the hands provided the catch is made before the ball strikes the ground again.

(i) Kickup and Overhead Dribble. A kickup is a ball that is so kicked by a player that he can catch it himself. A bounce from the ground may not be touched with the hand because it has touched the ground since being kicked. This rule prohibits the ordinary basketball dribble, but one overhead dribble (throwing the ball into the air and advancing to catch it before it hits the ground) is permitted.

(j) Out of Bounds. If a team causes the ball to go out of bounds over the sidelines, a free throw-in (any style) is given to the opposing team. When the ball goes over the end line without scoring, it is given to the opponents who may pass or kick from out of bounds at that point.

(k) Tie Ball. In case two players are contesting the possession of a held ball, even in the end zone, a tie ball is declared. The ball is tossed up between them.

(6) Scoring

(a) Field Goal(3 points).A soccer-type kick, in which a ground ball is kicked under the crossbar and between the goal post from the field of play or end zone, is a field goal. (A punt going straight through is not a field goal for it is not a ground ball. The ball must hit the ground first.) A drop kick from the field of play that goes under a crossbar does not count as a field goal. A drop kick from the end zone that goes under the crossbar counts as a field goal;if it goes over the crossbar, it is ruled as a touch back, and the other team takes possession of the ball.

(b) Dropkick(2 points). A scoring drop kick must be made from the field of play and go over the crossbar and between the uprights. The ball must hit the ground before it is kicked (usually with the instep).

(c) End goal(1 point). This is a ground ball which receives its impetus (kicked or legally propelled by the body) from any player, offensive or defensive, in the end zone and which passes over the end line but not between the goal posts.

(d) Penalty Kick(1 point).A ball kicked from the penalty mark that goes between the goal posts and under the crossbar is a penalty kick. The penalty mark is placed directly in front of the goal at the center of the goal line.

(e) Touchdown(1 point). A touchdown is a forward pass from the field of play completed in the end zone. The player must be entirely in the end zone. If the player is on the goal line or has one foot in the field of play and the other in the end zone, the ball is declared out of bounds. If a forward pass is missed, the ball continues in play, but it must be returned to the field of play before another forward pass or drop kick may be made.

(7) Substitutions. Substitutions may be made any time when the ball is not in play. A player who is withdrawn may not return during that same period.

(8) Timeout. Three legal timeouts of 2 minutes each are permitted each team during the game.

(9) Fouls

(a) Personal Foul. Kicking, tripping, charging, pushing, holding, blocking, and unnecessary roughness of any kind, such as running into an opponent from behind, are personal fouls. Kicking at a fly ball and thereby kicking an opponent is a personal foul. Four personal fouls disqualify a player from the game.

(b) Technical Foul. Illegal substitution, more than three timeouts in a game, unsportsmanlike conduct, and unnecessarily delaying the game are technical fouls.

(c) Violation. Traveling with the ball, touching a ground ball with the hands or arms, double overhead dribble, violating tie ball, and kicking or kneeing a fly ball before catching it are violations.

(d) Penalties. The offended player shall attempt the kick.

(e) Officiation of Fouls

1 Fouls in the field of play allow no follow-up while fouls in the end zone always allow follow-up.

2 On penalty kicks, with no follow-up, only the kicker and goalie are involved.

3 On penalty kicks, with a follow-up, the kicking side is behind the ball and the defending side behind the end line or in the field of play. No one is allowed in the end zone or between the goal posts except the goal guard. The kicker must make an actual attempt at goal and cannot play the ball again until after another player plays it.

c. Volleyball. Volleyball is a popular sport. The game entails much physical activity, yet it is not strenuous. It is, therefore, a game for young and old alike, for beginners and for skilled players. It may be played indoors or outdoors on any type of terrain. As an informal activity, volleyball can be played by any number; as an organized activity, it provides, as few other sports do, a game for 12 to play in a limited area. While volleyball requires no great skill to play, it permits a high degree of proficiency. Players naturally get more enjoyment when they know the game and play it well. For this reason, instruction in the basic skills should be provided.

(1) Place in the Program Volleyball may be used occasionally as a competitive activity during the sustaining stage. Though it is a year- round sport, it should be included in the physical training program only when a more strenuous activity is impractical. Volleyball is an excellent off-duty activity.

(2) Instruction. Usually during the first three or four classes, a 10- to 15- minute period of instruction, followed by scrimmage, is enough to teach the basic skills, rules, and techniques of volleyball. More time, if available, can be given to teaching basic skills, but emphasize competitive play rather than formal instruction. Divide the unit into 6-person teams, organized on the basis of ability .Teams should be as nearly equal as possible.

(3) Basic Skills

(a) Passing the High Ball. The chest pass is the most effective method of playing the ball. To receive the ball, the knees are fixed, and the body is tilted forward. The elbows are raised sideward to a point in line with the shoulders. The wrists are extended in line with the forearm. The arms, wrists, and hands are rotated inward. To pass the ball, the hands are chest-high, thumbs pointing inward. The fingers are flexed, forming a cup, allowing them to contact the ball. On contact with the ball, the wrists are snapped while the fingers and elbows are pushed upward, sending the ball upward. A high ball is much easier to handle than a low one.

(b) Passing the Low Ball. A ball that is lower than the waist is one of the easiest to hit, but it is also a frequent cause of fouls (holding or carrying the ball). The best position for handling a low ball is to have the knees flexed and arms flexed at the elbows and rotated so the thumbs are pointing outward, the palms up. When the fingers contact the ball, the entire body reacts in a lifting motion. The arms and hands swing upward in a scooping action. It is important that the fingers, not the palms, contact the ball, and that the ball is batted, not thrown.

(c) Underhand Serve. Take a position behind the back line facing the net, left foot forward, holding the ball in the palm of the left hand. The left knee is flexed; the right knee is straight. Swing the right arm back and at the same time move the left hand (holding the ball) across the body in line with the right hip. Then swing the right arm forward hitting the ball off of the left hand with the palm of the right hand, raising the hips and arching the back in the same motion. Be certain to swing the right arm in a straight line, or the ball will be difficult to control. When the opposition is in formation, the server should try to place the ball in the right or left back area, and not near the net.

(d) Setup. A setup is a ball hit into the air near the net by one player, so that a teammate may hit or "spike" it sharply downward into the opponent's court. The chest pass is the best pass to use. The ball is sent approximately 10 feet into the air toward the spiker so it will descend from 4 to 20 inches from the net.

(e) Spiking. The spike is a leap into the air and a sharp downward hitting of the ball into the opponent's court. A spiker must be able to spring easily from the floor, judge the movement of the ball, and strike it with a downward movement of the arm. To jump from the floor, step off with one foot and jump with the other. Stand with the right or left side to the net, facing the setup player. Much depends upon the setup player to place the ball in the proper position. The spiker jumps into the air and strikes the ball above its center so as to drive it downward. A snapping movement of the arm and wrist will drive the ball forward and downward with power and control. Aim for a weak spot in the opponent's defense.

(f) Blocking. The block is a defense technique used to prevent a spiker from driving the ball across the net. It is an attempt by one or more defensive players at the net to block a hard-hit shot by using the force of the ball to send it immediately back into the opponent's court. In an effective block, forwards on the defensive team spring into the air at the time of the spike, placing both hands and arms in the expected path of the ball. An effective block tends to upset the offense and presents another element for the spiker to worry about. To be effective, the blocker must anticipate the path of the ball and time the block with the spike.

(4) Abridged Rules

(a) The volleyball court is 30 feet wide by 60 feet long. (See fig. 5-18.)

(b) The top of the net is 8 feet high.

(c) A volleyball team consists of 6 players.

(d) A match consists of the best two out of three games.

(e) The first team scoring 15 points wins the game, provided that they have 2 points more than their opponents.

(f) A deuce game is a game in which both teams score 14 points. The game is continued until one team obtains a 2-point advantage over the other.

(g) Only the serving team can score. If the serving team commits a fault, it loses the serve to the opposing team.

(h) The team members who receive the served ball rotate one position in a clockwise direction.

(i) The ball is put into play by serving from behind the back line.

(j) A served ball touching the net results in the loss of the serve. At any other time during play, a ball touching the net is still in play.

Figure 5-18 is available in the illustration section at the end of the book.

(k) The ball is out of play when it touches the ground or goes outside one of the boundary lines.

(l) All line balls are good.

(m) The players must hit or bat the ball; they may not throw, lift, or grasp it.

(n) A player may not touch the ball with any part of the body below the knees.

(o) A player may not play the ball twice in succession. In receiving a hard-driven spike, a defensive player may make several contacts with the ball even if they are not simultaneous. All such contacts, however, must constitute one continuous play, and all must be above the knees.

(p) The ball may be touched no more than three times on one side of the net before being returned across the net to the opposing team.

(q) A player must not touch or reach across the net.

(r) A player may touch the line under the net, but may not cross it.

Chapter 6
EVALUATION OF PERFORMANCE DURING TRAINING

6001.OBJECTIVES OF COMBAT READINESS

The objectives of the Marine Corps physical fitness program are-

a. To contribute to the health and well-being of every Marine through regular exercise and health education.

b. To develop Marines who are physically capable of performing their duties in garrison and in combat.

c. To develop Marines with a reserve level of physical fitness that will help them win in combat.

d. To provide a medium for developing the self-confidence of the individual Marine and thereby enhancing overall discipline, morale, esprit de corps, unit efficiency, and the desire to excel within the Marine Corps.

6002. THE PHYSICAL FITNESS TEST

The physical fitness test (PFT) has distracted attention from the proper goal of the physical fitness program. Simply stated, the goal of the physical fitness program is the success of Marines in combat. Unit commanders must not allow the PFT to become an end in itself. The PFT should be used within the commander's program of combat readiness training as a standard, an index of the physical fitness of individual Marines at a given time and place. The PFT is not difficult, and it must be kept in perspective. All athletes should be first class, and first class is attainable for all Marines. The conduct of the PFT is described in Marine Corps Order 6100.3_.

6003. OTHER METHODS OF EVALUATION

The commander should use other ways of evaluation to provide variety and imagination. Intangible objectives, such as confidence and aggressiveness, are hard to measure. The use of either inspection or observation is necessary to evaluate objectives of this nature. The commander has several other methods available when evaluating the physical condition of the command. The additional methods of inspection, observation, medical examination, and testing are summarized as follows:

a. Formal Inspections. Formal inspections, using inspection officers and standardized rating criteria, may help evaluate unit physical fitness.

b. Daily Observation. Routine observation of physical performance and appearance can serve as an indicator of an individual's or unit's physical readiness. However, mere observation is not a totally reliable or accurate means of evaluation.

c. **Physical Examination.** Medical examination may detect any individual disability or detrimental physical condition. It may guide in application of remedial, therapeutic, or limited exercise programs.

d. Other Physical Fitness Tests. Some other tests are described in detail in this manual for information purposes. Commanders are cautioned, however, not to place too great an emphasis on testing.

6004. RESPONSIBILITY

The commander is responsible for the physical fitness of the command and for the measurement and evaluation of its physical readiness. The goal of combat readiness training remains to prepare Marines and units for the physical demands of combat. The bulk of the commander's combat readiness training program should be such activities as obstacle and confidence courses, endurance courses, and progressive load-bearing marches which will prepare Marines for combat. Group activities which are competitive and combative in nature are advantageous and add imagination and enthusiasm to the program. Activities such as orienteering, which is not only physically demanding but also competitive and which teaches essential subject skills, are excellent and motivate all concerned. A commander's goal should be to provide an integrated training program that is so aggressive and demanding of Marines and leaders alike that a PFT administered at any time will reflect that each and every Marine is both physically and mentally prepared for the demands of combat.

6005. COMMANDER'S PHYSICAL FITNESS PROGRAM

The Commandant of the Marine Corps has directed that the physical fitness test shall be the universal measure of adequate individual physical fitness. The focus of a commander's physical fitness training program, however, should not be to prepare Marines to pass the physical fitness test, but to prepare them for the physical rigors of combat.

a. Administration. The physical fitness program requires each Marine to spend a minimum of 3 hours per week in physical fitness training, to be tested semiannually, and to obtain a minimum standard of third class. The attainment of a higher score is a laudable individual goal which should be encouraged but should not become a unit objective. Overemphasis could be detrimental to training required to develop the complete Marine. All commanders should exercise sound judgment as well as positive, aggressive leadership in striving toward the training objective.

b. Variety of Activities. The program must include the following elements:

(1) Physical Readiness Training. physical readiness training is a complete physical training program which develops and maintains the strength, endurance, and physical skills needed to sustain the individual during combat.

(2) Remedial Physical Conditioning. Remedial physical conditioning is a process by which physically substandard Marines are conditioned to meet prescribed standards.

(3) Competitive Conditioning Activities. Competitive conditioning activities consist of teams or individuals competing against an opponent to win. This involves a combination of sports and military skills designed to foster competitive attitudes and develop unit pride and esprit de corps.

(4) Occupational Conditioning. Marching long distances, dry net training, etc., should not be ignored as a means to fulfill the time requirement.

c.Time. In order to develop the desired level of physical fitness, unit commanders are advised that a minimum program of 3 hours per week is mandatory. The program must encompass at least three exercise periods weekly. Each period should include calisthenics, running, or other forms of vigorous activity .The minimum desirable length for each exercise period is 60 minutes. Shorter periods may be authorized by the unit commander if dictated by the local training situation. Unit commanders are encouraged to use the normal working day to satisfy this training requirement. However, commanders are authorized to conduct required physical fitness training during off-duty hours when the mission, workload, personnel status or other significant factors preclude fulfilling it during duty hours.

d.Action. Marine Corps Order 6100.3_ requires that commanders-

o Establish and maintain an effective physical fitness program which ensures that all Marines maintain an acceptable level of physical fitness.

o Ensure that all individuals are medically qualified prior to participating in the physical fitness program.

o Establish a minimum physical fitness program of at least 3 hours per week. Physical training may be authorized on an individual basis at the discretion of the unit commander.

o Ensure that all Marines participate in physical fitness conditioning activities commensurate with their medical qualifications and limitations.

o Conduct physical fitness testing for all Marines.

o Place all Marines who fail to pass the PFT on a daily command-supervised remedial physical conditioning program until they pass the test.

o Ensure that results of physical fitness testing are entered on section A of the fitness report. Comments are placed in section C in connection with attainment of superior physical performance or a medical excuse from the PFT.

6006. OTHER FORMAL TESTS

Commanders must realize that the best index of combat readiness is personal observation of performance on conditioning marches of 15 miles while under load. This manual presents three additional indicators of unit fitness. These indices are the physical readiness test, which was formerly the Marine Corps standard and is now routinely administered at Officer Candidate School; the airborne trainee physical fitness test which many Marines who attend U.S. Army formal schools are required to take; and the battle fitness test, the ultimate test of the combat readiness of Marines. Commanders may wish to include these tests or portions thereof within their combat readiness programs.

6007. PHYSICAL READINESS TEST

a. Purpose. The purpose of this test is to measure the Marine's ability to meet the minimum standard of physical readiness for duty in the field. This test is currently administered to all male officer candidates at Officer Candidate School, MCCDC, Quantico, VA. Female candidates do not take this test.

b. Test Area. Administration of this test requires an area suitable for crawling, jumping, and running 160 yards to include a trench that is 8 feet wide.

c. Equipment. The equipment required for this test includes—
o A platform, rail on a ledge, or a box 18 inches high,
o A 20-foot climbing rope,
o A stopwatch.

d. Conduct of Test. Adequate timers and other supervisory personnel must be provided by the company staff. All events should be conducted in a single session of one morning or afternoon. The events may be run in any sequence.

e. Uniform. The uniform for the physical readiness test is as follows: full utilities, cartridge belt, belt suspender straps, two canteens (full of water), rifle, and helmet. Depending on the heat condition, the PT shirt may be worn in place of the utility jacket. All equipment will remain secure throughout the test. Except for emergencies, canteens are not to be emptied.

f. Events
(1) Event 1: Climbing Uphill (Step-ups)This event simulates marching uphill at a rapid and steady pace. The Marine stands in front of the step. On the command GO, the Marine places either foot on the step and steps up. Hand pressure on the knee may be used if desired. The Marine then stands erect and steps backward and down, on one foot at a time. The same leg may be used for each repetition or the legs may be alternated. For 100 points, Marines will complete 60 repetitions in the maximum allotted time of 1 minute and 20 seconds (80 seconds).
*(2) Event 2: Rope Climb. T*his event resembles entering and leaving a hovering helicopter, using ropes in house-to-house fighting, and ascending and descending landing nets. The Marine assumes a sitting position on the ground at the bottom of the rope with hands at the highest point that can be reached. On the command GO, the Marine jumps up and climbs to the top. The Marine must touch the 20-foot mark with one hand, then descend without sliding or dropping. Marines must reach the 20-foot mark in the maximum allotted time of 30 seconds or less.
(3) Event 3: Evacuation. This event simulates reaching and evacuating a wounded Marine under fire. The Marine assumes the prone position at a distance of 50 yards from a casualty who is of approximately equal weight. On the command GO, the Marine springs up and covers the distance in a sprint. The Marine lifts the casualty to a fireman's carry and returns to the starting point. Assistance by the casualty is permitted. The tested individual must carry all equipment; the casualty has none. Marines must negotiate the course in a maximum time of 47 seconds or less.
(4) Event 4.'Advance by Fire and Maneuven. This event simulates advancing and assaulting during an attack. The Marine assumes the prone position at the starting point. On the command GO, the Marine then alternately crawls, runs in a zigzag manner, rolls, jumps, and assumes different firing positions for the next 130yards. Before reaching the objective, the Marine must "hit the deck," roll over, and assume the firing position three times. While assuming the firing position, the Marine must pull back the charging handle, take aim, and squeeze the trigger before rising from the ground. After rising from the ground for the third time,the Marine continues running and jumps to clear the 8-foot trench in one leap. The Marine then finishes the course in

the maximum allotted time of 2 minutes. The Marine is allowed multiple attempts only within the specified time limit.

(5) Event 5: Forced March. This event is a combat run of 3 miles without halts and against time. The Marine must double-time. Marines must complete the 3-mile course in a maximum allotted time of 30 minutes or less.

g. Event Standards

(1) To successfully pass the physical readiness test, the Marine must pass each event with a minimum of 80 points for each. Points are then assessed for each event in accordance with the scoring tables. (See fig. 6-1.)

(2) If a Marine fails one event, the highest possible score that the Marine may receive for the entire test is 79 percent. (See example 1 in fig. 6-2.) If two events are failed, the highest possible score for the entire test is 69 percent, regardless of the total points accumulated. (See example 2 in fig. 6-2.) If three events are failed, the highest possible score is 59 percent. Failure in four or more events will result in a score of zero for the entire test.

(3) To convert the total points for all five satisfactorily passed events to a final percentage score, total the accumulated points for all five events and divide by five. (See example 3 in fig. 6-2.)

Figure 6-1 is available in the illustration section at the end of the book.

Figure 6-2 is available in the illustration section at the end of the book.

6008. AIRBORNE TRAINEE PHYSICAL FITNESS TEST

a. Use and Composition of the Test.

(1) Use. The airborne trainee physical fitness test is a means of determining the physical ability of the applicant for acceptance to and retention in the airborne course of instruction.

(2) Test Events The test battery consists of five events as follows: chinups, knee bender, pushups, situps, and an endurance run.

b. Method of Scoring and Standards

(1) Scoring. The Marine will be scored by a trained scorer who is thoroughly familiar with the minimum standards for the test events. The Marine will be scored on a pass or fail basis. The performance on each event may be recorded on the reverse side of DA Form 705 (Physical Fitness Testing Record).The examinee records the personal information on the face of the card as specified on the first, second, and third lines. The face of the card is identified by writing diagonally across the lower half of the card "Airborne Trainee PPT—See reverse side."Using line 7 on the reverse side of the scorecard, the scorer enters the test title in the block entitled "(Other (Specify))" and completes the required information on the remainder of that line. The "Remarks" section can be used to record the test event titles and the applicant's performance on each test event.

(2) Standards. To successfully pass the test, the examinee must reach the standard in each test event. The standards follow:

(a) Chinups

(b) Knee bender 80(2-minute period)

(c) Pushups 45(2-minute period)

(d) Situps45(2-minute period)

(e) Endurance runs:2 miles in 15:54 minutes or less in athletic gear and 4 miles in 32 minutes in utilities and running shoes.

c. Uniform for Testing

(1) Examinees. The prescribed uniform for test participation is boots and the work uniform of the season. No headdress is worn. When climatic conditions permit, jackets or outer shirts may be removed.

(2) Officials. Scorers and other test officials should be uniformly and distinctively dressed for contrast with Marines being tested.

d. Test Administration Procedure

(1) Preparation for Test. The administration of the test to a large group makes it mandatory that the test effort be organized and efficiently operated. All testing is not completed with large test groups; on certain occasions, individuals and small groups are tested. Care must be exercised to administer the test uniformly and to standardize the conduct of all elements of the test. Regardless of the size of the test group, the following elements of sound test administration should be included:

(a) An orientation to include the purpose, method of administration, scoring of the test, preparation of the score card, and required standards.

(b) A correct demonstration of each event to ensure that there is no misunderstanding of the proper form and required standards.

(c) Completion of all five test events in one test period with all Marines taking the events in the same sequence.

(d) Adequate rest periods between the test events to allow for recovery before the next event.

(2) Method of Administration with a Large Group. With a minimum of 12 lanes per test event, 14 officials can administer the test battery to 150 or 200 Marines in 2 hours. The officials are designated as follows: one officer in charge, one demonstrator,and 12 scorers. If more or fewer Marines are to be tested, a greater or lesser number of officials will be required. The following procedure is recommended:

(a) Conduct an orientation and ensure the examinees have properly completed their score cards.

(b) Assign Marines to lanes and caution them to remain in the same lane order throughout the test.

(c) Explain and demonstrate the chinup event, administer it, and score it. Then proceed to the knee bender and pushup events, and administer them in the same manner.

(d) Grant a 5- to 10-minute rest period after the pushup event. Advise against excessive consumption of water during the break period.

(e) Explain and demonstrate the situp event, administer it, and score it as prescribed. Then move to the run area, explain the running event, and complete it.

(f) Retain the scorecards at the completion of the running event.

(3) Method of Administration with a Small Group. A similar procedure is followed for the testing of individuals and small groups. The informality usually associated with small groups must not conflict with sound test administration. With fewer examinees, fewer officials are required.

e. Description and Explanation of Test Events

(1) Test Event 1:Chinups

(a) Purpose. This event is devised to test arm and shoulder flexor strength.

(b)Equipment There is one horizontal bar per lane, made of plumber's pipe or a gymnasium horizontal bar 1 1/2 inches in outside diameter. The bar must be rigidly supported at a height of 8 feet above the ground. The upright supports must be 5 feet apart. There must be a movable stand at each bar for short Marines to stand on to reach the bar.

(c) Officials There is one scorer per lane.

(d) Organization. The examinees, holding their scorecards, stand in order behind the restraining line in their respective lanes. The scorers take each scorecard when the Marine is called forward for the test.

(e) Starting Position The bar is grasped with the palms turned toward the face, the thumbs underneath the bar. The body is fully extended in a "dead" hanging position with the arms straight and the feet above the ground.

(f) Movement. Pull the body directly upward until the chin is placed over the bar. Lower the body until the elbows are completely straight and the body is again in the "dead" hanging position. Repeat as many times as required.

(g) Instructions. Explain and demonstrate the fully extended "dead" hanging position with the proper grasp. Show that the chin is placed over the bar at the top of the movement and that the arms are fully extended, the elbows completely straight, at the bottom of the movement (the hanging position). Explain that the body must be kept from swinging and that it is permissible to raise the legs and flex the hips when pulling up, but that any kicking, bicycling, or jerking motion with the trunk or legs is not acceptable. Inform the examinees that no penalty is exacted for hanging on the bar to rest in the bottom position but that this is not to their advantage. Tell them that half-completed chinups are not counted, and that the scorer will repeat the number of the last correct chinup when incorrect execution is detected.

(h) Administration and Scoring. Caution the examinees to assume the "dead" hanging position and wait for the scorer's command to begin. The scorer is at the examinee's left with a clear view of the bar. If the examinee begins to swing widely, the scorer should stop the swinging by extending the left arm across the front of the examinee's body, being sure not to hinder the execution of the chinup. The scorer counts aloud the number of chinups correctly executed. When a chinup is not correctly executed, the scorer repeats the number of the last correct one. The scorer records the number of correct chinups on the scorecard and returns the card to the examinee.

(2) Test Event 2:Knee Bender

(a) Purpose. This event measures the strength and endurance of the leg muscles.

(b) Equipment. None.

(c) Officials There is one scorer per lane.

(d) Organization. Marines stand in numerical order behind the restraining line in their respective lanes. The scorer takes each scorecard when the Marine is called forward for the test.

(e) Starting Position. The feet are spread less than shoulder width apart, hands on hips, thumbs in the small of the back, elbows back.

(f) Movement. Do a knee bend and at the same time bend slightly forward at the waist and thrust the arms between the legs until the extended fingers touch the ground. The hands are about 6 inches apart. The bend is approximately a three-quarters bend. From this knee bend position, recover to the starting position by moving the body upward, straightening the knees, and returning the hands to the hips. Repeat as many times as required.

(g) Instructions. Explain and demonstrate the correct starting position. Be certain examinees understand the correct knee bend and that only the tips of the fingers touch the ground. Tell them the scorer will repeat the number of the last correct knee bender when incorrect execution is detected. Some of the common errors are failure to correctly bend the knees, failure to touch the ground, and failure to assume the completely erect position after the bend has been executed.

(h) Administration and Scoring. The scorer stands to one side so as to see that the knees are properly bent and the fingers touch the ground as prescribed. From this position, the scorer can view the examinee to see that a properly erect position is assumed after each knee bend. The scorer counts aloud the number of correctly executed knee bends. When a knee bend is done incorrectly, the scorer repeats the number of the last correct one. The scorer records the number of correct knee bends on the scorecard and returns it to the examinee.

(3) Test Event 3:Pushups

(a) Purpose. Pushups measure arm and shoulder extensor strength.

(b) Equipment. None.

(c) Official. There is one scorer per lane.

(d) Organization. Marines stand behind the restraining line in their respective lanes until the scorer calls on them to perform. The scorer takes each scorecard when the Marine comes forward.

(e) Starting Position. The front-leaning rest position is the starting position. The body is straight from head to heels, palms are flat on the ground directly underneath the shoulders, and elbows are straight and locked. The body weight is supported on the hands and toes throughout the event.

(f) Movement. Bending only the elbows, lower the body in one straight plane until the chest touches the scorer's hand. Straightening and locking the elbows, raise the body in one straight plane, returning to the original front-leaning rest position. Repeat as many times as required, keeping the body in a straight line from head to heels.

(g) Instructions. Explain and demonstrate that the arms are straight at the beginning and completion of the movement and that the chest must touch the scorer's hand, but the stomach and thighs must not touch the ground. Also explain that the whole body must be maintained in a straight line as it is lowered and raised; that is, there is to be no breaking at the hips or shoulders so that any body part is lowered or raised in advance of the other or as a separate segment .Dipping or rolling through the shoulders is illegal, as is lowering or raising the body with one arm or shoulder at any time. Resting is not permitted during repetitions. Instruct

Marines that the scorer will repeat the number of the last correct pushup when incorrect execution is detected.

(h) Administration and Scoring. It is recommended that Marines assume a prone position while placing their feet and hands in the proper positions. This permits them to rest while the scorer gets into position and, at the same time, provides a feel of the body in a straight plane from head to heels. The scorer lies on the right hip and side to the right of the examinee. This gives the scorer a clear view of the examinee's body to see any errors. The palm of the scorer's right hand rests flat on the ground underneath the lowest part of the examinee's chest. By keeping the right forearm flat on the ground at an angle in front of the examinee's right arm, the scorer's position will not prevent the examinee from lowering the body completely. The scorer's left hand is free to test the straightening of the elbow at the completion of the movements and to point out body segments being lowered or raised separately. When in position and ready, the scorer has the examinee assume the starting position and begin doing pushups. The scorer counts aloud the repetitions done correctly and repeats the number of the last correct pushup if one is done incorrectly. There is no penalty if the contour of the examinee's body causes the hips to protrude slightly out of line, provided that the whole body is raised and lowered simultaneously. The scorer enters the number of repetitions on the scorecard and returns it to the examinee.

(4) Test Event 4: Situps

(a) Purpose. This event primarily measures abdominal strength.

(b) Equipment. None.

(c) Official. There is one scorer per lane.

(d) Organization Marines stand behind the restraining line in their respective lanes until the scorer calls on them to perform. The scorer takes each scorecard when the Marine comes forward.

(e) Starting Position. The examinee lies flat on the back with knees flexed, both feet flat on the ground. The correct angle of the thighs to the ground is 45 degrees. If the heels are too near the buttocks, the applicant will not be able to sit up. Legs are spread shoulder width apart. The examinee interlaces fingers and places them behind the head in contact with the ground. The feet are not held by another person.

(f) Movement. Bend forward at the waist and raise the upper body until the head is directly over the knees. Heels are not to leave the ground. Elbows remain in the same plane to the head and body throughout the event. The upper body is slowly lowered to the starting position until the head touches the ground. Repetitions are done at a slow cadence with no rest periods.

(g) Instructions. Explain and demonstrate the correct starting position and the proper execution of the situps to be sure that Marines understand the movement. Warn them that their knees must remain flexed during each situp, the heels cannot leave the ground at any time, and they may not roll up on one side and push up with one elbow. Tell them they must do the repetitions at a slow cadence with no rest periods. Instruct Marines that the scorer will repeat the number of the last correct situp when incorrect execution is detected.

(h) Administration and Scoring. When the performer is in position and ready, the scorer has the performer assume the starting position and begin doing situps. The scorer counts aloud the correct executions. When a situp is improperly done, the scorer repeats the number of the last correct one. No situp is credited if the hands are unclasped from behind the head, if the back is used to bounce up from the ground (which means the shoulders would not touch the ground), or if one shoulder or elbow is used to push up. The examinee is not penalized if the heels slide forward slightly so long as the knees remain flexed and the heels maintain contact with the ground. The scorer enters the number of repetitions on the scorecard and returns it to the examinee.

(5) Test Event 5:Endurance Run

(a) Purpose. This event measures cardiovascular endurance.

(b) Equipment One stopwatch or watch with a sweep second hand.

(c) Area. A large training field on which a quarter-mile track has been staked out or a level road over flat terrain may be used as a running surface. A 1-mile route is designated with wooden stakes marking the start point, finish point, and one-quarter mile intervals.

(d) Officials For large groups, a scorer times the event and controls the conduct of the run, and a guide runs with the group and sets the pace.

(e) Organization The run is conducted with groups of Marines in a column formation. Company-sized units may run at the same time with platoons serving as running groups. The scorer issues the command to assume double-time.

(f) Starting Position. Marines are assembled in proper column formation (column of twos, threes, or fours, as appropriate to the size of the group) with short men to the front. When all is ready, the column is moved forward a short distance before the running period is started.

(g) Movement. At the command DOUBLE-TIME, MARCH, the examinees retain their places in the column formation and execute the command. Length of steps is about 40 inches. The scorer has the group execute the run. The formation is maintained during therun.

(h) Instructions. Marines are instructed to maintain formation while running and are informed that the guide will set the proper pace. They will be instructed in the command used to control the column in the execution of the test. The scorer should announce the 4-minute, 2-minute, 1-minute, and 30-second remaining time intervals.

(i) Administration and Scoring. The event may be administered as previously prescribed to a large group, to several Marines, or to an individual. An individual examinee usually does not require a guide or pacer. If the event is administered on a training field, the scorer may stand in the center as the runner(s) circles about the field. This method of administration relieves the scorer of running with each group to be tested. Scoring is based on successful completion of the run as prescribed. The scorer should announce the remaining times as prescribed in (h).

6009. BATTLE FITNESS TEST

The battle fitness test is a physical fitness evaluation which is currently under development as part of the basic warrior training concept plan.

Chapter 7
THE HUMAN BODY

7001.GENERAL

To implement an effective physical conditioning program, a leader must understand how the human body functions and how exercise affects the body .This chapter provides only an introductory discussion of anatomy and functioning.

a. Body Functioning During the Stages of Conditioning. For more on the stages of conditioning, see chapter 1.

(1) During the toughening stage, the waste products of muscle activity (lactic acids) collect more rapidly than the blood can remove them. This acid waste builds up in the muscle tissue and irritates the nerves in the muscle fiber, causing pain and stiffness. As the exercise program continues, more blood is carried through the muscle, removing the waste materials more rapidly and eventually causing the soreness to disappear.

(2) During the slow improvement stage, the blood circulation in the muscles increases, and the body as a whole becomes a more efficient machine. The improvement is rapid in the first few weeks, but as a higher level of skill and conditioning is reached, the improvement becomes less noticeable. The body reaches its maximum level of performance after 6 to 10 weeks and should then be maintained at this peak.

(3) During the sustaining stage, it is possible to maintain this state of conditioning through 15 to 20 minutes of exercise a day, but the exercise must be quite strenuous.

b. Diet. Regular exercise increases the appetite. If the desire for greater amounts of food is satisfied by a balanced diet, the body benefits.

(1) There are two main types of foods: body-building and energy- producing.

(a) Body-building foods consist of proteins, which build up tissue and repair wear and tear.

(b) Energy-producing foods include carbohydrates and fats. Carbohydrates provide a quick source of energy, while fats act as a reserve store of energy.

(2) In addition, food contains vitamins, mineral salts, and water. During hot weather and strenuous training periods, the body requires greater fluid intake.

(3) Diet should be supplemented with proper rest to allow the digestive system to digest the lood.

(4) Occasionally, especially during early stages of conditioning, strenuous exercise may cause vomiting. Although vomiting is not a frequent occurrence, it is not usually a cause for concern.

7002. SYSTEMS OF THE BODY

The systems of the body include the skeletal, muscular, circulatory, respiratory, endocrine, digestive, genitourinary, and nervous systems. Each has a different function, but all must work in cooperation with one another to insure a sound body. Of these systems, the first four are the most affected by exercise and are discussed in succeeding paragraphs.

7003. THE SKELETON

a. Bones. The skeleton (see fig. 7-1) is composed of about 206 bones, which-

o Provide structure for the body and support for the attachment of muscles,

o Protect vital organs, such as the brain, lungs, and heart,

o Manufacture red blood cells, which carry oxygen through the body.

Figure 7-1 is available in the illustration section at the end of the book.

b. Joints. A joint is a place of union between two or more bones. Joints can be-

o Immovable, such as in the face and head.

o Slightly movable, such as in the vertebrae or sternum.

o Freely movable, such as in the shoulder, hip, knee, ankle, and elbow. The bones in a freely movable joint are held in place by ligaments. Freely movable joints are of the greatest importance in physical training because exercise improves their mobility and stability.

c. Cartilage and Ligaments. The joints in the body, where bones connect, are supported by cartilage and ligaments.

o Cartilage is a tough, elastic, translucent tissue that acts as a shock absorber or buffer between bones. Examples are the discs between the vertebrae, the tissue attaching the ribs to the sternum, and buffers in the knee joints. Damaged cartilage does not heal.

o Ligaments are connective tissue that bind bones together. They contract and extend but are not elastic. Consequently, ligaments can be torn or strained. Damaged ligaments will heal to some extent.

7004. THE MUSCLES

a. General. Muscles are tissue, or an organ composed of tissue, which contract or extend to produce movement. Muscles are of three classifications: involuntary, voluntary, and cardiac. The voluntary and cardiac muscles are the most affected by physical training; physical training has little effect on involuntary muscles.(Seefigs. 7-2, 7-3, 7-4.)

o People have no control over the involuntary muscles, an example of which is the diaphragm,

o Voluntary muscles are the larger skeletal muscles which cause movement and which the individual can control.

o Cardiac muscle is found only in the heart and for all practical purposes is an involuntary muscle. Physical training has a significant effect on the fitness of cardiac muscles.

b. Muscle Structure. Muscle is composed of cells which are grouped into bundles called fibers. Groups of fibers form muscles. These bundles of fibers are held in place by sarcolemma, a thin, sheath-like material that surrounds the muscle bundles and secretes a fluid that lubricates the muscle tissue. The fused ends of the sarcolemma form the tendons which attach the muscles to the bones.

c. Attachment of Muscles. The arrangement of muscles on the skeleton provides the proper angle of pull to make movement possible. Voluntary muscles are usually attached to the skeleton in two places.

d.Action of Muscles. To produce motion and to do work a muscle usually shortens its fibers. The movement may be flexion such as bending the arm at the elbow, or extension, such as the straightening the arm. Some muscles raise the arms or legs, others depress the raised limb. Some muscles have the primary function of rotating the trunk from side to side. In addition, muscles act as stabilizers as well as prime movers.

Figure 7-2 is available in the illustration section at the end of the book.

Figure 7-3 is available in the illustration section at the end of the book.

Figure 7-4 is available in the illustration section at the end of the book.

7005. FUNCTIONING OF THE SKELETON AND MUSCLES

a. General. The skeleton and muscles function in close coordination. Muscles move the body, and bones support the muscles.

b. Effect of Exercise on Bones. Continuous exercise, particularly among younger people, usually strengthens the bones, allowing them to withstand greater stress and strain. Bones which are not used lose much of their strength, a factor which should be considered when individuals return to physical training after a prolonged period of inactivity. People in this category should refrain from activities which might result in bone breakage before the bone is exercised back to normal condition. The condition known as "march fracture" is usually a result of this condition.

c. Muscular Strength. When exercised vigorously to improve strength, muscle grows in size. In general, the larger the muscle, the stronger the muscle. Furthermore, trained muscle functions more efficiently than untrained muscle. Trained muscles contract more vigorously and with less effort. Regular and strenuous exercise also toughens muscle, making it firmer and able to stand more strain.

d. **Muscular Endurance.** Muscles naturally become fatigued during continuous, repeated work. Through specific exercises, such as chinups or situps, local muscle groups become exhausted long before an individual fatigues. Training to develop muscular endurance enables people to continue a relatively heavy load of exercise over a long period of time. Lengthened exercise periods improve muscular endurance.

e. Muscular Coordination. Improved speed and strength result in part from improved muscular coordination. An unskilled individual may use irrelevant muscles to perform a particular activity, increasing the amount of work without increasing the mechanical output. This increase in skill is highly desirable, but it should be offset by greater effort in duration or intensity to compensate for the loss in overload due to increased skill.

f. Muscular Fatigue. When the rate of work is increased, the energy required is proportionately much greater than the increase in rate. For example, if an individual doubles running speed, the amount of power demanded to do this is increased eight times.

g. Circulation in Muscles Regular, strenuous exercise causes the creation of new capillaries and the opening of inactive, latent capillaries, increasing blood circulation within the muscle by as much as 400 percent. This increases the supply of food materials and oxygen to the muscle, improving its endurance. This process takes about 8 to 12 weeks of regular conditioning in young adults and longer as age increases.

7006. THE CIRCULATORY AND RESPIRATORY SYSTEMS

a. The Circulatory System. The circulatory system transports blood to all parts of the body, removes waste products for disposal, and delivers protecting and repairing substances where needed. The circulatory system includes the heart and blood vessels.

(1) The Heart. The heart is a pump which forces blood through the blood vessels to the parts of the body. The heart is a little larger than a fist and is located in the left center of the chest between the lungs.

(2) Blood Vessels. The vessels carrying blood away from the heart are the arteries. They eventually divide into capillaries, the very small vessels through which diffusion and osmosis take place. The capillaries gradually increase in size, forming the veins which carry blood back to the heart.

b. The Respiratory System. The respiratory system performs the function of breathing, whereby oxygen is inhaled and carbon dioxide is exhaled. The respiratory system consists of the mouth, nose, trachea, lungs, and diaphragm.

o Trachea. The trachea, or windpipe, is a hollow, tube-like structure that carries air from the mouth to the lungs.

o Lungs. The lungs are elastic bags in the chest where the exchange of oxygen and carbon dioxide takes place.

o Diaphragm. The diaphragm is a thin, sheet-like muscle just below the lungs. During inhalation, the diaphragm flattens out and lowers, allowing the lungs to expand and fill with air. During exhalation, the diaphragm raises into a dome shape, helping to expel air from the lungs.

7007. CARDIOVASCULAR FUNCTIONING

a. General. Cardiovascular functioning is the combined functioning of the circulatory and respiratory systems. The chief organs of these systems, the heart and lungs, function together to provide oxygen to the body.(See fig. 7-5.)

Figure 7-5 is available in the illustration section at the end of the book.

b. Heart Action. The heart, the organ which propels blood through the blood vessels, is the chief organ of cardiovascular endurance. A person tires quickly if the heart lacks the capacity to circulate the blood through the body. The heart is a muscular organ and is developed by exercise just like any other muscle. The heart cannot be exercised alone; any exercise that puts a sufficient load on the heart, such as running or cycling, will also exercise other body parts. A conditioned heart tends to beat more slowly but pump more blood when at rest. This is known as an increase in "stroke volume,"a desirable condition because it enables the heart to pump more blood with a slower contraction rate.

c. Functions of the Lungs. Strenuous and regular exercise can improve the efficiency with which the lungs transmit oxygen to the blood by as much as 25 percent In a poorly conditioned person, some of the alveoli (air sacs) within the lungs are closed or collapsed. The forced breathing created by exercise over a period of weeks causes the air sacs to slowly expand, increasing the ability to absorb oxygen.

d. Relationship of Heart and Lungs. The heart pumps carbon dioxide-laden blood through the pulmonary artery to the lungs. In the lungs the carbon dioxide is exchanged for oxygen and the purified blood is returned to the heart by way of the pulmonary vein. The heart then pumps the blood through the aorta for circulation throughout the body. Blood moving into the muscles exchanges oxygen for carbon dioxide, after which time it is ready for the return trip to the heart.

e. Cardiovascular Functioning in High Altitudes. Marines to be employed in high altitudes should be acclimated in a similar area for 10 to 14 days prior to employment. Persons not accustomed to the rarefied air of higher altitudes tire more quickly and may collapse after rapid physical exertion. Air is much less dense at high altitudes than at sea level. Persons accustomed to sea level inhale only about 80 percent of the oxygen they are used to no matter how hard they breathe. Further, people accustomed to sea level or moderate altitude simply do not have enough red corpuscles in the blood to fulfill their needs at high altitudes. The red blood cells will increase over time.

f. Symptoms of Cardiovascular Malfunctioning. The brain is the first organ to react to a lack of oxygen. Unconsciousness results when the brain is denied sufficient oxygen. "Blacking out" is actually a defense mechanism to keep the body alive.

APPENDIX A
INSTRUCTOR HINTS AND INSTRUCTOR TRAINING

1. General. This appendix informs instructors on administration of exercise periods, commands, the extended rectangular and circle formations, methods of instruction and conduct of exercise activities, explanation of basic positions associated with the various activities, and instructor training.

2. Aim of the Physical Fitness Program. The aim of the physical fitness program is to prepare Marines physically for war. This can be conveniently split into two phases.

a. Preparation Phase

(1) Development of all-around physical fitness.

(2) Development of purposeful physical skills.

(3) Development of mental alertness.

(4) Development of character and leadership.

b. Maintenance Phase. Once the preparation is completed, the program must progress to a maintenance program. This phase revolves around the need to maintain the trained Marine's physical fitness for a particular role. This will include—

(1) Maintenance of a high standard of all-around physical fitness.

(2) Application of purposeful physical skills to combat skills.

(3) Development of leadership and self-confidence.

(4) Development of mental toughness and endurance.

3. Physical and Mental Development

a. Marines need to be developed physically and mentally. The physical training (PT) instructor must ensure that the program accomplishes this.

b. The system of the body functions best in response to progressive exercise. This means that Marines should master the easier exercises before they attempt the more difficult ones.

c. The body and mind are linked inseparably and must be considered as one in physical fitness training. Physical exercises assist in developing the mind.

d. The confidence in oneself and in one's comrades attained by physical achievement helps to promote the indomitable will to win. The Marine must be taught to realize this. It will help during arduous training, which at all times must be imaginative, realistic, and filled with enthusiasm. All physical fitness training must be associated with developing the qualities required of a Marine.

4. Planning and Preparation of ProgramsThe following factors must be considered when compiling a program (see also ch. 2)~

a. Aim or desired skill level.

b. Duration of the complete training program and number of physical training lessons.

c. Purpose of included exercises,drills, or practical applications.

d. Beginning/inventory fitness levels.

e. Medical facilities, medical personnel, and Marine Corps or local command regulations governing their proximity to the training.
f. Number of Marines to be trained.
g. Availability of instructors, assistants, and facilities.
h. Apparatus and equipment available.
i. Alternative training areas for varying weather conditions,
j. Stages of progression and tests of achievement.

5. Five Basic Principles of Program Design
a. Regularity. Regularity of exercise is far more important than the amount of exercise performed on a given day. A Marine should exercise a minimum of three times per week. Human muscle begins to atrophy after 3 days without exercise.
b. Balance. The program must be balanced in terms of total physical fitness. Imbalance is created when, for example, an individual merely runs. While the cardiovascular system is improved, general upper body strength may be lacking. Conversely, a person who just does weight training may develop strength, but not cardiovascular endurance.
c. Overload. If total body fitness is to be developed, then the muscular and cardiovascular systems must be made to perform more work than previously. This is done by increasing the frequency, intensity, and duration of the exercise. The body adapts to the amount of stress placed upon it and becomes stronger and more efficient.
d. Progression. To be successful, any type of training program must be progressive. This is an area where untrained persons conducting PT can cause injury to Marines resulting in failure of the PT program. Close supervision must be made of untrained instructors. The result of progressing too fast is that the class begins to reject PT because it is no longer enjoyable or motivating. The program should start at a slow pace, and the pace should increase as proficiency improves. Average persons will encounter three stages of progressiom the toughening stage, the slow improvement stage, and the sustaining stage. (See ch. 1.)
e. Variety. Some programs fail because the training becomes boring. Variety is essential and perhaps presents the greatest challenge to the commander and staff.

6. Grouping. Providing for different levels of physical fitness is particularly recommended in the early stages of conditioning. Older Marines and those in poorer physical condition should be expected to attain a group level of fitness, but they should be given more time to do it.
a. Homogeneous Grouping One simple method of providing for the difference in levels is to group them according to their condition. A two-group classification would divide Marines into highly conditioned and average groups. A finer classification could be obtained by dividing them into three groups—a highly conditioned, a moderately conditioned, and an unconditioned group.
b. Performance or Age Grouping. The segregation of Marines into different exercise groups should be based on physical fitness test scores or on the level of fitness they demonstrate. They may also be grouped at first according to age. A common classification by ages is under 30, 30 to 34, and 35 and above.

7. Preparatory Commands and Commands of Execution. The prepreparatory command describes and specifies what is required. The command of execution calls into action what has been prescribed. All preparatory commands are given with a rising inflection. The interval between commands is long enough to permit the average Marine to understand the first one before the second one is given.

8. Extended Rectangular Formation. The traditional formation for carrying on many physical training activities is the extended rectangular formation (fig. A-l). This formation is more compact than any other. It is the best type to use for large numbers of Marines because it is simple and easy to assume.

a. For the formation of oneplatoon, the base man paces off five paces from the stand, faces left and moves five paces, halts, and again faces left. With the base man positioned facing the stand, the platoon leader then commands. **FALL OUT AND FALL IN ON THE BASE MAN**. At this command, all Marines run to the designated area and reform. This procedure is preferred to marching the unit into position. If more control is desired, the unit may march at double time to the vicinity of the base man and then be directed to fall out and fall in on him. Much valuable time is wasted in the field by needless maneuvering of Marines at quick time in an effort to position the platoon or unit on the exact spot for the exercises.

b. A company-size unit assumes the extended rectangular formation from a column of threes or fours at normal intervals between squads. This extension can also be executed from a company mass without interval between platoons. In extending either a platoon or company-size unit, take your place at the head of the column and command:

(1) TAKE INTERVAL TO THE LEFT, MARCH. At this command, the Marines in the right flank file stand fast with arms extended sideward. All others turn to the left and run forward at double time. After taking a sufficient number of steps, all Marines face the front with both arms extended sideward. The distance between finger-tips is about 12 inches, and dress is right.

Figure A-1 is available in the illustration section at the end of the book.

(2) ARMS DOWNWARD, MOVE. At this command, the arms are lowered smartly to the sides.

(3) LEFT, FACE

(4) EXTEND TO THE LEFT, MARCH. At this command, those in the right flank file stand fast with arms extended sideward. All others turn to the left and run forward at double time. Spacing is the same as in (1) above, and dress is right.

(5) ARMS DOWNWARD, MOVE. Same as in (2) above.

(6) RIGHT, FACE.

(7) FROM FRONT TO REAR, COUNT OFF. At this command, the leading Marine in each column turns the head to the right rear, calls of **ONE** and faces the front. Successive Marines in each column call off in turn **TWO, THREE, FOUR, FIVE** in the same manner.

(8) EVEN NUMBERS TO THE LEFT, UNCOVER. At this command, each even-numbered Marine stride jumps sideward to the left, squarely in the center of the interval. In doing this, each one swings the left leg sideward, jumps from the right foot to the left foot, and smartly brings the right into position against the left.

c. To assemble the unit, you commandASSEMBLE TO THE RIGHT, MARCHAt this command, all return to their original position in the column at double time and reform on the base man.

d.It is recommended that the area for grounding equipment and arms be at the edge of, or well away from, the area to be used for exercising. To conserve time and ensure proper position of the unit, the base man or, if the unit is composed of several platoon-sizegroups, the various base men may precede the unit and establish their positions in relation to the instructor's stand.

9. Circle Formation.The circle formation is effective for conducting various exercise activities (fig. A-2). This formation has an advantage in that supervision of all Marines is facilitated and a moving formation is available which provides control. Guerrilla exercises, grass drills, and some forms of running are examples of activities which are more easily conducted in the circle formation than in the extended rectangular formation.

a. It is not advisable to have more than 60 Marines in a circle. When more must be accommodated, separate circles should be used.Concentric circles may be employed to reduce the size of the circle or to accommodate more persons. If concentric circles are formed, the different circles are made by designating squads for each circle.Each additional circle requires more Marines than the one inside it.For example, one squad of a platoon may form the inner circle and theremaimng three squads the outer circle.

b. When a platoon is to form a circle, the commands ai€IRCLEFORMATION, MARCH, FOLLOW MEUpon this command, the right flank squad of the column moves forward at double time with the leader of the platoon group gradually forming a circle in a counterclockwise direction. Each succeeding file falls in behind that on the right. After the rough outline of the circle is formed, the leader commandsPICK UP A **5**-YARD INTERVAL.

Figure A-2 is available in the Illustration section at the end of the book.

c.The group may be halted and faced toward the center, or, if instruction is not necessary, the exercise activity may be executed without stopping the platoon.

lO.Leadership Techniques

a. Unless you experience all the exercises, you cannot appreciate how arduous they are, what movements are most strenuous and difficult, where the errors in performance are likely to occur, and what the proper cadence should be.

b. You must give everyone careful supervision and participate in the exercises to show that you can do them. When you participate, your assistant instructors should supervise because it is difficult for you to supervise and exercise at the same time.

c. Marines should never be kept too long in one position, especially a constrained one. They should never have to perform an exercise more times than they can do it without losing the proper form. You should insist upon proper form in the execution of all exercise activities.Even slight deviations from the proper form reduce the value of the exercise.

d. Avoid long explanations.As a rule, it should be necessary to give a full explanation of new exercises only. Explain the most essential features of an exercise first; add details later. Too many details at one time are more likely to confuse than to assist them.Minor corrections should be made to the entire class while the exercise is in progress (for exampHEADS **UP,** KNEES STRAIGHT). If necessary, follow this direction by the name of the person who is particularly at fault. If a Marine requires special attention, givethat person separate instruction by an assistant instructor to avoidwasting the time of the group.

e. The heavy demand on your voice can be lightened by training assistant instructors to assume some of the instruction. Using mass cadence is also an effective method of lessening the demand on your voice.

“Illustrations”

DAY	TIME	ACTIVITY
MONDAY	60 MIN	RUNNING ACTIVITY
TUESDAY	60 MIN	COMBATIVES
WEDNESDAY	60 MIN	CONDITIONING & LOG DRILLS
THURSDAY	60 MIN	CONDITIONING & RELAYS
FRIDAY	60 MIN	RUNNING & CONDITIONING
SATURDAY	60 MIN	RUNNING ACTIVITY

Figure 2-1. Sample A- -Single Activity.

DAY	TIME	ACTIVITY	REMARKS
MONDAY	20 MIN 20 MIN 20 MIN	RIFLE DRILL COMBATIVES RUNNING	PROGRESSIVE ACTIVITY SYSTEM
TUESDAY	30 MIN	GUERRILLA EXERCISES TABLE I	SINGLE ACTIVITY
WEDNESDAY	30 MIN	RIFLE DRILL	SINGLE ACTIVITY
THURSDAY	30 MIN	CIRCUIT INTERVAL TRAINING	SINGLE ACTIVITY
FRIDAY	20 MIN 20 MIN 20 MIN	STRENGTH CIRCUIT CONDITIONING DRILL CROSS-COUNTRY RUN	ROTATION ACTIVITY
SATURDAY	60 MIN	WARM UP RUN AND GRASS DRILL	SINGLE ACTIVITY

Figure 2-2. Sample B- -Various Time Periods, Activities, and Systems.

DAY	TIME	ACTIVITY
MONDAY	45 MIN	RUNNING – CROSS-COUNTRY
TUESDAY	75 MIN	TEAM ATHLETICS – TOUCH FOOTBALL
WEDNESDAY	45 MIN	RUNNING – CROSS-COUNTRY
THURSDAY	20 MIN 40 MIN	LOG DRILL DUAL COMBATIVES
FRIDAY	90 MIN	SPEED MARCH
SATURDAY	90 MIN	TEAM CONTESTS

Figure 2-3. Sample C--Sustaining Activities.

Route	Kilometers from SP	Recommended rate of march (kmph)	Remarks
SP: RJ 8th Div Rd—Superhighway	......	4	Bridge; hard surface; two guides.
RJ Jamestown—Hersey Rd	2,25	4	Traffic heavy; two guides; Jamestown Rd bears to right.
RJ Jamestown—Yankee Rd	7.10	4	Two guides
RJ Jamestown—Lightning Rd	10.50	4	Two guides
RJ Jamestown—Sunshine Rd	12.40	4	Light traffic; one guide Sunshine Rd; gravel, poor traction when wet.
RP; RJ Sunshine—Sedan Rd	14.80	4	Two guides

Figure 3-1.Example of Route Reconnaissance Report.

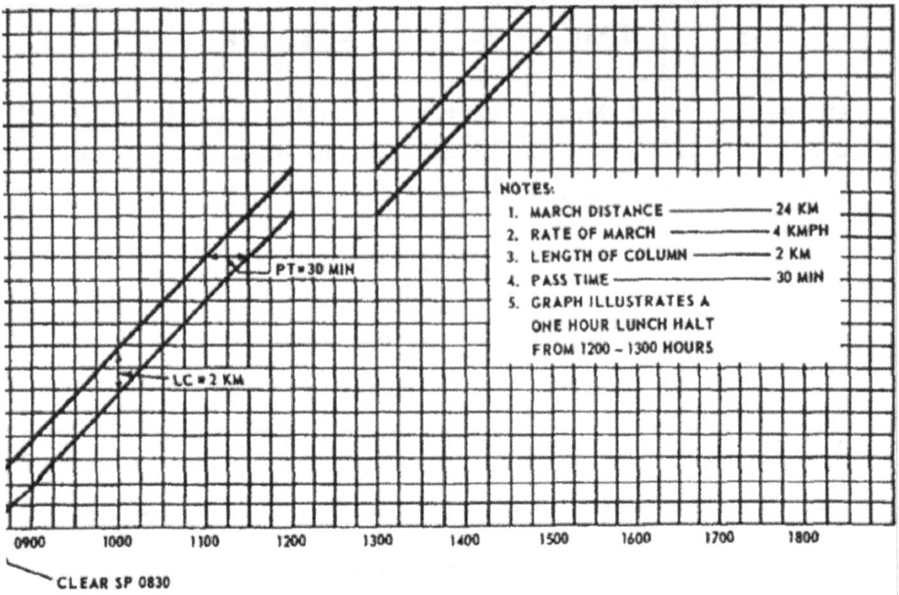

Figure 3-2.Road Movement Graph.

Figure 3-3.The Pace-Setter.

	SINGLE FILE	COLUMN OF TWO
2 m/MAN APART	2 4	1 2
5 m/MAN APART	5 4	2 7

GAP DISTANCES			
SERIAL:	(3-1) × 100 m	=	200 m
MARCH UNIT:	(12-1-2-) × 50 m	=	450 m
	TOTAL COLUMN GAP		650 m

Figure 3-4.Conditioning Drill One.

Figure 3-5.Conditioning Drill Two.

Figure 3-6.Conditioning Drill Three.

Figure 3-7. Rifle Drill.

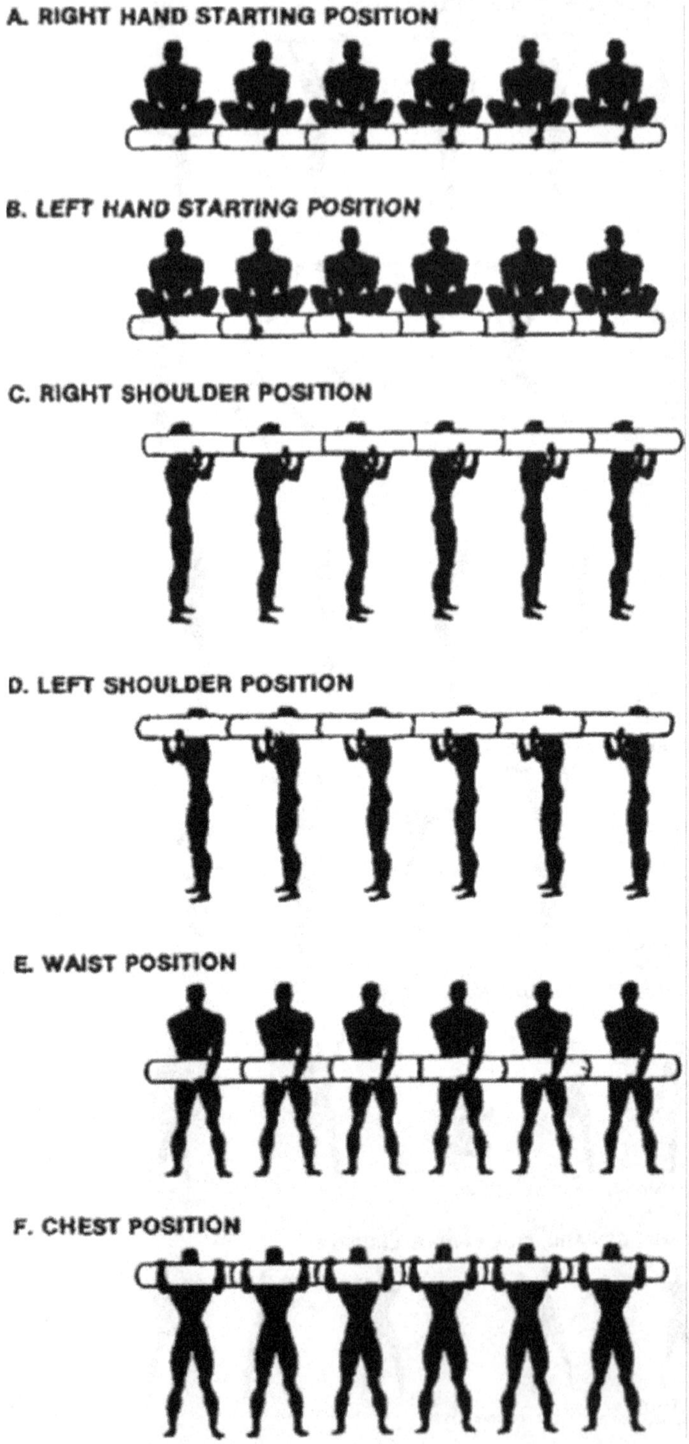

Figure 3-8. Starting Positions.

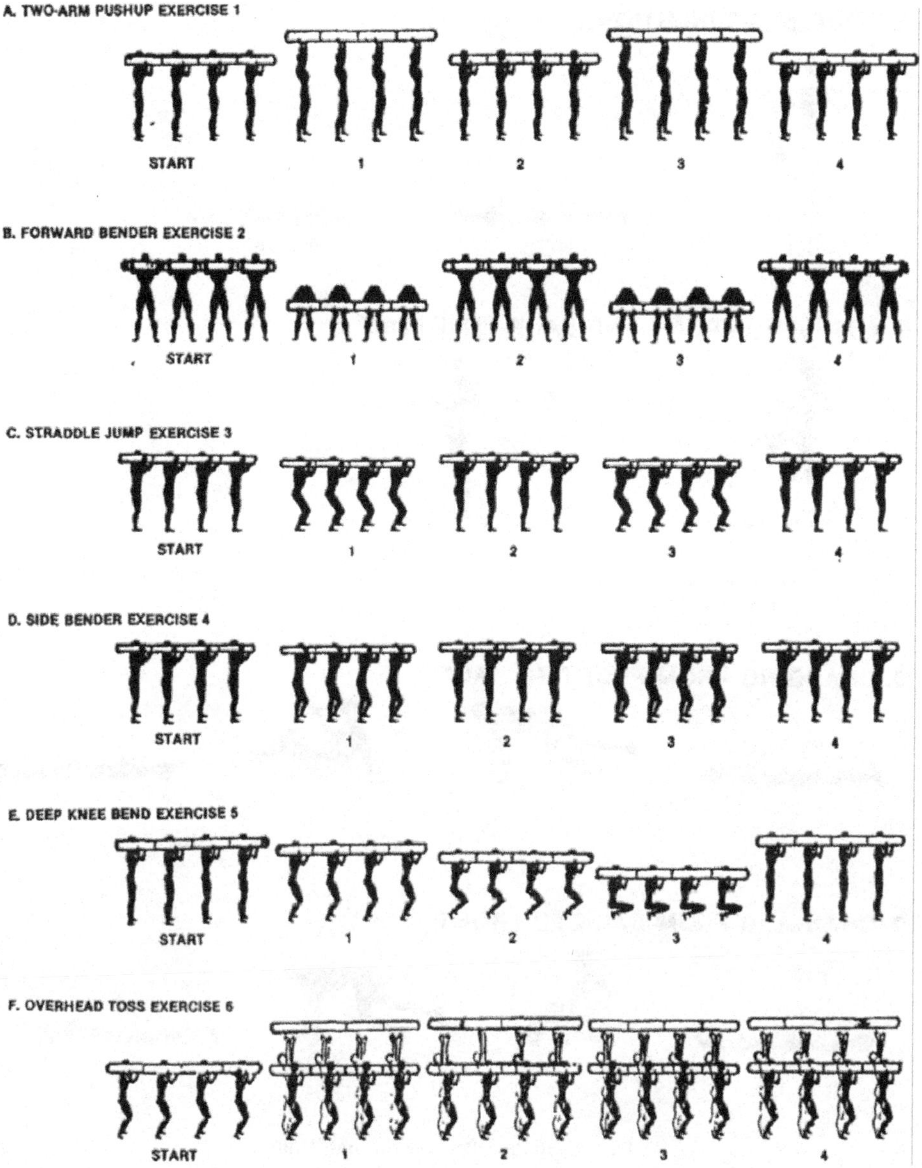

Figure 3-9. Log Drill

Figure 3-10.Basic Positions for Grass Drill.

Figure 3-11.Grass Drill One.

Figure 3-12.Grass Drill Two.

Figure 3-13.Guerrilla Table I.

Figure 3-14.Guerrilla Table II.

Figure 3-15.Proper Running Form.

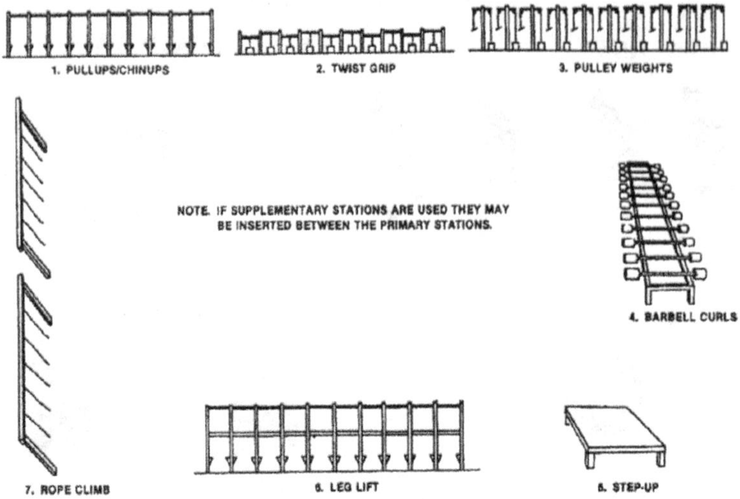

Figure 3-16. Fixed Strength Circuit.

STATION	ITEM	NO.	SPECIFICATION
1	BARBELL		1¼-INCH PIPE 5 FEET LONG WITH CONCRETE FILLED NO. 10 CANS
2	JUMP ROPE		¼- OR ⅜-INCH ROPE, 10 FEET LONG
3	TWIST GRIP		HANDLE 12 INCHES LONG, ROPE 4 FEET LONG, NO. 10 CAN CONCRETE FILLED.
4	INCLINE PLANE		¾-INCH PLYWOOD PLATFORM 2 FEET WIDE AND 6 FEET, 6 INCHES LONG, ELEVATED 10 INCHES AT ONE END. STRAP TO HOLD FEET DOWN.
5	WAR CLUB		HEAD IS 6 BY 12 INCHES, HANDLE IS 14 INCHES LONG BY 1¼ INCHES IN DIAMETER, ABOUT 20 POUNDS.
6	BICYCLE RIDE		PLYWOOD BOARD OR PLATFORM 2 BY 3 FEET WITH 2 BY 2 RUNNERS.
7	STEP – UP		A BOX OR STURDY PLATFORM 18 INCHES HIGH, 18 INCHES WIDE, 24 INCHES LONG.
8	ISOMETRIC PULL		TWO HANDLES 12 INCHES LONG WITH 4 FEET (BETWEEN HANDLES) OR LIGHT WIRE CABLE OR ¼-INCH ROPE

Figure 3-17.Movable Strength Circuit.

Figure. 3-18.Stations of Movable Strength Circuit.

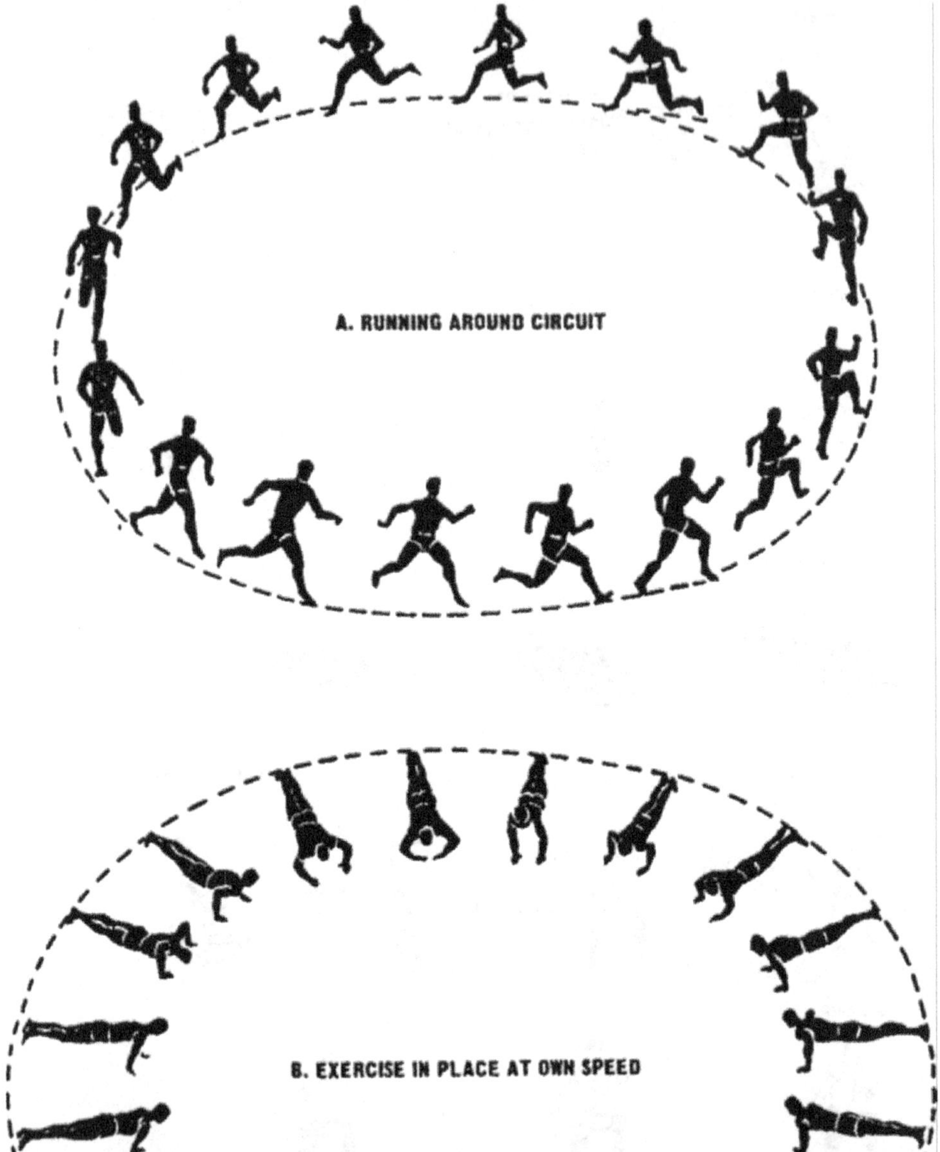

Figure 3-19. Circuit-IntervalTable.

Figure 3-20. Jumping.

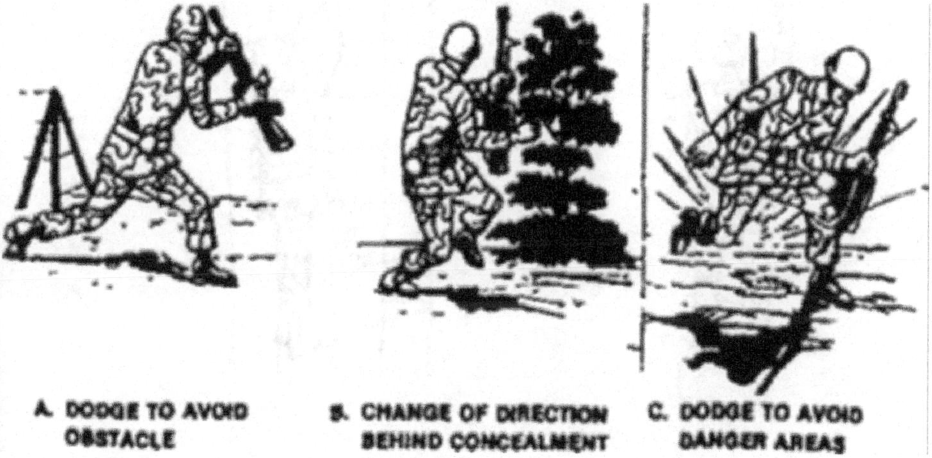

Figure 3-21.Dodging.

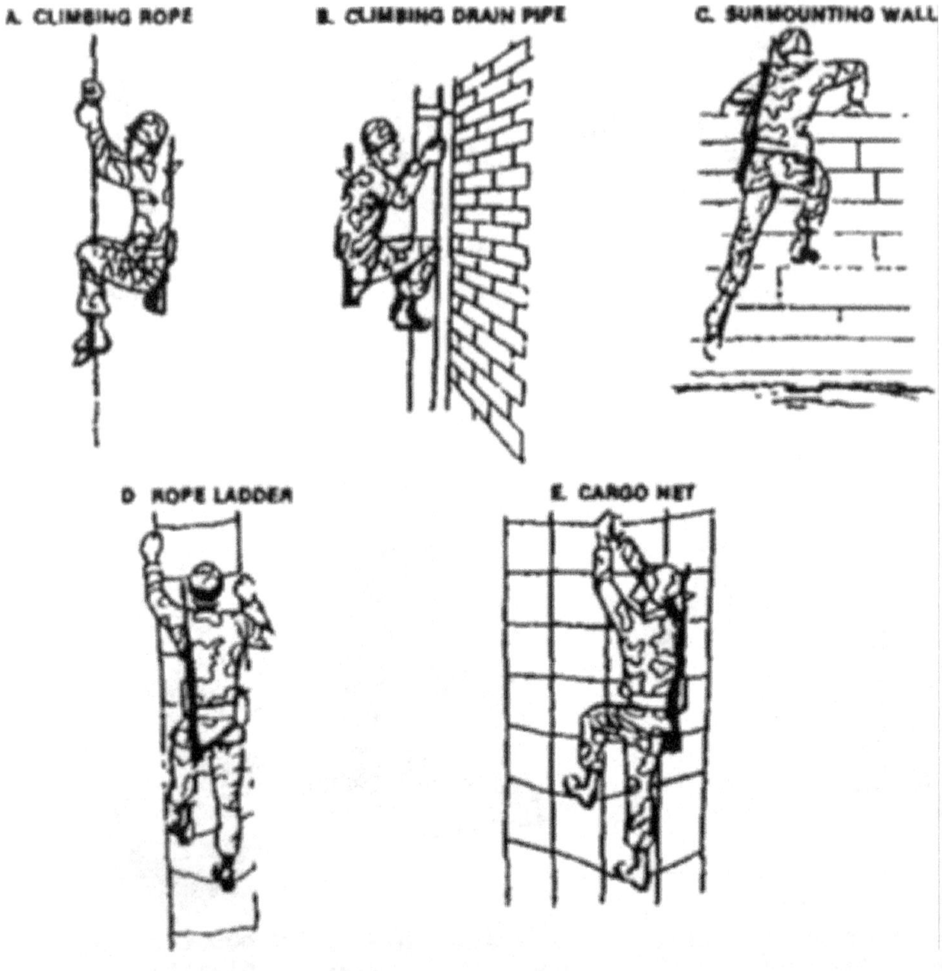

Figure 3-22.Climbing and Surmounting Vertical Objects.

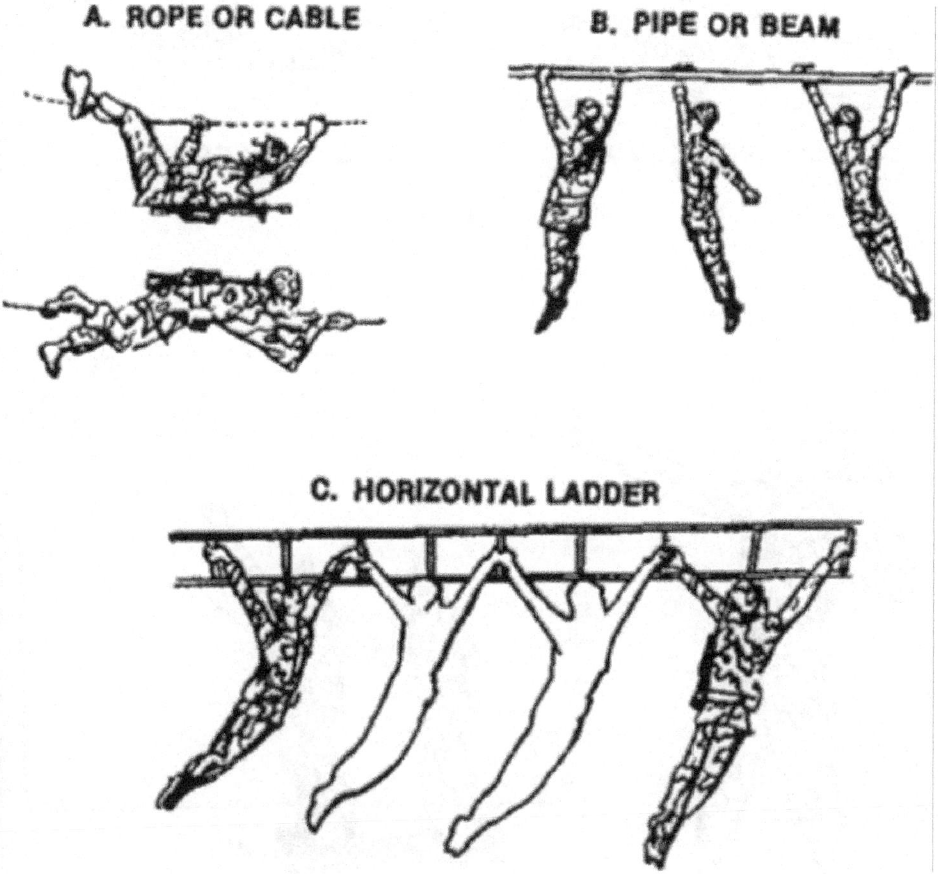

Figure 3-23. Traversing Horizontal Objects.

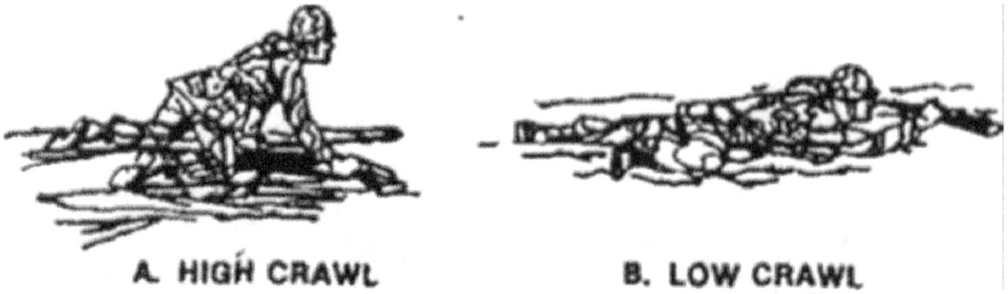

Figure 3-24.Crawling.

Figure 3-25.Throwing.

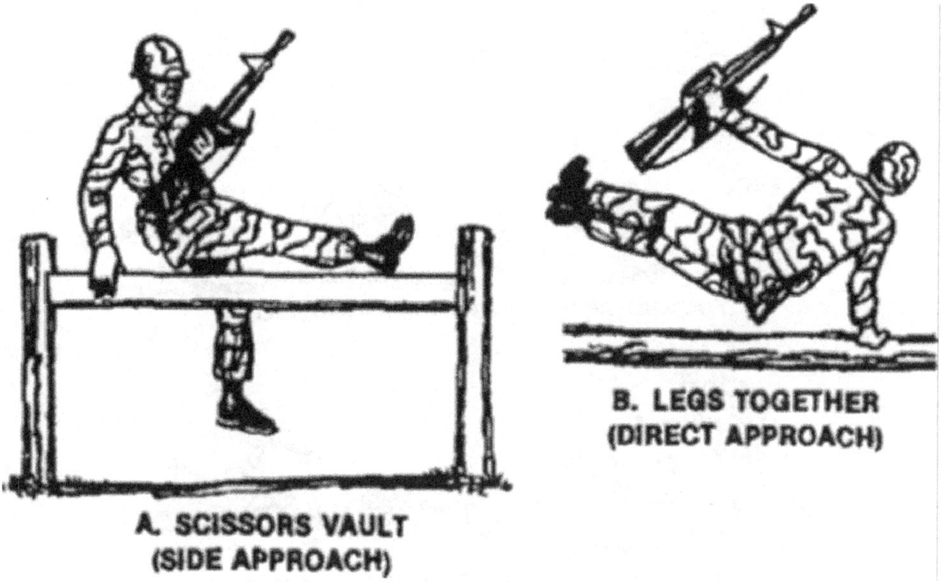

Figure 3-26. Vaulting

Figure 3-27.Balancing.

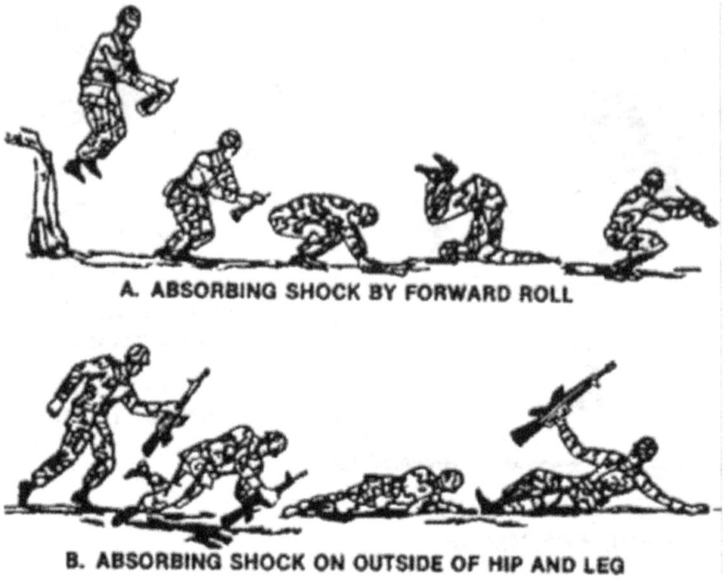

Figure 3-28.Falling.

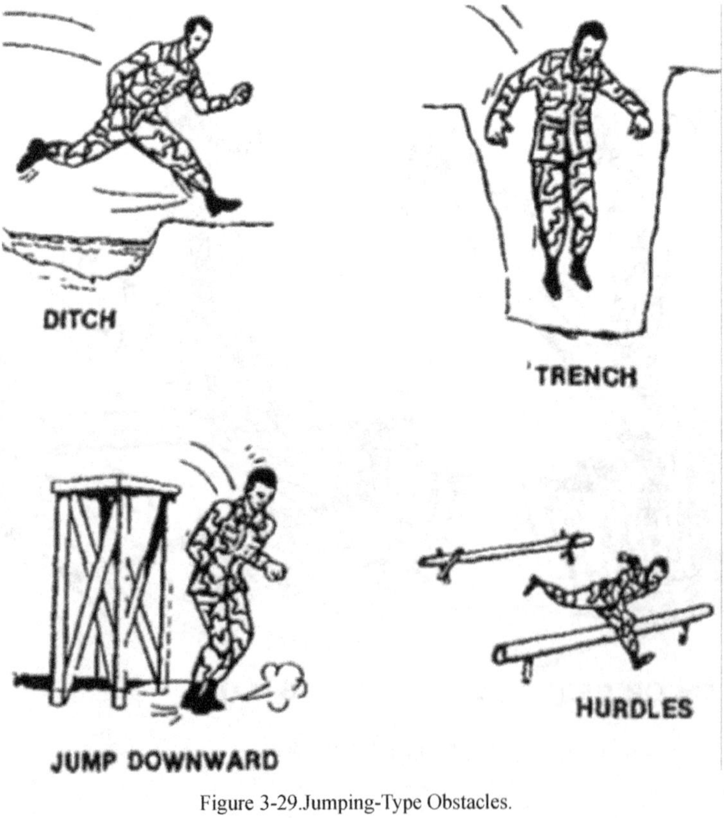

Figure 3-29.Jumping-Type Obstacles.

LANES TO GUIDE CHANGE OF DIRECTION

MAZES TO CAUSE CHANGE OF DIRECTION

Figure 3-30.Dodging-Type Obstacles.

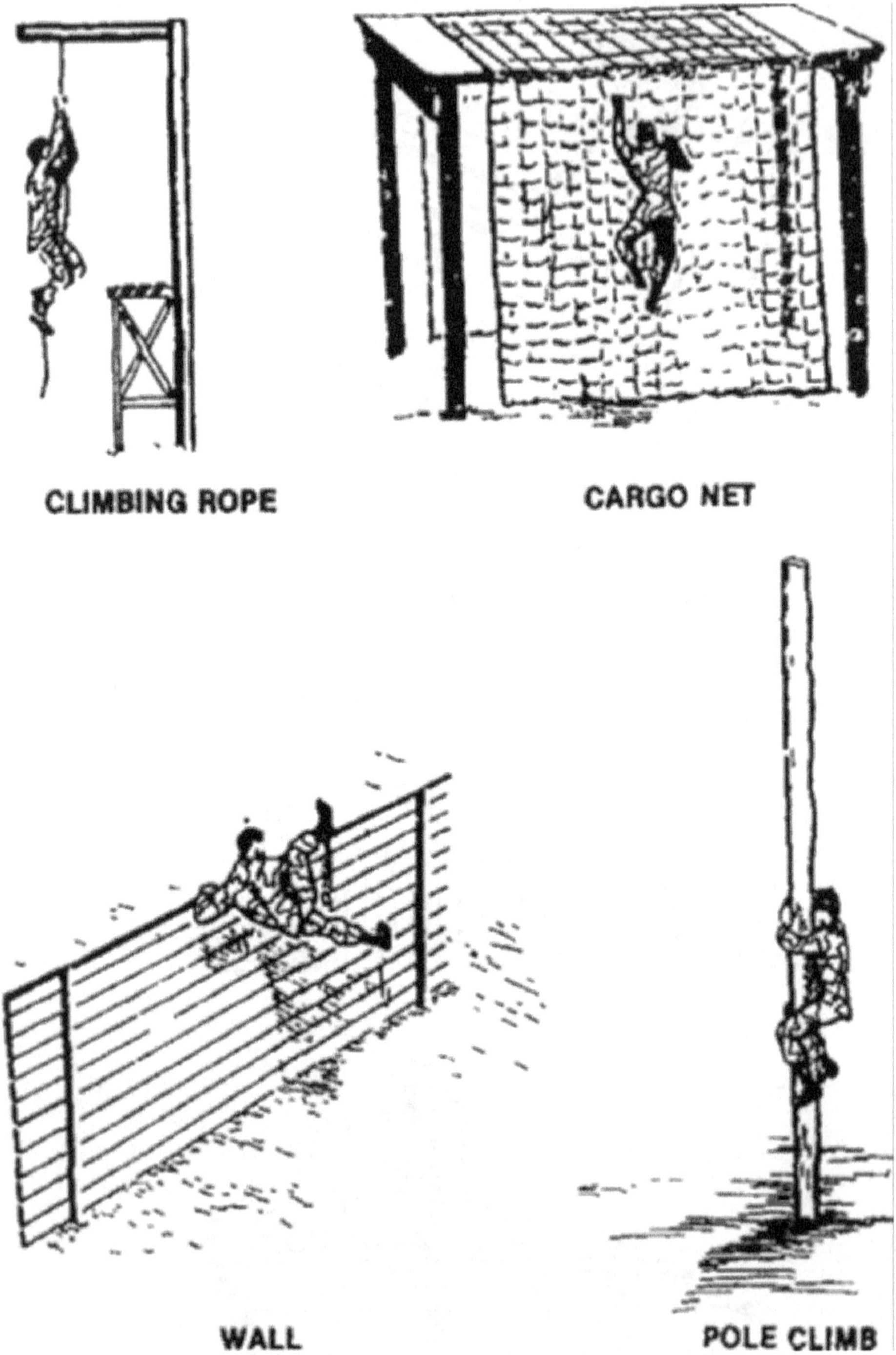

Figure 3-31. Vertical Climbing Obstacles.

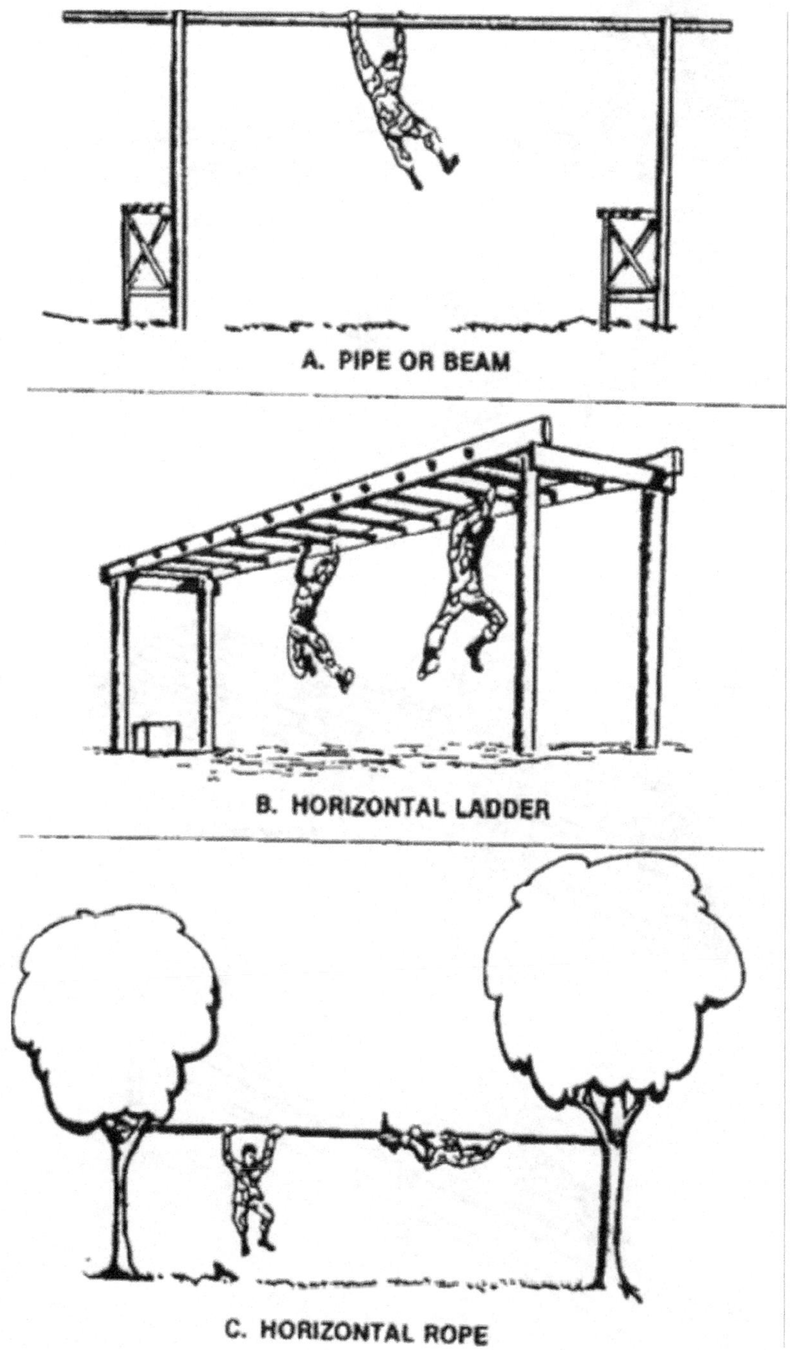

Figure 3-32.Horizontal Type Obstacles.

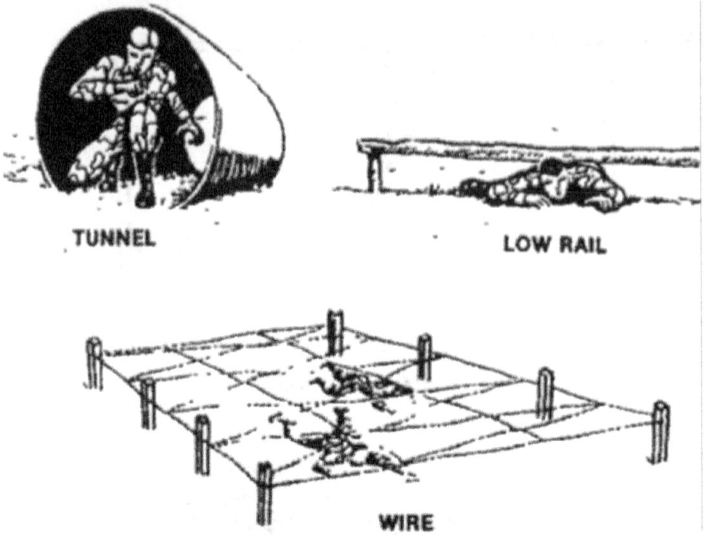

Figure 3-33.Crawling-Type Obstacles.

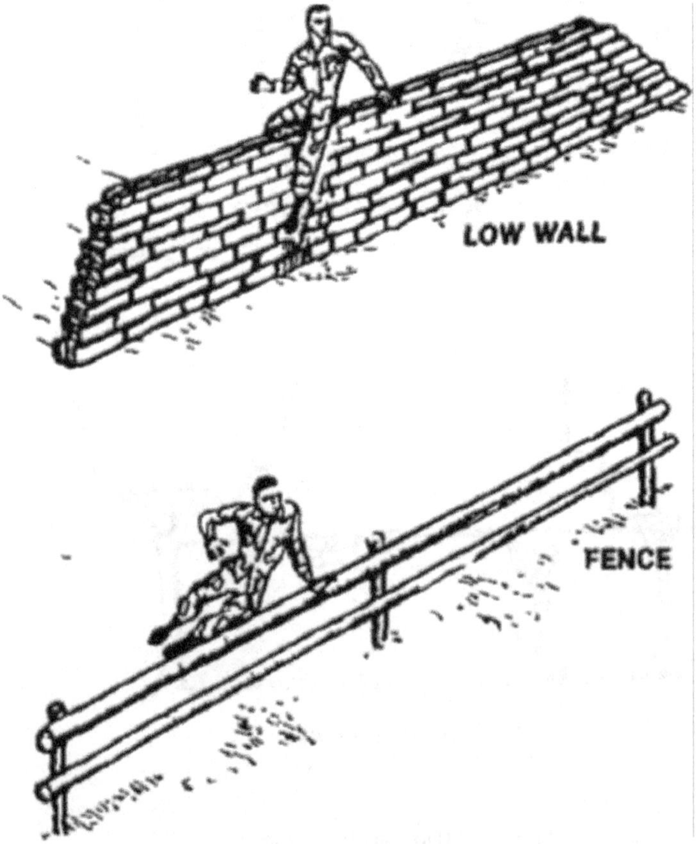

Figure 3-34. Vaulting Obstacles.

Figure 3-35.Balance-Type Obstacles.

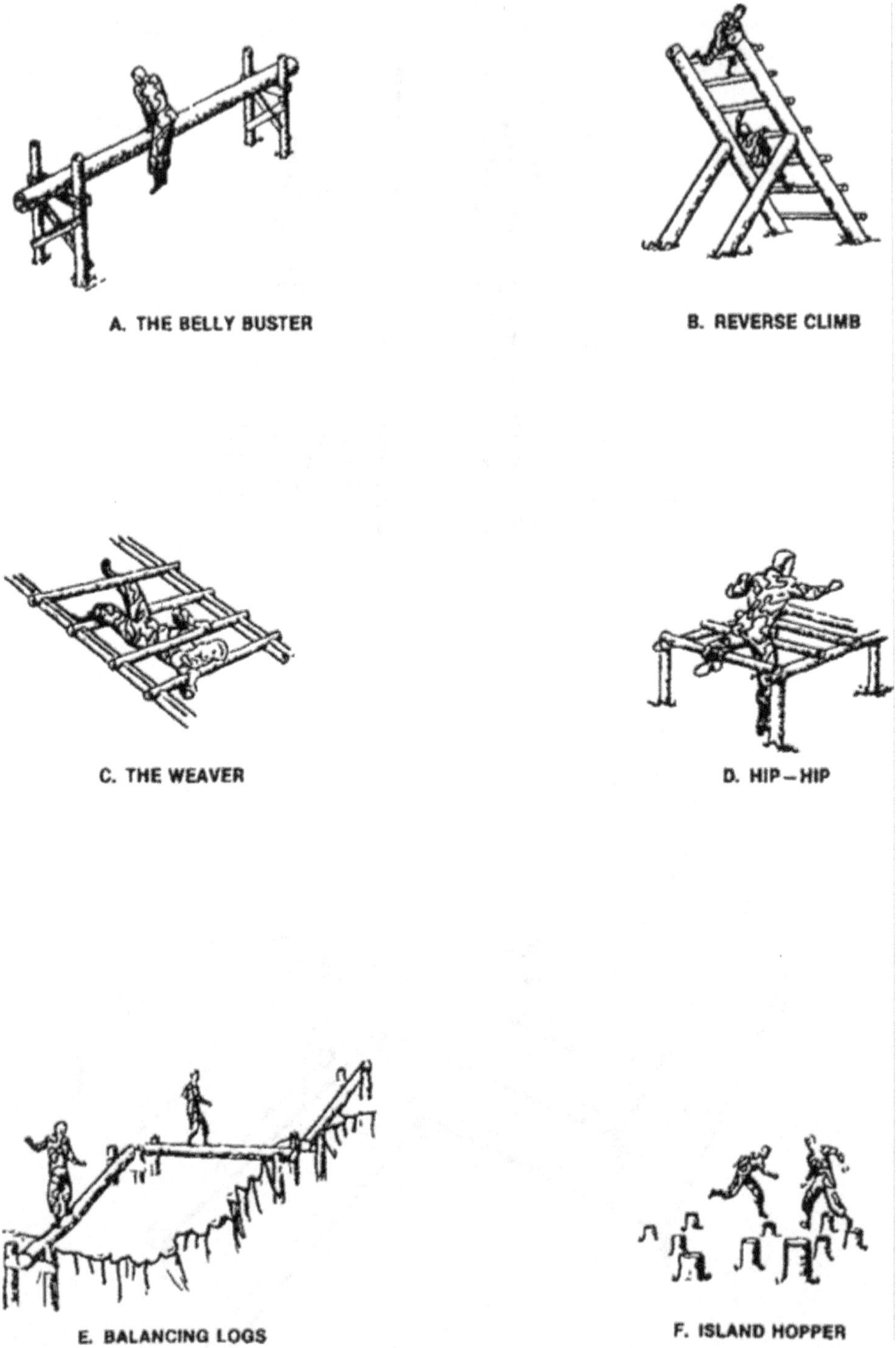

Figure 3-36.Red Group.

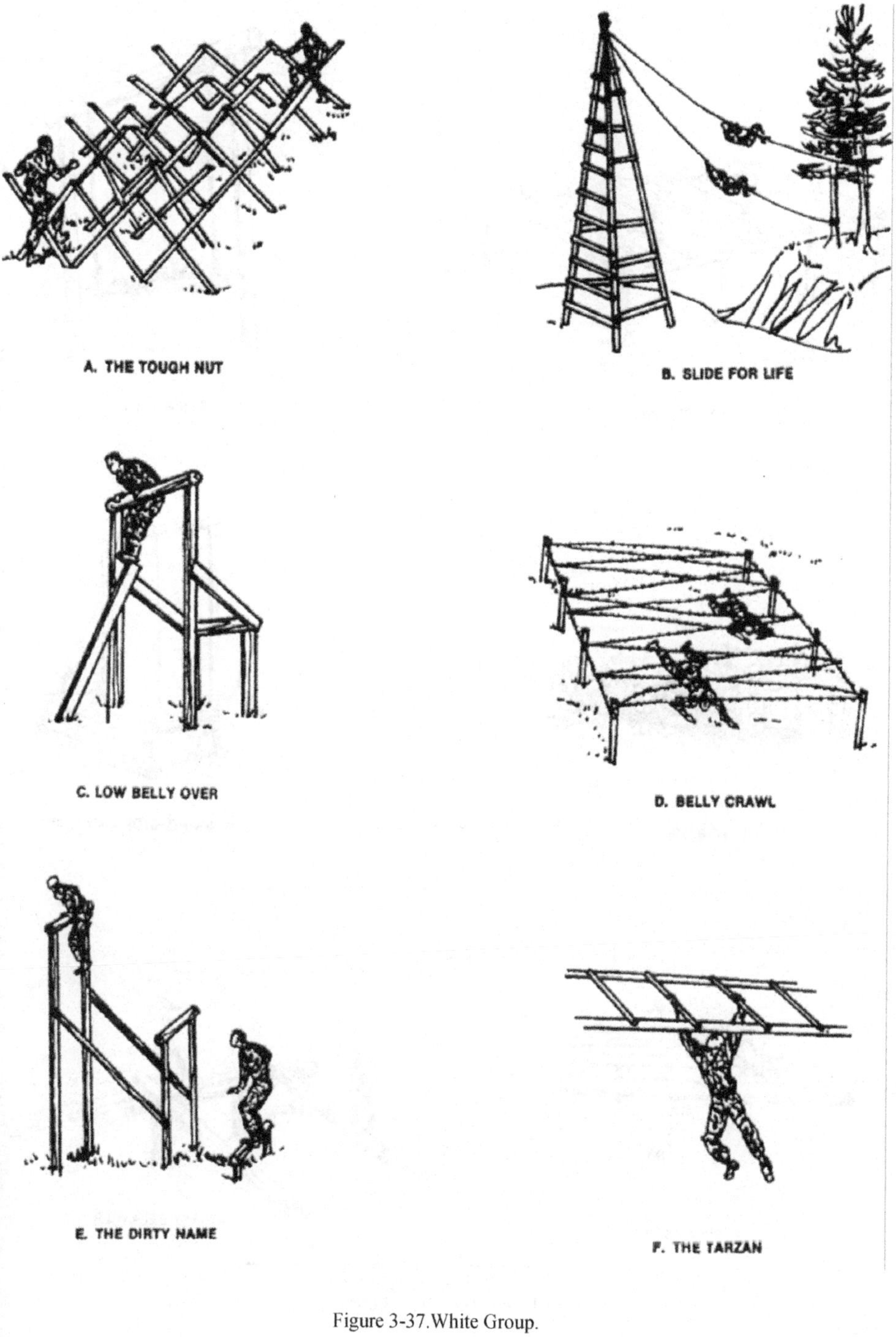

Figure 3-37.White Group.

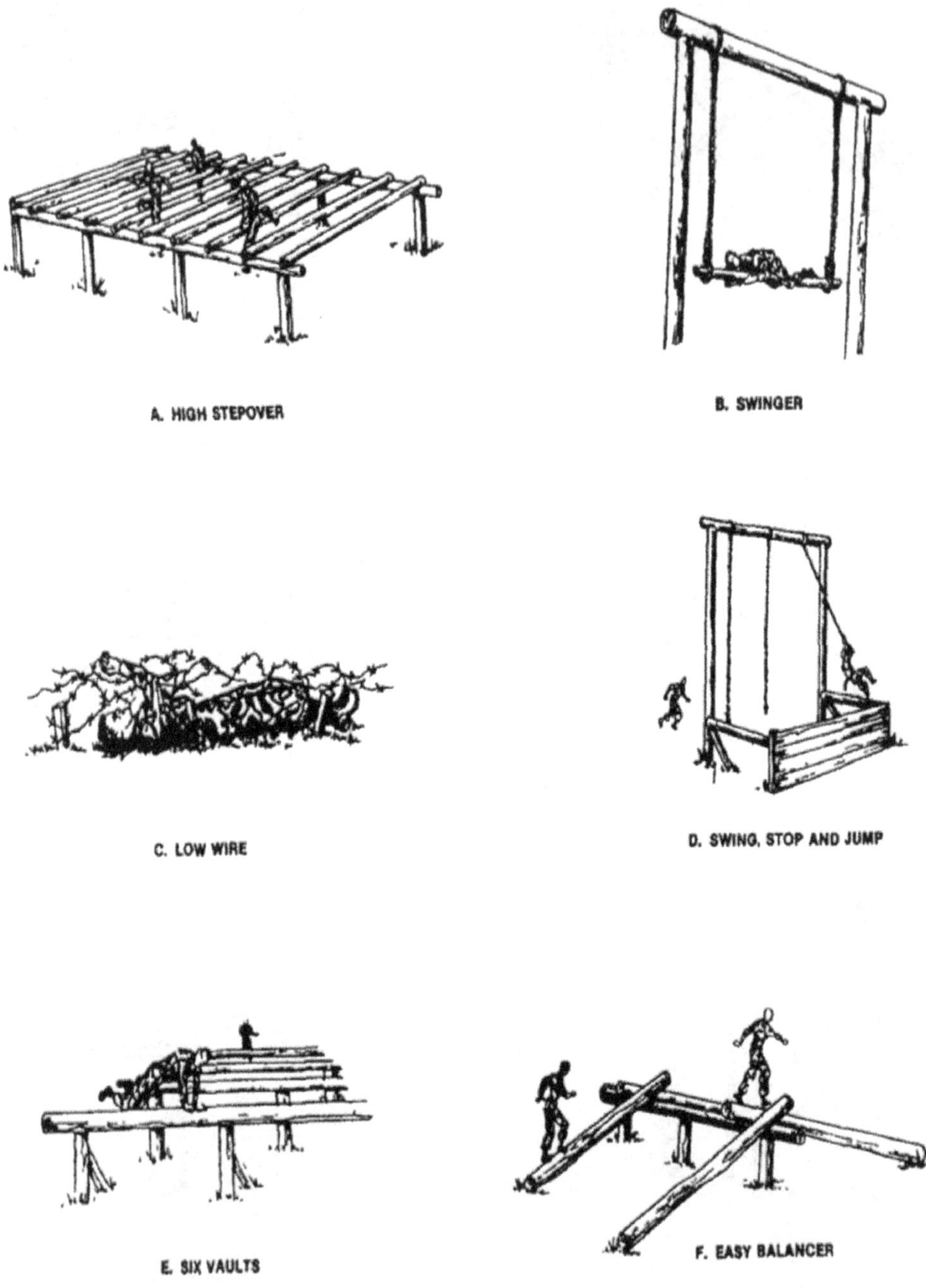

Figure 3-38.Blue Group.

A. INCLINE WALL

B. SKYSCRAPER

C. JUMP AND LAND

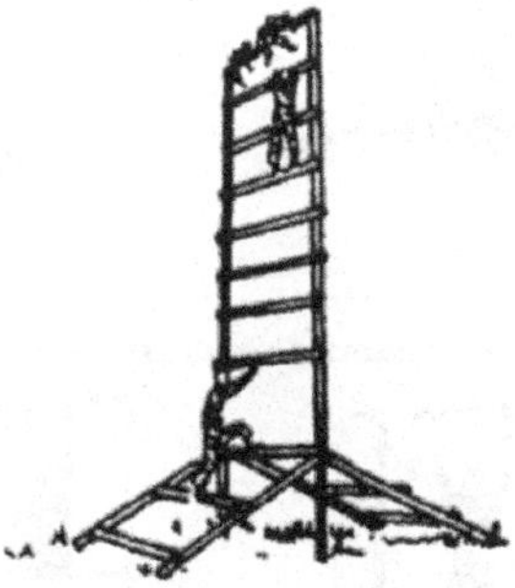

D. CONFIDENCE CLIMB

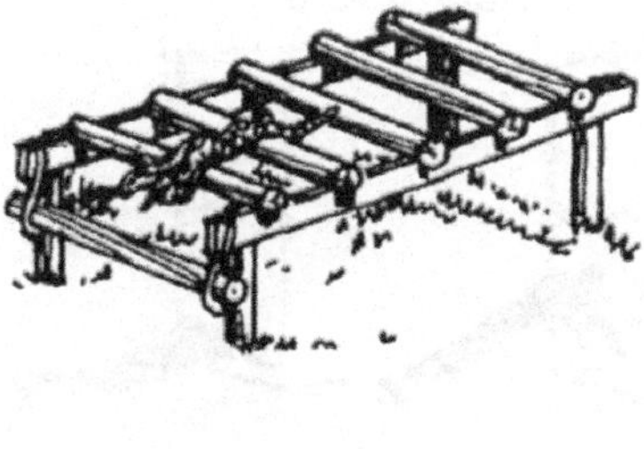

E. BELLY ROBBER

F. THE TOUGH ONE

Figure 3-39.Black Group.

Figure 3-40. Bench Conditioning, Table I.

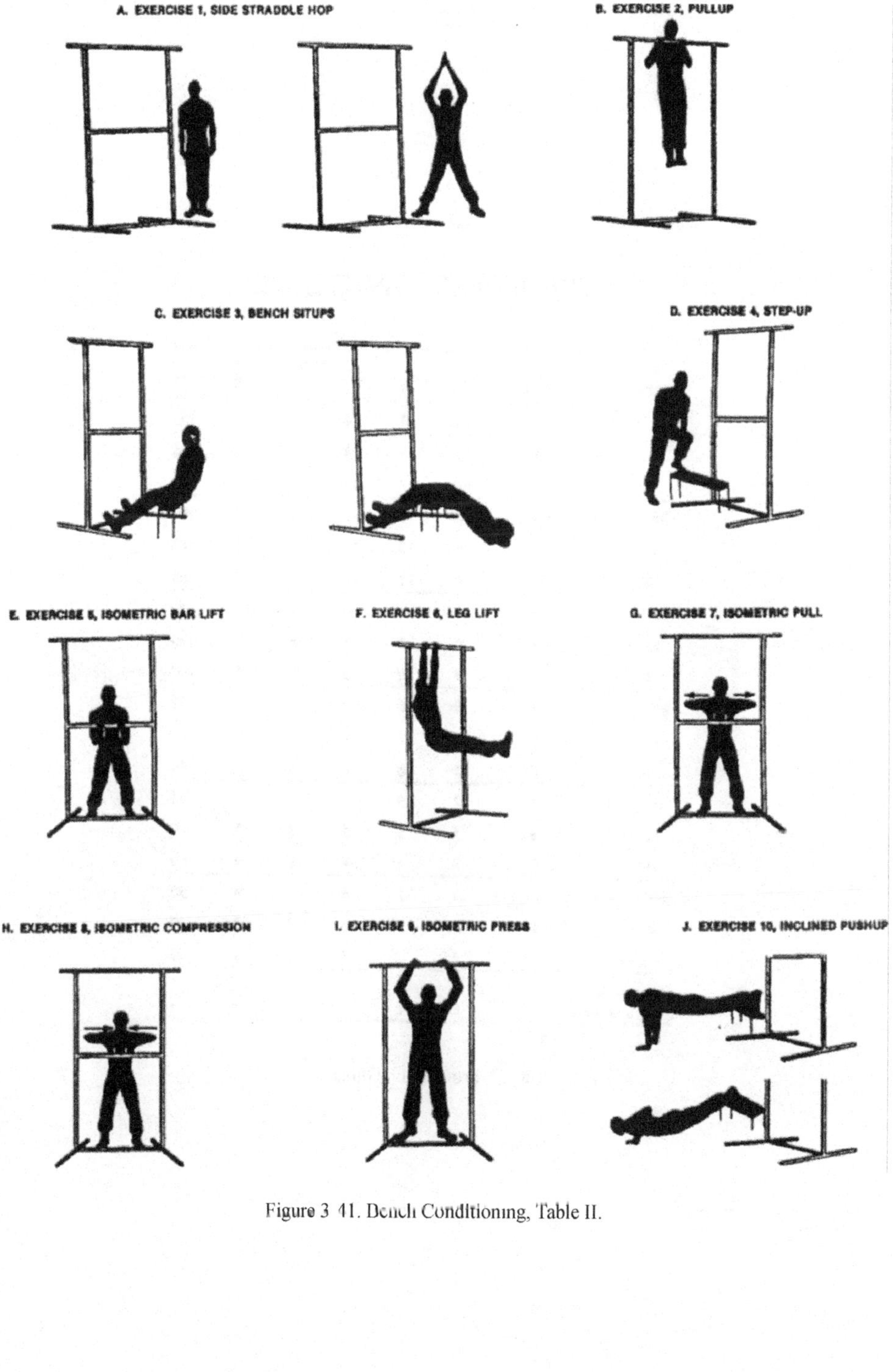

Figure 3-41. Bench Conditioning, Table II.

PROGRESSION GUIDE

AGE GROUP	LEVEL	EXERCISES 1	2	3	4	5	6
17	A	15	18	14	15	15	250
to	B	13	16	13	13	13	235
29	C	11	14	12	11	11	[illegible]
30	A	13	14	12	13	13	200
to	B	11	13	11	11	11	185
39	C	9	12	10	9	9	165
40	A	11	11	10	11	11	150
to	B	9	10	9	9	9	135
44	C	7	9	8	7	7	120
45	A	9	8	8	9	9	100
to	B	7	7	7	7	7	90
49	C	5	6	6	5	5	80
50	A	7	6	6	7	7	75
to	B	5	5	5	5	5	70
59	C	3	4	4	3	3	60
60	A	4	5	4	4	4	50
and	B	3	4	3	3	3	40
over	C	2	3	2	2	2	30
Minutes for each exercise		2	1	1	1	2	5

Table I. Progression Guide.

Table I. Exercises.

PROGRESSION GUIDE

AGE GROUP	LEVEL	EXERCISES 1	2	3	4	5	6
17	A	17	17	17	9	19	300
to	B	15	15	15	8	17	270
29	C	13	1[illegible]	13	7	15	245
30	A	15	15	15	8	17	235
to	B	13	13	13	7	15	210
39	C	11	11	11	6	13	190
40	A	13	13	13	7	15	175
to	B	11	11	11	6	13	155
44	C	9	10	9	5	11	135
45	A	11	11	11	6	13	125
to	B	9	9	9	5	11	110
49	C	7	7	7	4	9	100
50	A	9	9	9	5	11	95
to	B	7	7	7	4	9	85
59	C	5	5	5	3	7	75
60	A	6	7	7	4	9	70
and	B	5	5	5	3	7	60
over	C	4	4	4	2	5	50
Minutes for each exercise		1	1	1	1 1/2	1 1/2	6

Table II. Progression Guide.

Table II. Exercises.

PROGRESSION GUIDE

AGE GROUP	LEVEL	EXERCISES 1	2	3	4	5	6
17	A	10	19	19	16	10	350
to	B	9	17	17	15	9	315
29	C	8	15	15	14	8	280
30	A	9	17	17	14	9	270
to	B	8	15	15	13	8	240
39	C	7	13	13	12	7	210
40	A	8	15	15	12	8	200
to	B	7	13	13	11	7	180
44	C	6	11	11	10	6	160
45	A	7	13	13	10	7	150
to	B	6	11	11	9	6	135
49	C	5	9	9	8	5	120
50	A	6	11	11	8	6	115
to	B	5	9	9	7	5	105
59	C	4	7	7	6	4	95
60	A	5	9	9	7	5	90
and	B	4	7	7	6	4	80
over	C	3	5	5	4	3	70
Minutes for each exercise		1½	1	1	1½	1	6

Table III. Progression Guide.

Table III. Exercises.

PROGRESSION GUIDE

AGE GROUP	LEVEL	EXERCISES 1	2	3	4	5	6
17	A	12	9	12	24	25	400
to	B	11	8	11	22	23	380
29	C	10	7	10	21	21	360
30	A	11	8	11	23	23	305
to	B	10	7	10	21	21	290
39	C	9	6	9	20	20	275
40	A	10	7	10	20	21	225
to	B	9	6	9	18	18	215
44	C	8	5	8	16	16	205
45	A	8	6	8	16	16	175
to	B	7	5	7	14	14	165
49	C	6	4	6	12	12	155
50	A	6	5	6	13	13	135
to	B	5	4	5	11	11	130
59	C	4	3	4	10	10	120
60	A	5	4	5	10	10	100
and	B	4	3	4	9	9	95
over	C	3	2	3	8	8	90
Minutes for each exercise		1	2	1	1	1	6

Table IV. Progression Guide.

Table IV. Exercises.

PROGRESSION GUIDE

AGE GROUP	LEVEL	EXERCISES 1	2	3	4	5	6
17	A	14	13	28	14	30	450
to	B	13	12	27	13	28	430
29	C	12	11	26	12	26	410
30	A	12	12	25	12	26	350
to	B	11	11	24	11	24	330
39	C	10	10	23	10	22	310
40	A	11	11	23	11	23	250
to	B	10	10	21	10	21	240
44	C	9	9	19	9	19	230
45	A	9	9	20	9	20	200
to	B	8	8	18	8	18	190
49	C	7	7	16	7	16	180
50	A	7	7	16	7	16	170
to	B	6	6	14	6	14	155
59	C	5	5	12	5	12	140
60	A	6	6	12	6	12	115
and	B	5	5	11	5	10	110
over	C	4	4	9	4	9	105
Minutes for each exercise		2	1	1	2	1	5

Table V. Progression Guide.

Table V. Exercises.

PROGRESSION GUIDE

AGE GROUP	LEVEL	EXERCISES 1	2	3	4	5	6
17	A	17	15	32	32	35	500
to	B	16	14	30	30	33	480
29	C	15	13	28	28	31	460
30	A	15	13	30	30	31	400
to	B	14	12	28	28	29	380
39	C	13	11	26	26	27	360
40	A	13	10	27	27	27	310
to	B	12	9	25	25	25	285
44	C	11	8	23	23	23	265
45	A	11	9	23	23	23	250
to	B	10	8	21	21	21	230
49	C	9	7	19	19	19	210
50	A	9	8	19	19	19	200
to	B	8	7	17	17	17	190
59	C	7	6	15	15	15	175
60	A	8	7	15	15	17	140
and	B	7	6	13	13	15	130
over	C	5	5	10	10	12	120
Minutes for each exercise		2	1	1	1	1	6

Table VI. Progression Guide.

Table VI. Exercises.

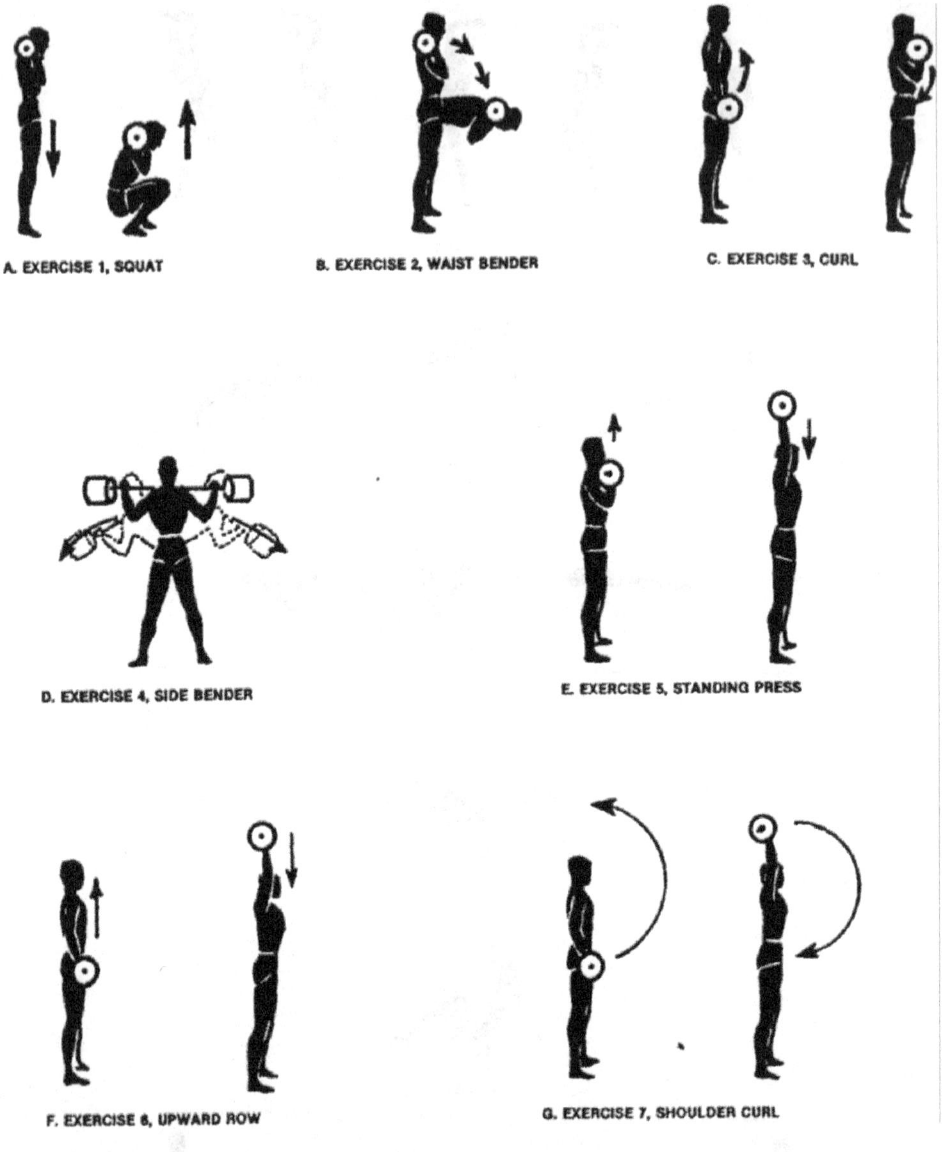

Figure 3-42. Weight Training.

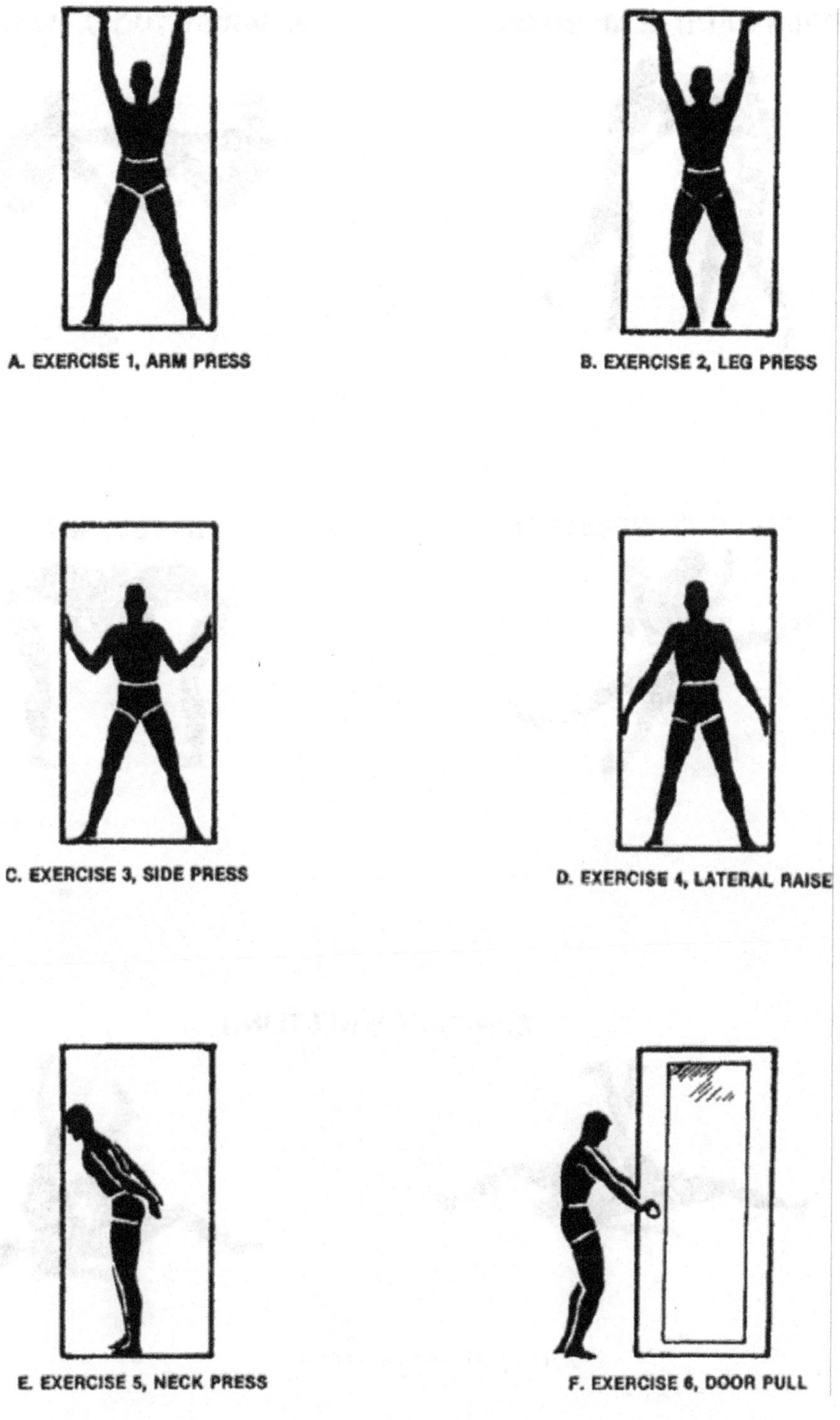

Figure 3-43. Isometric Exercises.

Figure 5-1.Combatives Table I.

Figure 5-2.Combatives Table II.

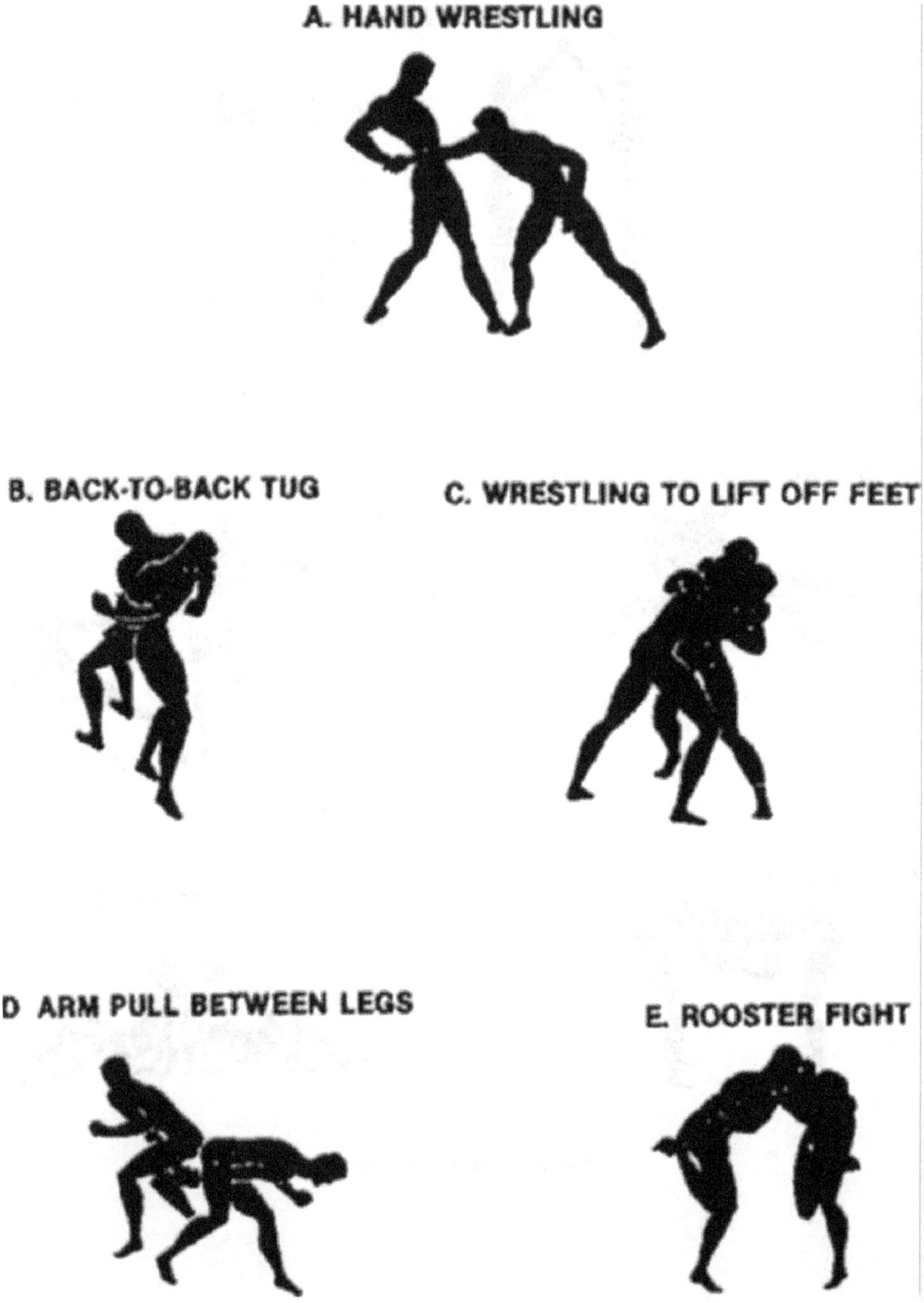

Figure. 5-3.Combatives Table III.

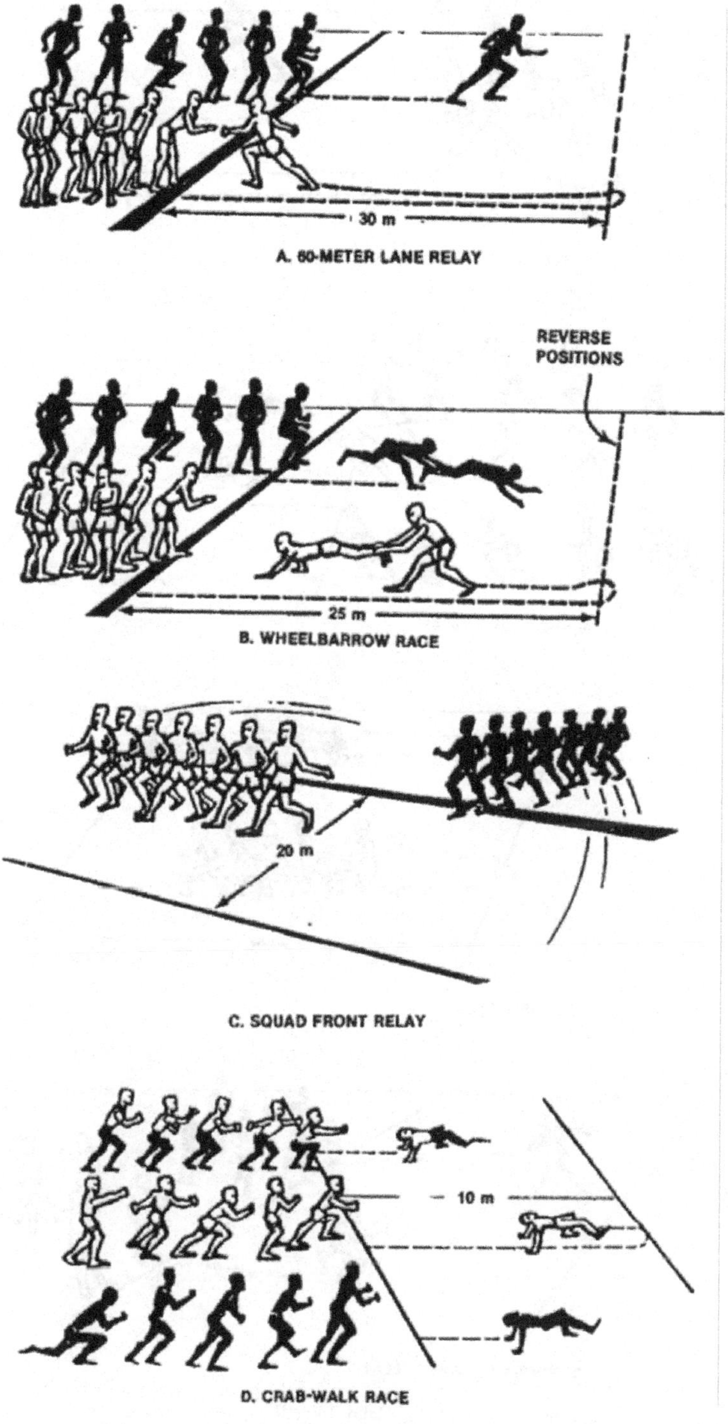

Figure 5-4.Relay Table I.

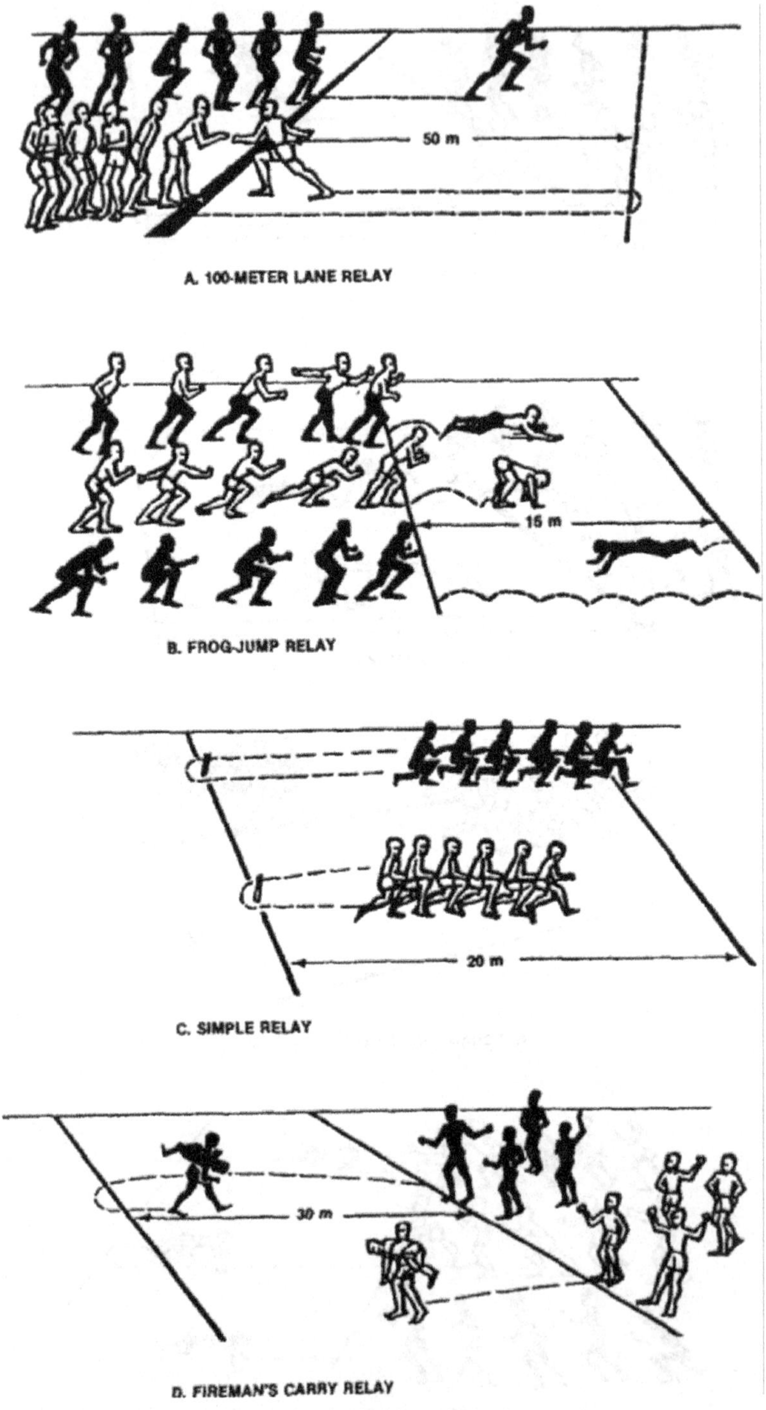

Figure 5-5.Relay Table II.

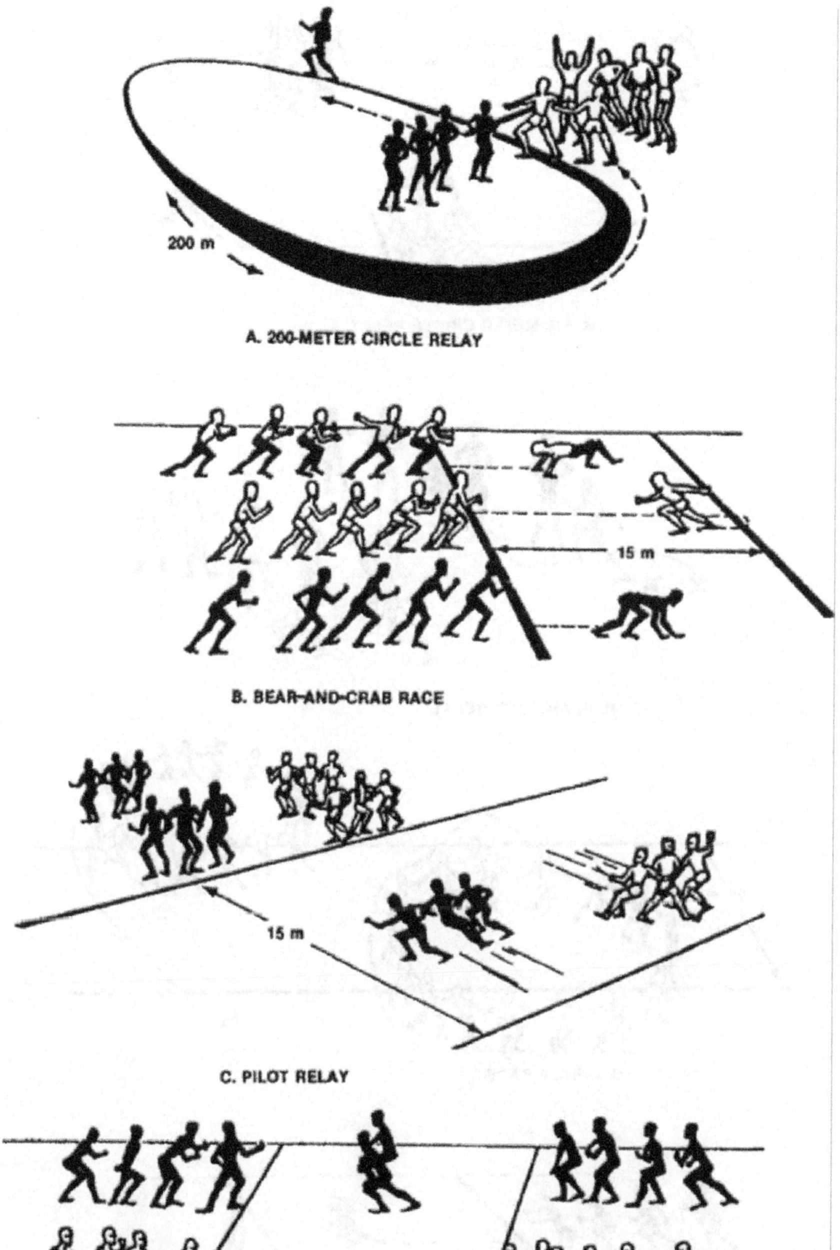

Figure 5-6.Relay Table III.

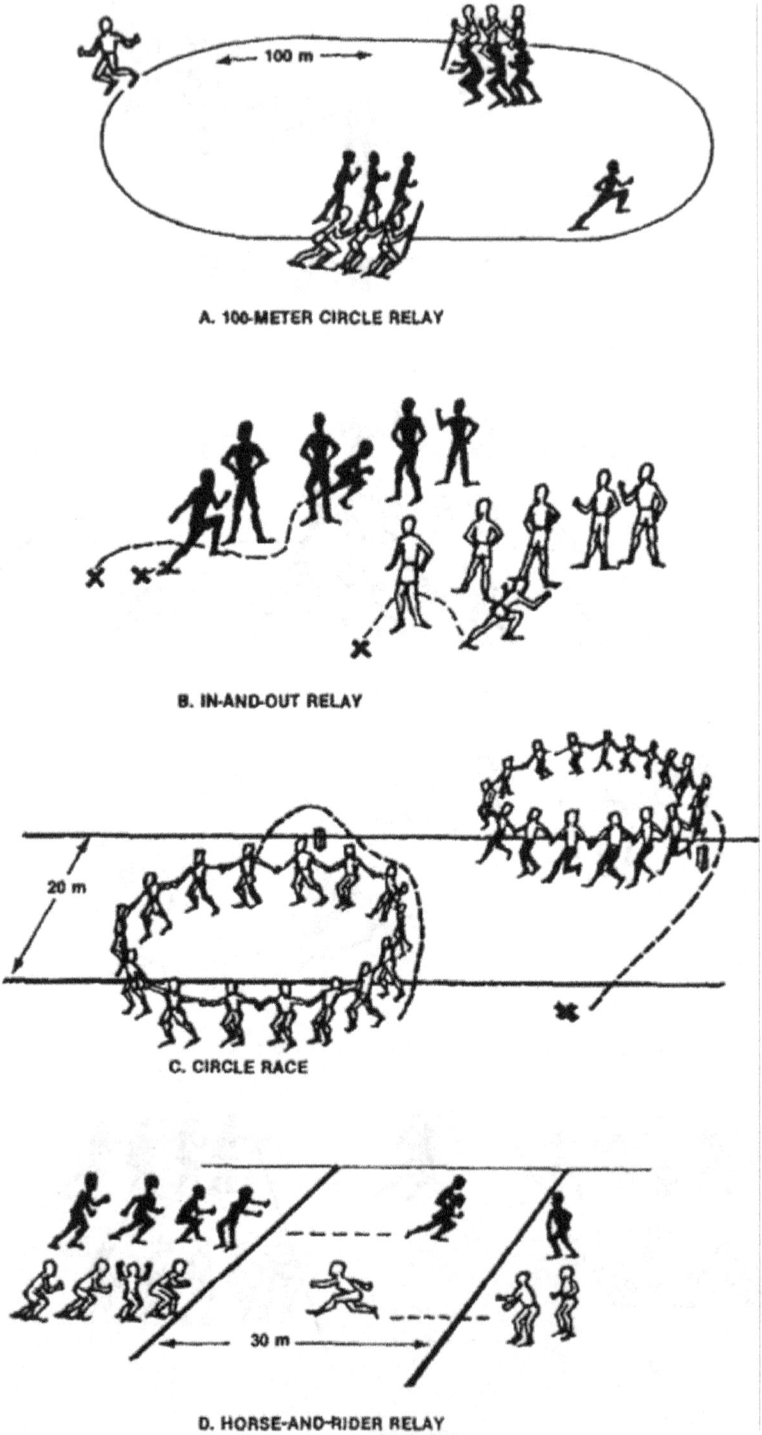

Figure 5-7.Relay Table IV.

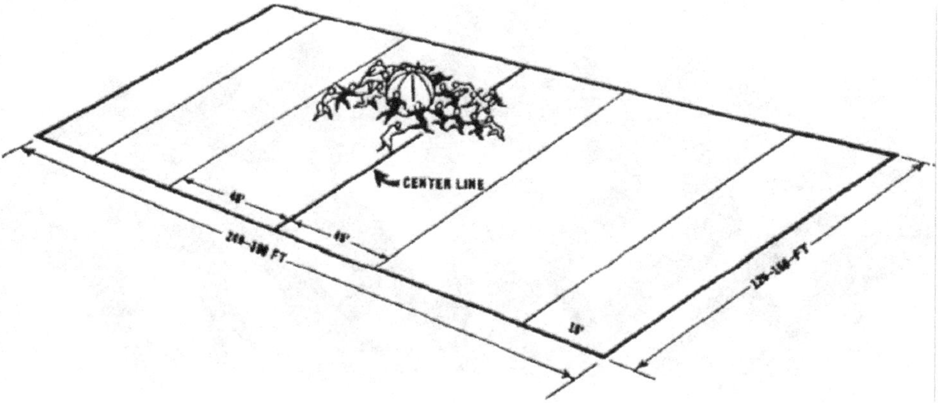

Figure 5-8.Pushball.

Figure 5-9.Line Rush.

Figure. 5-10. Human Tug-of-War.

Figure 5-11. Master of the Ring.

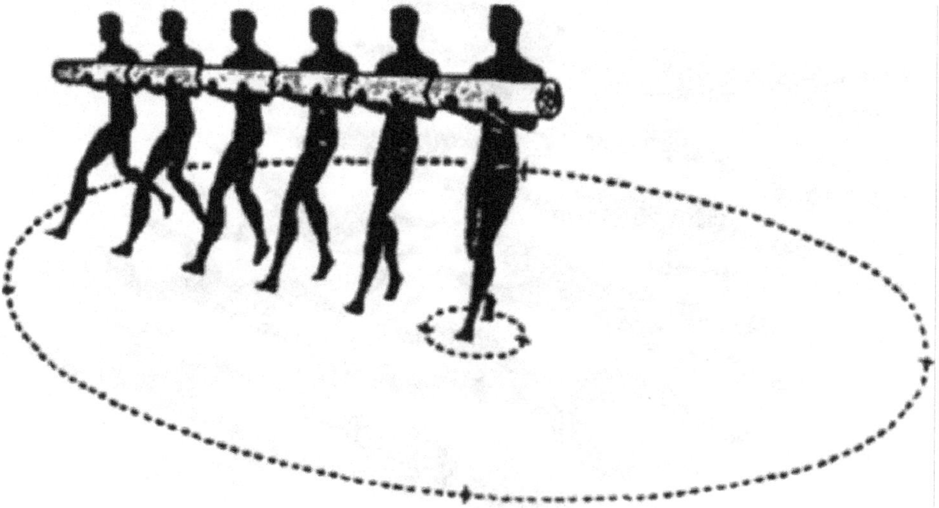

Figure 5-12.Log Pivot Circle.

Figure 5-13.Log-Rolling Race.

Figure 5-14. Prone Pushing Contest.

Figure 5-15. Shuttle Relay Race.

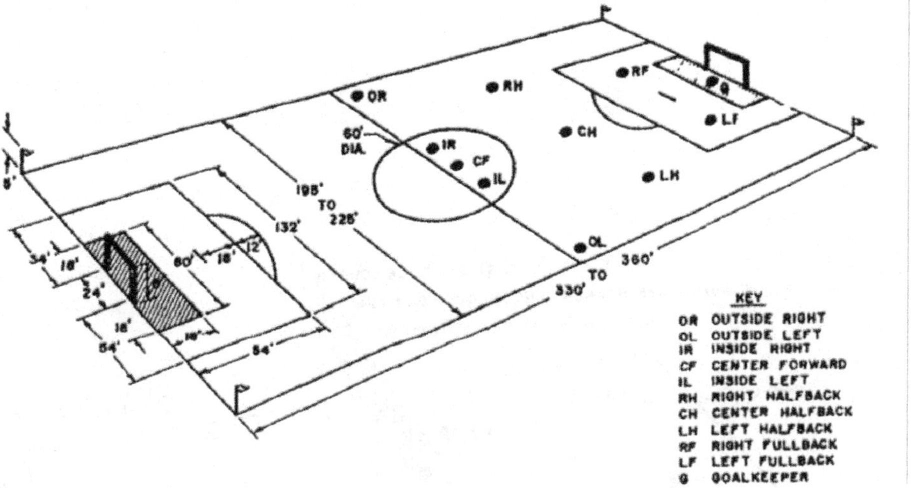

Figure 5-16. Soccer.

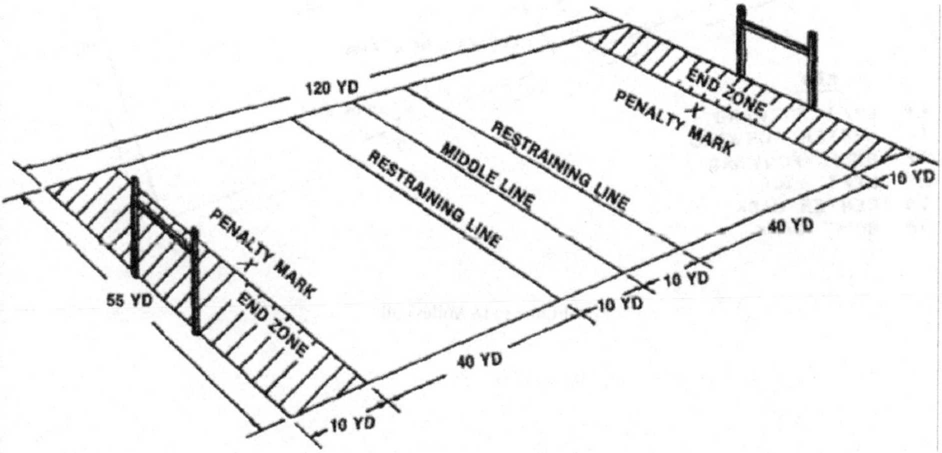

Figure 5-17.Speedball Field.

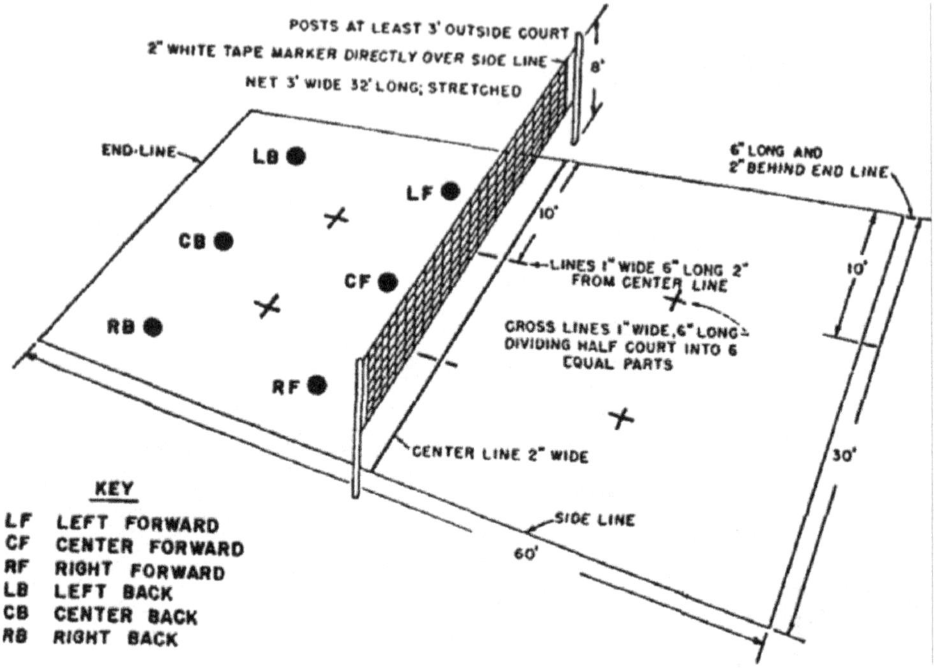

Figure 5-18. Volleyball.

PTS	UPHILL CLIMB	ROPE CLIMB	EVACUATION	FIRE AND MANEUVER	FORCED MARCH
100	1:20-BELOW	0:10-BELOW	0:25-BELOW	1:20-BELOW	26:00-BELOW
99	1:21-1:22	0:11		1:21-1:22	26:01-26:12
98	1:23-1:24	0:12	0:26	1:23-1:24	26:13-26:24
97	1:25-1:26	0:13		1:25-1:26	26:25-26:36
96	1:27-1:28	0:14	0:27	1:27-1:28	26:37-26:48
95	1:29-1:30	0:15		1:29-1:30	26:49-27:00
94	1:31-1:32	0:16	0:28	1:31-1:32	27:01-27:12
93	1:33-1:34	0:17		1:33-1:34	27:13-27:24
92	1:35-1:36	0:18	0:29	1:35-1:36	27:25-27:36
91	1:37-1:38	0:19		1:37-1:38	27:37-27:48
90	1:39-1:40	0:20	0:30	1:39-1:40	27:49-28:00
89	1:41-1:42	0:21	0:31	1:41-1:42	28:01-28:12
88	1:43-1:44	0:22	0:32	1:43-1:44	28:13-28:24
87	1:45-1:46	0:23	0:33	1:45-1:46	28:25-28:36
86	1:47-1:48	0:24	0:34	1:47-1:48	28:37-28:48
85	1:49-1:50	0:25	0:35	1:49-1:50	28:49-29:00
84	1:51-1:52	0:26	0:36	1:51-1:52	29:01-29:12
83	1:53-1:54	0:27	0:37	1:53-1:54	29:13-29:24
82	1:55-1:56	0:28	0:38	1:55-1:56	29:25-29:36
81	1:57-1:58	0:29	0:39	1:58-1:58	29:37-29:48
80	1:59-2:00	0:30	0:40	1:59-2:00	29:49-30:00
79	2:01-2:02	0:31	0:41	2:01-2:02	30:01-30:12
78	2:03-2:04	0:32	0:42	2:03-2:04	30:13-30:24
77	2:05-2:06	0:33	0:43	2:05-2:06	30:25-30:36
76	2:07-2:08	0:34	0:44	2:07-2:08	30:37-30:48
75	2:09-2:10	0:35	0:45	2:09-2:10	30:49-31:00
74	2:11-2:12	0:36	0:46	2:11-2:12	31:01-31:12
73	2:13-2:14	0:37	0:47	2:13-2:14	31:13-31:24
72	2:15-2:16	0:38	0:48	2:15-2:16	31:25-31:36
71	2:17-2:18	0:39	0:49	2:17-2:18	31:37-31:48
70	2:19-2:20	0:40	0:50	2:19-2:20	31:49-32:00
69	2:21-2:22	0:41	0:51	2:21-2:22	32:01-32:12
68	2:23-2:24	0:42	0:52	2:23-2:24	32:13-32:24
67	2:25-2:26	0:43	0:53	2:25-2:26	32:25-32:36
66	2:27-2:28	0:44	0:54	2:27-2:28	32:37-32:48
65	2:29-2:30	0:45	0:55	2:20-2:30	32:49-33:00

Figure 6-1.Physical Readiness Test Scoring Table for Male.

	EXAMPLE 1	EXAMPLE 2	EXAMPLE 3
Event #1	100 pts	100 pts	99 pts
Event #2	100 pts	99 pts	86 pts
Event #3	85 pts	75 pts (F)	80 pts
Event #4	90 pts	79 pts (F)	100 pts
Event #5	79 pts	85 pts	90 pts
Pts	454 pts	438 pts	455 pts
Final Score	79.0% (F)	69.0% (F)	91.0%
(F) = Failure			

Figure 6-2.Scoring Examples.

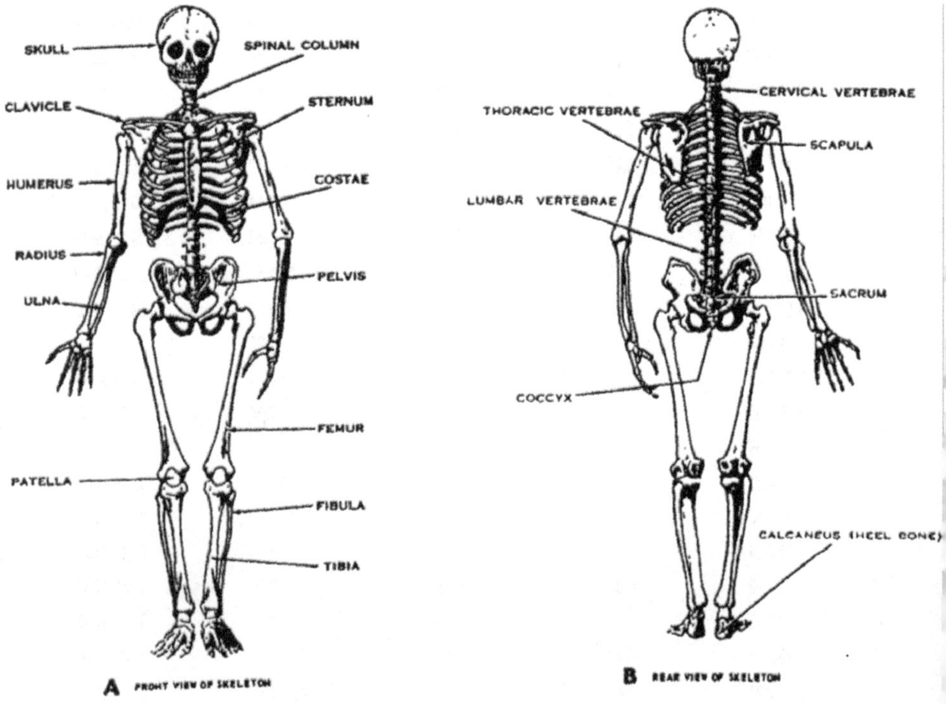

Figure 7-1.Front and Rear Views of Skeleton.

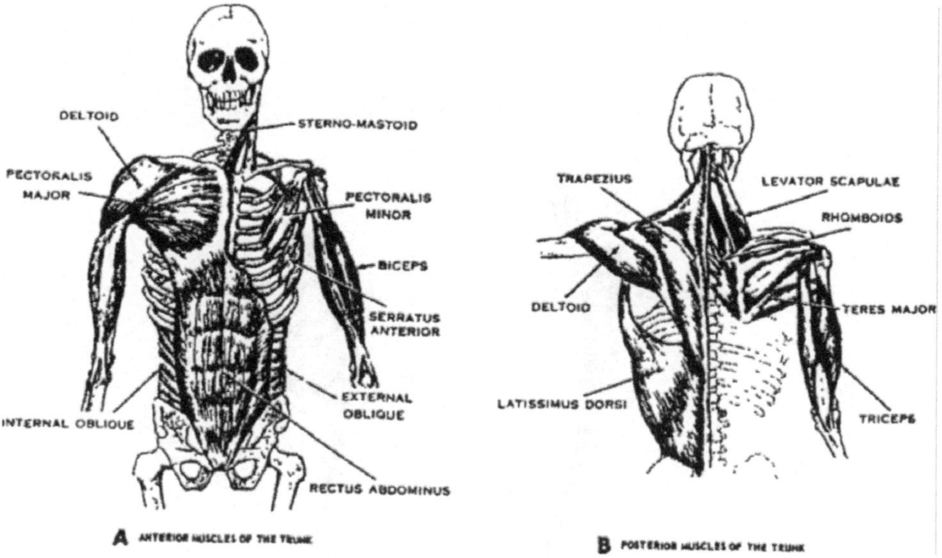

Figure 7-2.Muscles of the Trunk.

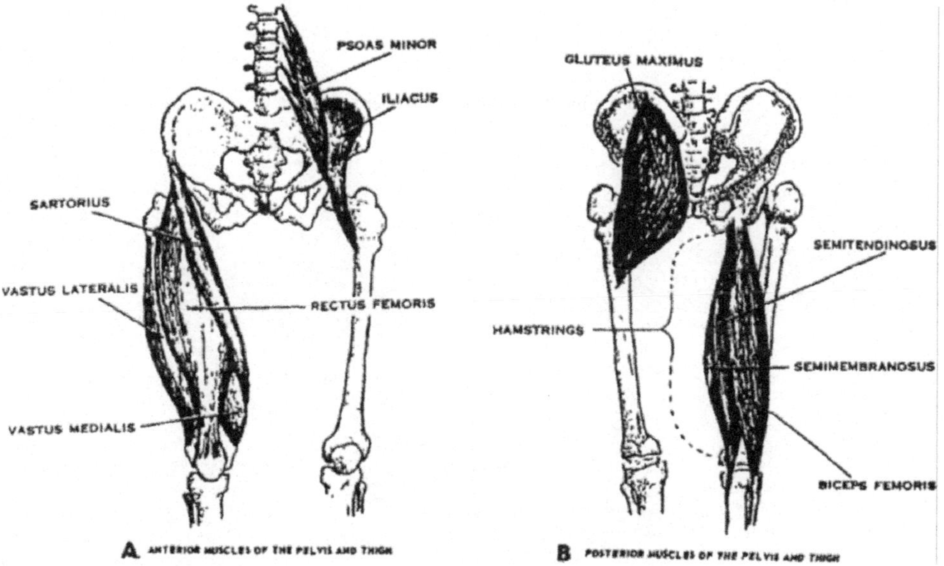

Figure 7-3.Anterior and Posterior Muscles of the Pelvisand Thigh.

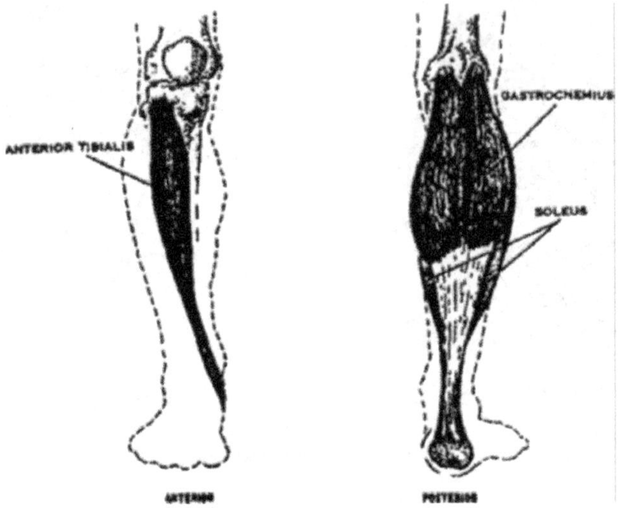

Figure 7-4. Muscles of Lower Leg.

Figure 7-5.Circulation of Blood from Heart to the Body.

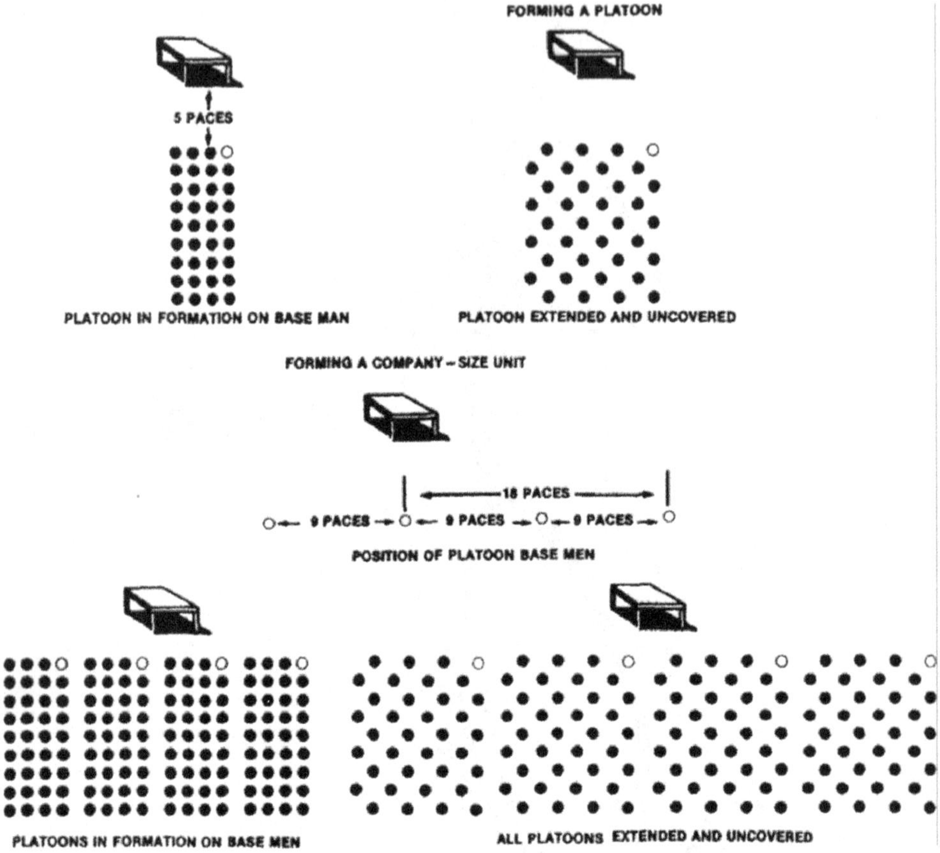

Figure A-l. Forming the Extended Rectangular Formation.

Figure A-2.Circle Formation.

Mountain Operations

Department of Defense

Pentagon
Publishing

Preface

FM 3-97.6 describes the tactics, techniques, and procedures that the United States (US) Army uses to fight in mountainous regions. It is directly linked to doctrinal principles found in FM 3-0 and FM 3-100.40 and should be used in conjunction with them. It provides key information and considerations for commanders and staffs regarding how mountains affect personnel, equipment, and operations, it also assists them in planning, preparing, and executing operations, battles, and engagements in a mountainous environment.

Army units do not routinely train for operations in a mountainous environment. Therefore, commanders and trainers at all levels should use this manual in conjunction with TC 90-6-1. Army Training and Evaluation Program (ARTEP) mission training plans, and the training principles in FM 7-0 and FM 7-10 when preparing to conduct operations in mountainous terrain.

The proponent of this publication is Headquarters TRADOC. Send comments and recommendations on DA Form 2028 directly to Commander, US Army Combined Arms Center and Fort Leavenworth, ATTN: ATZL-SWW. Fort Leavenworth, Kansas 66027-6900.

Unless this publication states otherwise, masculine nouns and pronouns do not refer exclusively to men.

Field Manual
No. 3-97.6

***FM 3-97.6**

Headquarters
Department of the Army
Washington, DC, 28 November 2000

Mountain Operations

Contents

Page

DISTRIBUTION RESTRICTION: Approved for public release; distribution unlimited

This publication supersedes FM 90-6, 30 June 1980.

Introduction

The US Army has a global area of responsibility and deploys to accomplish missions in both violent and nonviolent environments. The contemporary strategic environment and the scope of US commitment dictate that the US Army be prepared for a wide range of contingencies anywhere in the world, from the deserts of southwest Asia and the jungles of South America and southeast Asia to the Korean Peninsula and central and northern Europe. The multiplicity of possible missions makes the likelihood of US involvement in mountain operations extremely high. With approximately 38 percent of the world's landmass classified as mountains, the Army must be prepared to deter conflict, resist coercion, and defeat aggression in mountains as in other areas.

Throughout the course of history, armies have been significantly affected by the requirement to fight in mountains. During the 1982 Falkland Islands (Malvinas) War, the first British soldier to set foot on enemy-held territory on the island of South Georgia did so on a glacier. A 3.000-meter (10,000-foot) peak crowns the island, and great glaciers descend from the mountain spine. In southwest Asia, the borders of Iraq, Iran, and Turkey come together in mountainous terrain with elevations of up to 3,000 meters (10,000 feet).

Mountainous terrain influenced the outcome of many battles during the Iran-Iraq war of the 1980s. I n the mountains of Kurdistan, small Kurdish formations took advantage of the terrain in an attempt to survive the I raqi Army's attempt to eliminate them. I n the wake of the successful United Nations (UN) coalition effort against Iraq, US forces provided humanitarian assistance to Kurdish people suffering from the effects of the harsh mountain climate.

Major mountain ranges, which are found in desert regions, jungles, and cold climate zones, present many challenges to military operations. Mountain operations may require special equipment, special training, and acclimatization. Historically, the focus of mountain operations has been to control the heights or passes. Changes in weaponry, equipment, and technology have not significantly shifted this focus. Commanders should understand a broad range of different requirements imposed by mountain terrain, including two key characteristics addressed in this manual: (1) the significant impact of severe environmental conditions on the capabilities of units and their equipment, and (2) the extreme difficulty of ground mobility in mountainous terrain.

Field Manual
No. 3-97.6

***FM 3-97.6**

Headquarters
Department of the Army
Washington, DC, 28 November 2000

Mountain Operations

Contents

DISTRIBUTION RESTRICTION: Approved for public release; distribution unlimited

This publication supersedes FM 90-6, 30 June 1980.

Chapter 1

Intelligence

Before they can understand how to fight in mountainous environment, commanders must analyze the area of operations (AO), understand its distinct characteristics, and understand how these characteristics affect personnel and equipment. This chapter provides detailed information on terrain and weather necessary to conduct a thorough intelligence preparation of the battlefield (IPB). however, the IPB process remains unaffected by mountains (see FM 2-01.3 for detailed information on how to conduct IPB).

SECTION I - THE PHYSICAL ENVIRONMENT

1-1. The requirement to conduct military operations in mountainous regions presents commanders with challenges distinct from those encountered in less rugged environments and demands increased perseverance, strength. will, and courage. Terrain characterized by steep slopes, great variations in local relief, natural obstacles, and lack of

CONTENTS

accessible routes restricts mobility, drastically increases movement times, limits the effectiveness of some weapons, and complicates supply operations. The weather, variable with the season and time of day, combined with the terrain, can greatly affect mobility and tactical operations. Even under non-violent conditions, operations in a mountainous environment may pose significant risks and dangers.

TERRAIN

1-2. Mountains may rise abruptly from the plains to form a giant barrier or ascend gradually as a series of parallel ridges extending unbroken for great distances. They may consist of varying combinations of isolated peaks, rounded crests, eroded ridges, high plains cut by valleys, gorges, and deep ravines. Some mountains, such as those found in desert regions, are dry and barren, with temperatures ranging from extreme heal in the summer to extreme cold in thewinter. In tropical regions, tush jungleswith heavy seasonal rains and little temperature variation frequently cover mountains. High, rocky crags with glaciated peaks and year-round snow cover exist in mountain ranges at most latitudes along the western portion of the Americas and in Asia. No matter what form mountains take, their common denominator is rugged terrain.

MOUNTAINOUS REGIONS

1-3. The principal mountain ranges of the world lie along the broad belts shown in Figure 1-1. Called Cordillera, after the Spanish word for rope, they encircle the Pacific basin and then lead westward across Eurasia into North Africa. Secondary, though less rugged, chains of mountains lie along the Atlantic margins of America and Europe.

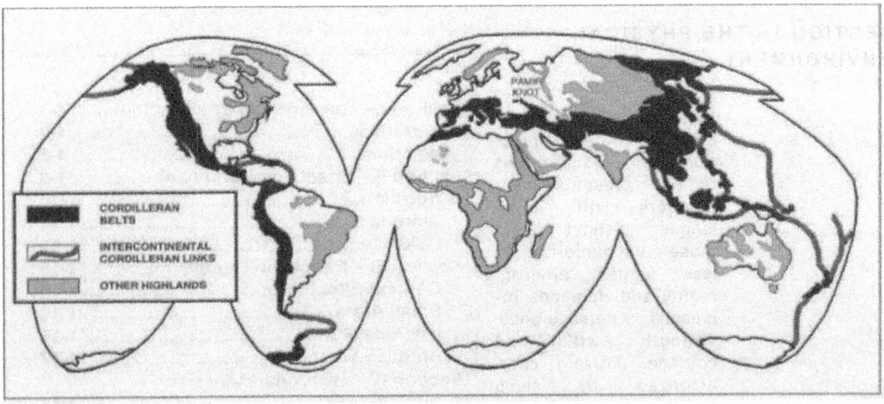

Figure 1-1. Mountain Regions of the World

1-4. A broad mountainous region approximateiy 1,600 kilometers wide dominates northwestern North America. It occupies much of Alaska, more than a quarter of Canada and the US, and all but a small portion of Mexico and Central America The Rocky Mountain Range includes extensive high plains and basins. Numerous peaks in this belt rise above 3,000 meters (10,000 feet). Its climate varies from arctic cold to tropical heat, with the full range of seasonal and local extremes.

1-5, Farther south, the Andes stretch as a continuous narrow band along the western region of South America, Narrower than its counterpart in the north, this range is less than 800 kilometers wide. However, it continuously exceeds an elevation of 3,000 meters (10,000 feet) for a distance of 3,200 kilometers.

1-6. In its western extreme, the Eurasian mountain belt includes the Pyrenees, Alps, Balkans, and Carpathian ranges of Europe. These loosely linked systems are separated by broad low basins and are cut by numerous valleys. The Atlas Mountains of North Africa are also a part of this belt Moving eastward into Asia, this system becomes more complex as it reaches the extreme heights of the Hindu Kush and the Himalayas. Just beyond the Pamir Knot on the Russian-Afghan frontier, it begins to fan out across all parts of eastern Asia. Branches of this belt continue south along the rugged island chains to New Zealand and northeast through the Bering Sea to Alaska.

MOUNTAIN CHARACTERISTICS

1-7. Mountain slopes generally vary between 15 and 45 degrees. Cliffs and other rocky precipices may be near vertical, or even overhanging. Aside from obvious rock formations and other local vegetation characteristics, actual slope surfaces are usually found as some type of relatively firm earth or grass. Grassy slopes may include grassy clumps known as tussocks, short alpine grasses, or tundra (the latter more common at higher elevations and latitudes). Many slopeswill be scattered with rocky debris deposited from the higher peaks and ridges. Extensive rock or boulder fields are known as talus. Slopes covered with smaller rocks, usually fist-sized or smaller, are called scree fields. Slopes covered in talus often prove to be a relatively easy ascent route. On the other hand, climbing a scree slope can be extremely difficult, as the small rocks tend to loosen easily and giveway. However, this characteristic often makes scree fields excellent descent routes. Before attempting to descend scree slopes, commanders should carefully analyze the potential for creating dangerous rockfall and take necessary avoidance measures.

1-8. In winter, and at higher elevations throughout the year, snow may blanket slopes, creating an environment with its own distinct affects. Some snow conditions can aid travel by covering rough terrain with a consistent surface. Deep snow, however, greatly impedes movement and requires soldiers well-trained in using snowshoes, skis, and over-snow vehicles. Steep snow covered terrain presents the risk of snow avalanches as well. Snow can pose a serious threat to soldiers not properly trained and equipped for movement under such conditions. Avalanches have taken the lives of more soldiers engaged in mountain warfare than all other terrain hazards combined,

1-9. Commanders operating in arctic and subarctic mountain regions, as well as the upper elevations of the world's high mountains, may be confronted with vast areas of glaciation. Valleys in these areas are frequently buried under massive glaciers and present additional hazards, such as hidden crevices and ice and snow avalanches. The mountain slopes of these peaks are often glacialed and their surfaces are generally composed of varying combinations of rock, snow, and ice. Although glaciers have their own peculiar hazards requiring special training and equipment, movement over valley glaciers is often the safest route through these areas (TC 90-6-1 contains more information on avalanches and glaciers, and their effects on operations).

MOUNTAIN CLASSIFICATIONS

1-10. Thereisnosimple system available to classify mountain environments. Soil composition, surface configuration, elevation, latitude, and climatic patterns determine the specific characteristics of each major mountain range. When alerted to the potential requirement to conduct mountain operations, commanders must carefully analyze each of these characteristics for the specific mountain region in which their forces wi 11 operate. H owever, mountains are generally classified or described according to their local relief; for military purposes, they may be classified according to operational terrain levels and dismounted mobility and skill requirements.

Local Relief

1-11. Mountains are commonly classified as low or high, depending on their local relief and, to some extent, elevation. Low mountains have a local relief of 300 to 900 meters (1,000 to 3,000 feet) with summits usually below the limberline. High mountains have a local relief usually exceeding 900 meters (3,000 feet) and are characterized by barren alpine zones above the timberline. Glaciers and perennial snow cover are common in high mountains and usually present commanders with more obstacles and hazards to movement than do low mountains.

Operational Terrain Levels

1-12. Mountain operations are generally carried out at three different operational terrain levels (see Figure 1-2). Level I terrain is located at the bottom of valleys and along the main lines of communications. At this level, heavy forces can operate, but maneuver space is often restricted. Light and heavy forces are normally combined, since vital lines of communication usually follow the valley highways, roads, and trails.

Level	Description
I	The bottoms of valleys and main lines of communications
II	The ridges, slopes, and passes that overlook valleys
III	The dominant terrain of the summit region

Figure 1-2. Operational Terrain Levels

1-13. Level II terrain lies between valleys and shoulders of mountains. Generally, narrow roads and trails, which serve as secondary lines of communication, cross this ridge system. Ground mobility is difficult and light forces will expend great effort on these ridges, since they can easily influence operations at Level I. Similarly, enemy positions at the next level can threaten operations on these ridges.

1-14. Level MI includes the dominant terrain of summit regions. Although summit regions may contain relatively gentleterrain, mobility in Level III is usually the most difficult to achieve and maintain Level III terrain, however, can provide opportunities for well-trained units to attack the enemy from the flanks and rear At this terrain level, acclimatized soldiers with advanced mountaineering training can infiltrate to attack lines of communication, logistics bases, air defense sites, and command infrastructures

Dismounted Mobility Classification

1-15. When conducting mountain operations, commanders must clearly understand the effect the operational terrain level has on dismounted movement. Therefore, in addition to the general mobility classification contained in FM 2-01.3 (unrestricted, restricted, severely restricted), mountainous terrain may be categorized into five classes based on the type of individual movement skill required (see Figure 1-3), Operations conducted in class 1 and 2 terrain require little to no mountaineering skills. Operations in class 3, 4, and 5 terrain require a higher level of mountaineering skills for safe and efficient movement. Commanders should plan and prepare for mountain operations based, in large part, on this type of terrain analysis.

Class	Terrain	Mobility Requirements	Skill Level Required*
1	Gentler slopes/ trails	Walking techniques	Unskilled (with some assistance) and Basic Mountaineers
2	Steeper/rugged terrain	Some use of hands	
3	Easy climbing	Fixed ropes where exposed	Basic Mountaineers (with assistance from assault climbers)
4	Steep/exposed climbing	Fixed ropes required	
5	Near vertical	Technical climbing required	Assault Climbers

* See Chapter 2 for a discussion of mountaineering skill levels

Figure 1-3. Dismounted Mobility Classification

WEATHER

1-16. In general, mountain climates tend to be cooler, wetter versions of the climates of the surrounding lowlands. Most mountainous regions exhibit at least two different climatic zones — a zone at low elevations and another at elevations nearer the summit regions. In some areas, an almost endless variety of local dimates may exist within a given mountainous region. Conditions change markedly with elevation, latitude, and exposure to atmospheric winds and air masses. In addition, the climatic patterns of two ranges located at the same latitude may differ radically.

1-17. Like most other landforms, oceans influence mountain dimates. Mountain ranges in dose proximity to oceans and other large bodies of water usually exhibit a maritime dimate. Maritime dimates generally produce milder temperatures and much larger amounts of rain and snow. Their relatively mild winters produce heavy snowfalls, while their summer temperatures rarely get excessively hot. Mountains farther inland usually display a more continental climate Winters in this type climate are often bitterly cold, while summers can be extremely hot. Annual rain- and snowfall here is far less than in a maritime climate and may be quite scarce for long periods. Relatively shallow snow-packs are normal during a continental dimale's winter season.

1-18. Major mountain ranges force air masses and storm systems to drop significant amounts of rain and snow on the windward side of the range. Asair masses pass over mountains, the leeward slopes receive far less precipitation than the windward slopes. It is not uncommon for the dimate on the windward side of a mountain range to be humid and the dimate on the leeward side arid. This phenomenon affeds coastal mountains, as well as mountains farther inland. The deepest winter snow-packs will almost always be found on the windward side of mountain ranges. As a result, vegetation and forest characteristics may be markedly different between these two areas. Prevailing winds and storm patterns normally determine the severity of these effect s.

1-19. Mountain weather can be erratic, varying from strong winds to calm, and from extreme cold to relative warmth within a short time or a minor shift in locality. The severity and variance of the weather require soldiers to be prepared for alternating periods of heat and cold, as well as conditions ranging from dry to extremely wet. At higher elevations, noticeable temperature differences may exist between sunny and shady areas or between areas exposed to wind and those protected from it. This greatly increases every soldier's clothing load and a unit's overall logistical requirements. Figure 1-4 summarizes the effects of mountain weather discussed below. FM 2-33,201 and FM 3-97 22 contain additional information on how weather affects operation s,

TEMPERATURE

1-20. Normally, soldiers encounter a temperature drop of three to five degrees Fahrenheit per 300-meter (1,000-foot) gain in elevation. In an atmosphere containi ng considerable water vapor, the temperature drops about one degree Fahrenheit for every 100-meter (300-foot) increase. In very dry air. it drops about one degree Fahrenheit for every 50 meters (150 feet). However, on cold, dear, and calm mornings, when a troop movement or dimb begins from a valley, soldiers may encounter higher temperatures as they gain elevation This reversal of the normal situation is called temperature inversion Additionally, during winter months, the temperature is often higher during a storm than during periods of dear weather. However, the dampness of predpitation and penetration of the wind may still cause soldiers to chill faster This is compounded by the fad that the cover afforded by vegetation often does not exist abovethe trt?e-line. Under these conditions, com ma riders must weigh the tactical advantage of retaining positions on high ground against seeking shelter and warmth at lower elevations with reduced visibility.

1-21, At high elevations, there may be differences of 40 to 50 degrees Fahrenheit between the temperature in the sun and that in the shade. This is similar in magnitude to the day-to-night temperature fluctuations experienced in some deserts (see FM 3-97.3). Besides permitting rapid heating, the dear air at high altitudes also results in rapid cooling at night. Consequently. temperatures ri se swiftly after sunrise and drop quickly after sunset. M uch of the chilled air drains downward so that the differences between day and night temperatures ere greater in valleys than on slopes.

Weather Condition	Flat to Moderate Terrain Effects	Added Mountain Effects
Sunshine	Sunburn Snow blindness Temperature differences between sun and shade	Increased risk of sunburn and snow blindness Severe, unexpected temperature variations between sun and shade Avalanches
Wind	Windchill	Increased risk and severity of windchill Blowing debris or driven snow causing reduced visibility Avalanches
Rain	Reduced visibility Cooler temperatures	Landslides Flash floods Avalanches
Snow	Cold weather injuries Reduced mobility and visibility Snow blindness Blowing snow	Increased risk and severity of common effects Avalanches
Storms	Rain/snow Reduced visibility Lightning	Extended duration and intensity greatly affecting visibility and mobility Extremely high winds Avalanches
Fog	Reduced mobility/visibility	Increased frequency and duration
Cloudiness	Reduced visibility	Greatly decreased visibility at higher elevations

Figure 1-4. Comparison of Weather Effects

WIND

1-22. In high mountains, the ridges and passes are seldom calm. By contrast, strong winds in protected valleys are rare. Normally, wind velocity increases with altitude and is intensified by mountainous terrain. Valley breezes moving up-slope are more common in the morning, while descending mountain breezes are more common in the evening. Wind speed increases when winds are forced over ridges and peaks (orographic lifting), or when they funnel through narrowing mountain valleys, passes, and canyons (Venturi effect). Wind may blow with great force on an exposed mountainside or summit. As wind speed doubles, its force on an object nearly quadruples

1-23, Mountain winds cause rapid temperature changes and may result in blowing snow, sand, or debris that can impair movement and observation. Commanders should routinely consider the combined cooling effect of ambient temperature and wind (windchill) experienced by their soldiers (see Figure 1-5 on page 1-8). At higher elevations, air is considerably dryer than air at sea level. Due to this increased dryness, soldiers must increase their fluid intake by approximately one-third. However, equipment will not rust as quickly, and organic matter will decompose more slowly.

WIND SPEED		COOLING POWER OF WIND EXPRESSED AS "EQUIVALENT CHILL TEMPERATURE"																				
KNOTS	MPH	TEMPERATURE (° F)																				
CALM	CALM	40	35	30	25	20	15	10	5	0	-5	-10	-15	-20	-25	-30	-35	-40	-45	-50	-55	-60
		EQUIVALENT CHILL TEMPERATURE																				
3-6	5	35	30	25	20	15	10	5	0	-5	-10	-15	-20	-25	-30	-35	-40	-45	-50	-55	-60	-70
7-10	10	30	20	15	10	5	0	-10	-15	-20	-25	-35	-40	-45	-50	-60	-65	-70	-75	-80	-90	-95
11-15	15	25	15	10	0	-5	-10	-20	-25	-30	-40	-45	-50	-60	-65	-70	-80	-85	-90	-100	-105	-110
16-19	20	20	10	5	0	-10	-15	-25	-30	-35	-45	-50	-60	-65	-75	-80	-85	-95	-100	-110	-115	-120
20-23	25	15	10	0	-5	-15	-20	-30	-35	-45	-50	-60	-65	-75	-80	-90	-95	-105	-110	-120	-125	-135
24-28	30	10	5	0	-10	-20	-25	-30	-40	-50	-55	-65	-70	-80	-85	-95	-100	-110	-115	-125	-130	-140
29-32	35	10	5	-5	-10	-20	-30	-35	-40	-50	-60	-65	-75	-80	-90	-100	-105	-115	-120	-130	-135	-145
33-36	40	10	0	-5	-15	-20	-30	-35	-45	-55	-60	-70	-75	-85	-95	-100	-110	-115	-125	-130	-140	-150
WINDS ABOVE 40 HAVE LITTLE ADDITIONAL EFFECT		LITTLE DANGER					INCREASING DANGER (Flesh may freeze within 1 minute)						GREAT DANGER (Flesh may freeze within 30 secs)									

Figure 1-55. Windchill Chart

PRECIPITATION

1-24. The rapid rise of air masses over mountains creates distinct locaf weather patterns. Precipitation in mountains increases with elevation and occurs more often on the windward than on the leeward side of ranges. Maximum cloudiness and precipitation generally occur near 1,800 meters (6,000 feet) elevation in the middle latitudes and at lower levels in the higher latitudes. Usually, a heavily wooded belt marks the zone of maximum precipitation.

Rain and Snow

1-25. Both rain and snow are common in mountainous regions. Rain presents the same challenges as at lower elevations, but snow has a more significant influence on all operations. Depending on the specific region, snow may occur at anytime during the year at elevations above 1,500 meters (5,000 feet). Heavy snowfall greatly increases avalanche hazards and can force changes to previously selected movement routes. In certain regions, the intensity of snowfall may delay major operations for several months. Dry, flat riverbeds may initially seem to be excellent locations for assembly areas and support activities, however, heavy rains and rapidly thawing snow and ice may create flash floods many miles downstream from the actual location of the rain or snow.

Thunderstorms

1-26. Although thunderstorms are local and usually last only a short time, they can impede mountain operations. Interior ranges with continental climates are more conducive to thunderstorms than coastal ranges with maritime climates. In alpine zones, driving snow and sudden wind squalls often accompany thunderstorms. Ridges and peaks become focal points for lightning strikes, and the occurrence of lightning is greater in the summer than the winter. Although statistics do not show lightning to be a major mountaineering hazard, it should not be ignored and soldiers should take normal precautions, such as avoiding summits and ridges, water, and contact with metal objects.

Traveling Storms

1-27. Storms resulting from widespread atmospheric disturbances involve strong winds and heavy predpitation and are the most severe weather condition that occurs in the mountains. If soldiers encounter a traveling storm in alpine zones during winter, they should expect low temperatures, high winds, and blinding snow. These conditions may last several days longer than in the lowlands. Specific conditions vary depending on the path of the storm. However, when colder weather moves in, dearing at high elevations is usually slow.

1-28. The effects of fog in mountains are much the same as in other terrain. However, because of the topography, fog occurs more frequently in the mountains The high inddence of fog makes it a significant planning consideration as it restricts visibility and observation complicating reconnaissance and surveillance. However, fog may help fadlitate covert operations such as infiltration. Routes in areas with a high occurrence of fog may need to be marked and charted to fadlitate passage.

SECTION II - EFFECTS ON PERSONNEL

1-29. The mountain environment is complex and unforgiving of errors. Soldiers conducting operations anywhere, even under the best conditions, become cold, thirsty, tired, and energy-depleted. In the mountains however, they may become paralyzed by cold and thirst and incapacitated due to utter exhaustion. Conditions such as high elevations, rough terrain, and extremely unpredictable weather require leaders and soldiers who have a keen understanding of environmental threats and what to do about them.

1-30. A variety of individual soldier charaderistics and environmental conditions influence the type, prevalence, and severity of mountain illnesses and injuries (see Figure 1-6 on page 1-10). Due to combinations of these characteristics and conditions, soldiers often succumb to more than one illness or injury at a time, increasing the danger to life and limb. Three of the most common, cumulative, and subtle fadors affeding soldier ability under these variable conditions are nutrition (to indude water intake), decreased oxygen due to high altitude, and cold. Preventive measures, early recognition, and rapid treatment help minimize nonbattle casualties due to these conditions (see Appendix A for detailed information on mountain-specific illnesses and injuries).

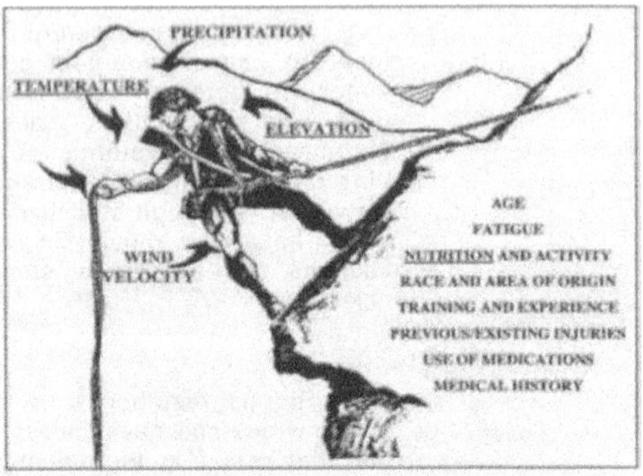

Figure 1-6. Environmental and Soldier Conditions Influencing Mountain Injuries and Illnesses

NUTRITION

1-31. Poor nutrition contributes to illness or injury, decreased performance, poor morale, and susceptibility to cold injuries, and can severely affect military operations. Influences at high altitudes that can affect nutrition include a dulled taste sensation (making food undesirable), nausea, and lack of energy or motivation to prepare or eat meals.

1-32, Caloric requirements increase in the mountains due to both the altitude and the cold A diet high in fat and carbohydrates is important in helping the body fight the effects of these conditions. Fats provide long-term, slew caloric release, but are often unpalatable to soldiers operating at higher altitudes. Snacking on high-carbohydrate foods is often the best way to maintain the calories necessary to function.

1-33. Products that can seriously impact soldier performance in mountain operations include:

- Tobacco. Tobacco smoke interferes with oxygen delivery by reducing the blood's oxygen-carrying capacity. Tobacco smoke in dose, confined spaces increases the amounts of carbon monoxide. The irritant effect of tobacco smoke may produce a narrowing of airways, interfering with optimal air movement. Smoking can effectively raise the"physiological altitude" as much as several hundred meters.
- Alcohol. Alcohol impairs judgement and perception, depresses respiration, causes dehydration, and increases susceptibility to cold injury
- Caffeine. Caffeine may improve physical and mental performance, but it also causes increased urination (leading to dehydration) and, therefore, should be consumed in moderation.

1-34. Significant body water is lost at higher elevations from rapid breathing, perspiration, and urination. Depending upon level of exertion, each soldier should consume about four to eight quarts of water or other decaffeinated fluids per day in low mountains and may need ten quarts or more per day in high mountains. Thirst is not a good indicator of the amount of water lost.

and in cold climates sweat, normally an indicator of loss of fluid, goes unnoticed. Sweat evaporates so rapidly or is absorbed so thoroughly by clothing layers that it is not readily apparent. When soldiersbecomelhirsty.they are already dehydrated. Loss of body water also plays a major role in causing altitude sickness and cold injury. Forced drinking in the absence of thirst, monitoring the deepness of the yellow hue in the urine, and watching for behavioral symptoms common to altitude sickness are important factors for commanders to consider In assessing the water balance of soldiers operating in the mountains.

1-35, In the mountains, as elsewhere, refilling each soldier's water containers as often as possible is mandatory. No matter how pure and clean mountain water may appear, water from natural sources should always be purified or chemically sterilized to prevent parasitical illnesses (giardiasis). Commanders should consider requiring the increased use of individual packages of powdered drink mixes, fruit, and juices to help encourage the required fluid intake.

ALTITUDE

1-36. As soldiers ascend in altitude, the proportion of oxygen in the air decreases. Without proper acclimatization, this decrease in oxygen saturation can cause altitude sickness and reduced physical and mental performance [see Figure 1-7). Soldiers cannot maintain the same physical performance at high altitudethat they can at low altitude, regardless of their fitness level.

Altitude	Meters	Feet	Effects
Low	Sea Level – 1,500	Sea Level – 5,000	None.
Moderate	1,500 – 2,400	5,000 – 8,000	Mild, temporary altitude sickness may occur
High	2,400 – 4,200	8,000 – 14,000	Altitude sickness and decreased performance is increasingly common
Very High	4,200 – 5,400	14,000 – 18,000	Altitude sickness and decreased performance is the rule
Extreme	5,400 – Higher	18,000 - Higher	With acclimatization, soldiers can function for short periods of time

Figure 1-7. Effects of Altitude

1-37. The mental effects most noticeable at high altitudes indude decreased perception, memory, judgement, and attention. Exposure to ait it udes of over 3,000 meters (10,000 feet) may also result in changes in senses, mood, and personality. Within hours of ascent, many soldiers may experience euphoria, joy, and excitement that are likely to be accompanied by errors i n judgement, leading to mistakes and accidents. After a period of about 6 to 12 hours, euphoria decreases, often changing to varying degrees of depression. Soldiers may become irritable or may appear listless. U sing the buddy system during this early exposure helps to identify soldiers who may be more severely affected. High morale and esprit instilled before deployment and reinforced frequently help to mini mize the i mpact of negative mood changes.

1-38. Trie physical effect most noticeable at high altitudes includes vision. Vision is generally the sense most affected by altitude exposure and can potentially affect military operations at higher elevations. Night vision is significantly reduced, affecting soldiers at approximately 2.400 meters (8,000 feet) or higher. Some effects occur early and are temporary, while others may persist after acclimatization or even for a period of time after descent. To compensate for loss of functional abilities, commanders should make use of tactics, techniques, and procedures that trade speed for increased accuracy. By allowing extra time lo accomplish tasks, commanders can minimize errors and injuries.

HYPOXIA-RELATED ILLNESSES AND EFFECTS

1-3g Hypoxia, a deficiency of oxygen reaching the tissues of the body, has been the cause of many mountain illnesses, injuries, and deaths. It affects everyone, but some soldiers are more vulnerable than others. A soldier may be affected at one time but not at another. Altitude hypoxia is a killer, but it seldom strikes alone. The combination of improper nutrition, hypoxia, and cold is much more dangerous than any of them alone. The three most significant altitude-related illnesses and their symptoms, which are essentially a series of illnesses associated with oxygen deprivation, are:

- Acute Mountain Sickness (AMS). Headache, nausea, vomiting, fatigue, irritability, and dizziness.

  High Altitude Pulmonary Edema (HAPE), Coughing, noisy breathing, wheezing, gurgling in the airway, difficulty breathing, and pink frothy sputum (saliva). Ultimately coma and death will occur without treatment,
- High Altitude Cerebral Edema (HACE). HACE is the most severe illness associated with high altitudes. Its symptoms often resemble AMS {severe headache, nausea, vomit i ng), often with more dram at ic signals such as a swaying of the upper body, especially when walking, and an increasingly deteriorating mental status. Early mental symptoms may include confusion, disorientation, vivid hallucinations, and drowsiness. Soldiers may appear to be withdrawn or demonstrate behavior generally associated with fatigue or anxiety. Like HAPE, coma or death wi 11 occur without treatment.

OTHER MOUNTAIN-RELATED ILLNESSES

1-40. Other illnesses and effects related to the mountain environment and higher elevations are:

- Subacute mountain sickness. Subacute mountain sickness occurs in some soldiers during prolonged deployments (weeks/months) to elevations above 3,600 meters (12,000 feet). Symptoms include sleep disturbance, loss of appetite, weight loss, and fatigue. This condition reflects a failureto acclimatize adequately.
- Carbon monoxide poisoning. Carbon monoxide poisoning is caused by the inefficient fuel combustion resulting from the low oxygen content of air and higher usage of stoves, combustion heaters, and engines in enclosed, poorly ventilated spaces.

Sleep disturbances. High altitude has significant harmful effects on sleep. The most prominent effects are frequent periods cf apnea (temporary suspension of respiration) and fragmented sleep Sleep disturbances may last for weeks at elevations less than 5.400 meters (18.000 feet) and may never stop at higher elevations. These effects have even been reported as low as 1,500 meters (5,000 feet).

Poor wound healing. Poor wound healing resulting from lowered immune functions may occur at higher elevations. Injuries resulting from burns, cuts, or other sources may require descent for effective treatment and healing.

ACCLIMATIZATION

1-41. Altitude acclimatization involves physiological changes that permit the body to adapt to the effects of low oxygen saturation in the air. It allows soldiers to achieve the maximum physical work performance possible for the altitude to which they are acclimatized. Once acquired, acclimatization is maintained as long as the soldier remains at that altitude, but is lost upon returning to lower elevations. Acclimatization to one altitude does not prevent altitude illnesses from occurring if ascent to higher altitudes is too rapid.

1-42. Getting used to living and working at higher altitudes requires acclimatization. Figure 1-8 shows the four factors that affect acclimatization in mountainous terrain. These factors are similar to those a scuba diver must consider, and the consequences of an error can be just as severe. In particular. high altitude dimbing must be carefully paced and staged in the same way that divers must pace and stage their ascent tothe surface.

- **Altitude**
- **Rate of Ascent**
- **Duration of Stay**
- **Level of Exertion**

Figure 1-8. Factors Affecting Acclimatization

1-43. For most soldiers at high to very high altitudes. 70 to 80 percent of the respiratory component of acclimatization occurs in 7 to 10 days, 80 to 90 percent of overall acclimatization is generally accomplished by 21 to 30 days, and maximum acdimatization may take several months to years. However, some soldiers may acdimatize more rapidly than others, and a few soldiers may not acclimatize at all. There is no absolute way to identify soldiers who cannot acdimatize, except by their experience during previous altitude exposures.

1-44. Commanders must be aware that highly fit, motivated individuals may go too high too fast and become victims of AMS. HAPE, or HACE. Slow and easy dimbing. limited activity, and long rest periods are critical to altitude acclimatization. Leaves that involve soldiers descending to lower altitudes and then returning should be limited. Acdimalization may be accomplished by either a staged or graded ascent. A combination of the two is the safest and most effective method for prevention of high altitude illnesses.

- Staged Ascent. A staged ascent requires soldiers to ascend to a moderate altitude and remain there for 3 days or more to acdimatize before ascending higher (the longer theduration, the more effedive and thorough the aoclimatization to that altitude). When possible, soldiers should make several slops for staging during ascent to allow a greater degree of acclimatization.

Graded Ascent. A graded ascent limits the daily altitude gain to allow partial acclimatization. The altitude at which soldiers sleep is the critical element in this regard. Having soldiers spend two nights at 2,700 meters (9,000 feet) and limiting the sleeping altitude to no more than 300 meters per day (1,000 feet) above the previous night's sleeping altitude will significantly reducethe incidence of altitude sickness,

1-45. In situations where there is insufficient time for a staged or graded ascent, commanders may consider using the drug acetazolamide to help accelerate acdimatization; however, commanders must ensure soldiers are acclimatized before they are committed to combat. When used appropriately, it will prevent symptoms of AMS in nearly all soldiers and reduce symptoms in most others. It has also been found to improve sleep quality at high altitudes. However, commanders should consult physidans trained in high-altitude or wilderness medidne concerning doses, side effects, and screening of individuals who may be allergic. Asa non-pharmacological method, high carbohydrate diets (whole grains, vegetables, peas and beans, potatoes, fruits, honey, and refined sugar) are effedive in aiding acdimatization.

COLD

1-46. After illnesses related to not being acdimatized. cold injuries, both freezing and nonfreezing, are generally the greatest threat. Temperature and humidity decrease with increasing altitude. Reviewing cold weather injury prevention, training in shelter construction, dressing in layers, and using the buddy system are critical and may preclude large numbers of debilitating injuries. Figure 1-9 lists the cold and snow injuries most common to mountain operations. See FM 3-97.11 and FM 4-25.11 for information regarding causes, symptoms, treatment, and prevention.

- **Frostbite (freezing)**
- **Hypothermia (nonfreezing)**
- **Trench/immersion Foot (nonfreezing)**
- **Snow Blindness**

Figure 1-9. Common Cold Weather Injuries

1-47. Altitude sickness and cold injuries can occur simultaneously, with signs and symptoms being confused with each other. Coughing, stumbling individuals should be immediately evacuated to medical support at lower levels to determine their medical condition. Likewise, soldiers in extreme pain from cold injurieswhodo not respond to normal pain medications, require evacuation. Without constant vigilance, cold injuries may significantly limit the number of deployable troops and drastically reduce combat power. However, with command emphasis and proper equipment, dothing, and training, all cold-weather injuries are preventable.

SECTION III - EFFECTS ON EQUIPMENT

1-48. No manual can cover the effects of terrain and weather on every weapon and item of equipment within the Army inventory. Although not all-encompassing, the list at Figure 1-10 contains factors that commanders should take into account when considering the effect the mountainous environment may have on their weapons and equipment. Of these, the most important factor is the combined effects of the environment on the soldier and his subsequent ability to operate and maintain his weapons and equipment. Increasingly sophisticated equipment requires soldiers that are mentally atert and physically capable. Failure to consider this important factor often results in severe injury. lowered weapons and equipment performance, and mission failure. The information provided within this manual, combined with the information found in weapon-specific field manuals (FMs) and technical manuals (TMs), provides the information necessary to know how to modify tactics, techniques, and procedurestowin on themountain battlefield

- Operator/Maintenance Personnel
- Line-of-Sight
- Range
- Thermal Contrast
- Ballistics and Trajectory
- Target Detection and Acquisition
- First Round Hit Capability
- Camouflage and Concealment/Noise
- Mobility
- Wear and Maintenance
- Aerodynamics and Lift
- Functioning and Reliability
- Positioning/Site Selection

Figure 1-10. Weapons and Equipment Factors Affected by the Environment

GENERAL EFFECTS

1-49. In a mountainous environment, the speed and occurrence of wind generally increase with elevation, and the effects of wind increase with range (depending on the speed and direction). Due to these factors, soldiers must be taught the effects of wind on ballistics and how to compensate for them. In cold weather, firing weapons often creates ice fog trails. These ice fog trails obscure vi sion and, at thesametime, al low the enemy to more easily discern the location of primary positions and the overall structure of a unit's defense. Th is situation increases the importance of alternate and supplementary firing positions,

1-50. Range estimation in mountainous terrain is difficult. Depending upon the type of terrain in the mountains, soldiers may either over- or underestimate range. Soldiers observing over smooth terrain, such as sand, water, or snow, generally underestimate ranges. This results in attempting to engage targets beyond the maximum effective ranges of their weapon systems. Looking downhill, targets appear to be farther away and looking uphill, they appear to be closer. This illusion, combined with the effects of gravity, causes the soldier shooting downhill to fire high, while it has the opposite effect on soldiers shooting uphill.

1-51. Higher elevations generally afford increased observation but low-hanging clouds and fog may decrease visibility, and the rugged nature of mountain terrain may produce significant dead space at mid-ranges. These effects mean that more observation posts are necessary to cover a given frontage in mountainous terrain than in non-mountainous terrain. They also require the routine designation of supplementary firing positions for direct fire weapons. Rugged terrain also makes ammunition resupply more difficult and increases the need to enforce strict fire control and discipline. Finally, the rugged environment creates compartmented areas that may preclude mutual support and reduce supporting distances.

SMALL ARMS

1-52 In rocky mountainous terrain, the effectiveness of small arms fire increases by the splintering and ricocheting when a bullet strikesa rock. M203 and M K-19 grenade launchers are useful for covering close-in dead space in mountainous terrain. Hand grenades are also effective Although it may seem intuitive, soldiers must still be cautioned against throwing grenades uphill where they are likely to roll back before detonation. Grenades (as well as other explosive munitions) lose much of their effectiveness when detonated under snow, and soldiers should be warned that hand grenades may freeze to wet gloves.

1-53. As elevation i ncreases. ai r pressure and air density decrease. At higher elevations, a round is more efficient and strikes a target higher, due to reduced drag. This effect does not significantly influence the marksmanship performance of most soldiers, however, designated marksmen and snipers should re-zero their weapons after ascending to higher elevations. (See FM 3-25.9 and FM 3-23,10 for further information on ballistics and weather effects on small arms.)

MACHINE GUNS

1-54. Machine guns provide long-range fire when visibility is good. However, grazing fire can rarely be achieved in mountains because of the radical changes in elevation. When grazing fire can be obtained, the ranges are normally short. More often, plunging fire is the result (see Figure 1-11 and FM 3-21.7). In mountainous terrain, situations that prevent indirect fire support from protecting advancing forces may arise. When

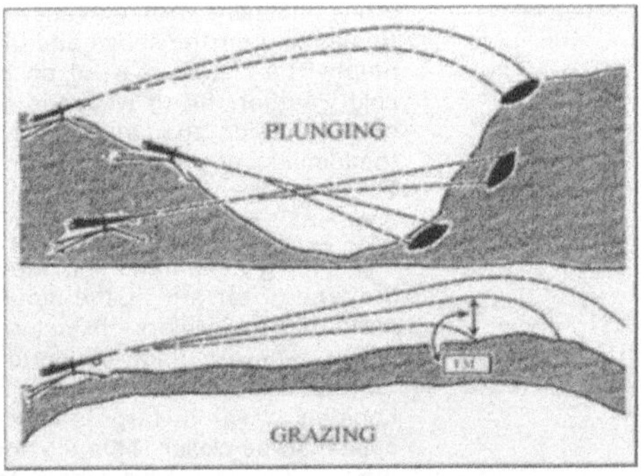

Figure 1-11. Classes of Fire with Respect to the Ground

these occur, the effects of machine-guns and other direct fire weapons must be concentrated to provide adequate supporting fires for maneuvering elements. Again, supplementary positions should be routinely prepared to cover different avenues of approach and dead space.

ANTITANK WEAPONS

1-55. The AT4 is a lightweight antitank weapon ideally suited for the mountainous environment and for direct fire against enemy weapon em placemen ts. Anti-tank guided missiles (ATGMs), such as the Javelin and the tube-launched, optically tracked, wire-guided, heavy antitank missile system (TOW), tend to hinder dismounted operations because of their bulk and weight. In very restrictive mountainous terrain, the lack of armored avenues of approach and suitable targets may limit their utility. If an armored or mechanized threat is present, TOWs are best used in long-range, antiarmor ambushes, while the shorter-range Javelin, with its fire-and-forget technology, is best used from restrictive terrain nearer the kill zone. However, their guidance systems may operate stiffly and sluggishly in extreme cold weather.

SECTION IV - RECONNAISSANCE AND SURVEILLANCE

RECONNAISSANCE

1-56. During operations in a mountainous environment, reconnaissance is as applicable to the maneuver of armies and corps as it is to tactical operations. Limited routes, adverse terrain, and rapidly changing weather significantly increase the importance of reconnaissance operations to focus fires and maneuver. Failure to conduct effective reconnaissance will result in units being asked to achieve the impossible or in missed opportunities for decisive action.

1-57. As in all environments, reconnaissance operations in a mountainous area must be layered and complementary in order to overcome enemy attempts to deny critical information to the friendly commander. In order to gather critical and timely information required by the commander, the activities of reconnaissance assets must be closely coordinated. Strategic reconnaissance platforms set the stage by identifying key terrain, as well as the general disposition and composition of enemy forces. Operational level commanders compare the information provided by strategic assets with their own requirements and employ reconnaissance assets to fill in the gaps that have not been answered by strategic systems and achieve the level of detail they require.

1-53. At the beginning of a campaign in a mountainous environment, reconnaissance requirements will be answered by aerial or overhead platforms, such as satellites, joint surveillance, target attack radar systems (JSTARSs), U2 aircraft, and unmanned aerial vehicles (UAVs). In a mountain AO, it may often be necessary to commit ground reconnaissance assets in support of strategic and operational information requirements. Conversely, strategic and operational reconnaissance systems may be employed to identify or confirm the feasibility of employing ground reconnaissance assets. Special reconnaissance (SR) and long-range surveillance (LRS) teams may be inserted to gather information thai cannot be collected by overhead systems, or to verify data that has already been collected. In this instance, satellite imagery is used to analyze a specific area for insertion for the team. The potential hide positions for the teams are identified using imagery and. terrain and weather permitting, verified by UAVs. See FM 3-100.55 for detailed information on combined arms reconnaissance.

1-59. In harsh mountain terrain, ground reconnaissance operations are often conducted dismounted. Commanders must assess the slower rate of ground reconnaissance elements to determine its impact on theentire reconnaissance and collection process. They must develop plans that account for this slower rate and initiate reconnaissance as early as possible to provide additional time for movement. Commanders may also need to allocate more forces, including combat forces, to conduct reconnaissance, reconnaissance in force missions, or limited objective attacks to gain needed intelligence. Based upon mission, enemy, terrain and weather, troops and support available, time available, civil considerations (METT-TC), commanders may need to prioritize collection assets, accept risk, and continue with less information from their initial reconnaissance efforts. In these cases, they must use formations and schemes of maneuver that provide maximum security and flexibility, to include robust security formations, and allow for the development of the situation once in contact.

1-60. Although reconnaissance patrols should normally use the heights to observe the enemy, it may be necessary to send small reconnaissance teams into valleys or along the low ground to gain suitable vantage points or physically examine routes that will be used by mechanized or motorized forces. In mountainous environments, reconnaissance elements are often tasked to determine:

- The enemy's primary and alternate lines of communication.
- Locations and directions from which the enemy can attack or counterattack.
- Heights that allow the enemy to observe the various sectors of terrain.
- Suitable observation posts for forward observers.
- Portions of the route that provide covert movement.
- Level of mountaineering skill required to negotiate routes (dismounted mobility classification) and sections of the route that require mountaineering installations.
- Suitability of routes for sustained combat service support (CSS) operations.
- Trails, routes, and bridges that can support or can be improved by engineers in order to move mechanized elements into areas previously thought to be impassable.
- Bypass routes
- Potential airborne and air assault drop^pick-up zones and aircraft landing areas.

RECONNAISSANCE IN FORCE

1-61. The compart merited geography and inherent mobility restrictions of mountainous terrain pose significant risk for reconnaissance in force operations. Since the terrain normally allows enemy units to defend along a much broader front with fewer forces, a reconnaissance in force may be conducted as a series of smaller attacks to determine the enemy situation at selected points. Commanders should carefully consider mobility restrictions that may affect plans for withdrawal or exploitation. Commanders should also position small reconnaissance elements or employ surveillance systems throughout the threat area of operations to gauge the enemy's reaction to friendly reconnaissance in force operations and alert the force to possible enemy counterattacks. In the mountains, the risk of having at least a portion of the force cut off and isolated is extremely high. Mobile reserves and preplanned fires must be available to reduce the risk, decrease the vulnerability of the force, and exploit any success as it develops.

ENGINEER RECONNAISSANCE

1-62. Engineer reconnaissance assumes greater significance in a mountainous environment in order to ensure supporting engineers are properly task organized with specialized equipment for quickly overcoming natural and reinforcing obstacles. Engineer reconnaissance teams assess the resources required for clearing obstacles on precipitous slopes, constructing crossing sites at fast-moving streams and rivers, improving and repairing roads, erecting fortifications, and establishing barriers during the conduct of defensive operations. Since the restrictive terrain promotes the widespread employment of point obstacles, engineer elements should be integrated into all mountain reconnaissance operations,

1-63. In some regions, maps may be unsuitable for tactical planning due to inaccuracies, limited detail, and inadequate coverage. En these areas, engineer reconnaissance should precede, but not delay operalions. Because rugged mountain terrain makes ground reconnaissance time-consuming and dangerous, a combination of ground and aerial or overhead platforms should be used for the engineer reconnaissance effort. Data on the terrain, vegetation, and soil composition, combined with aerial photographs and multispectral imagery, allows engineer terrain intelligence teams to provide detailed information that may be unavailable from other sources.

AERIAL AND OVERHEAD RECONNAISSANCE

1-64. During all but the most adverse weather conditions, aerial or overhead reconnaissance may be the best means to gather information and cover large areas that are difficult for ground units to traverse or observe. Airborne standoff intelligence collection devices, such as side-looking radar, provide excellent terrain and target isolation imagery. Missions must be planned to ensure that critical areas are not masked by terrain or other environmental conditions. Additionally, aerial or overhead photographs may compensate for inadequate maps and provide the level of detail needed to plan operations. Infrared imagery and camouflage detection film can be used to determine precise locations of enemy positions, even at night. Furthermore, AH-64 and

OH-58D helicopters can provide commanders with critical day or night video reconnaissance, utilizing television or lorward-lcoking infrared.

1-65. Terrain may significantly impact the employment of overhead reconnaissance platforms using radar systems to detect manmade objects. These systems may find themselves adversely impacted by the masking effect that occurs when the mountain terrain blocks the radar beam. Thus, the radar coverage may not extend across the reverse slope of a steep ridge or a valley floor. Attempts to reposition the overhead platform to a point where it can "see" the masked area may merely result in masking occurring elsewhere. This limitation does not preclude using such systems; however, the commander should employ manned or unmanned aerial reconnaissance when available, in conjunction with overhead reconnaissance platforms in order ta minimize these occurrences. The subsequent use of ground reconnaissance assets to verify the data that can be gathered by overhead and electro-optical platforms wi 11 ensure that commanders do not fall prey to del i berate enemy deception efforts that capitalize on the limited capabilities of some types of overhead platforms in this environment.

SURVEILLANCE

1-66. I n the mountains, surveillance of vulnerable flanks and gaps between units is accomplished primarily through well-positioned observation posts (OPs). These OPs are normally inserted by helicopter and manned by small elements equipped with sensors, enhanced electro-optical devices, and appropriate communications. Commanders must develop adequate plans that address not only their insertion, but their continued support and ultimate extraction. The considerations of METT-TC may dictate that commanders provide more personnel and assets than other types of terrain to adequately conduct surveillance missions. Commanders must also ensure that surveillance operations are fully integrated with reconnaissance efforts in order to provide a3dequate coverage of the AO,

1-67. Long-range surveillance units (LRSUs) and snipers trained in mountain operations also contribute to surveillance missions and benefit from the restrictive terrain and excellenl line-of-sight. Overhead platforms and air cavalry may also be used for surveillance missions of limited duration. However, weather may impede air operations, decrease visibility for both air and ground elements, and reduce the ability of ground surveillance elements to remain hidden for prolonged periods without adequate logistical support. As with overhead reconnaissance, terrain may mask overhead surveillance platforms.

Chapter 2

Command and Control

In the mountains, major axes of advance are limited to accessible valleys and often separated by restrictive terrain. The compart merited nature of the terrain makes it difficult to switch the effort from one axis to another or to offer mutual support between axes. The battle to control the major lines of communications of Level I develops on the ridges and heights of Level II. In turn, the occupation of the dominating heights in Level II may leave a force assailable from the restrictive terrain of Level III. Each operational terrain level influences the application of tactics, techniques, and procedures n ecessar y for su ccessf u l operations.

CONTENTS

In mountainous terrain, it is usually difficult to oonduct a coordinated battle. Engagements tend to be isolated, march columns of even small elements extremely long, and mutual support difficult to accomplish. Command and control of all available assets is best achieved if command posts are well forward. However, the mountainous environment decreases the commander's mobility. Therefore, commanders must be able to develop a clear vision of how the battle will unfold, correctly anticipate the decisive points on the battlefield, and position themselves at these critical points.

The success of a unit conducting mountain operations depends on how well leaders control their units. Control is limited largely to a well-thought-out plan and thorough preparation. Boundaries require careful planning in mountain operations. Heights overlooking valleys should be included in the boundaries of units capable of exerting the most influence over them. These boundaries may be difficult to determine initially and may require subsequent adjustment.

During execution, leaders must be able to control direction and speed of movement, maintain proper intervals, and rapidly start, stop, or shift fire. In the mountains, soldiers focus mainly on negotiating difficult terrain. Leaders, however, must ensure that their soldiers remain alert for, understand, and follow signals and orders Although in most instances audio, visual, wire, physical signals, and messengers are used to maintain control, operations may be controlled by time as a secondary means. However, realistic timetables must be based on thorough reconnaissance and sound practical knowledge of the mountain battlefield.

Commanders must devote careful consideration to the substantial effect the mountain environment may have on systems that affect their ability to collect, process, store, and disseminate information. Computers, communications, and other sophisticated electronic equipment are usually susceptible to jars, shocks, and rough handling associated with the rugged mountain environment. They are also extremely sensitive to the severecold often associated with higher elevations. Increased precipitation and moisture may damage electronic components, and heavy amounts of rain and snow, combined with strong surface winds, may generate background electronic interference that can reduce the efficiency of intercept/direction finding antennas and ground surveillance radars. Localized storms with low sustained cloud cover reduce the effectiveness of most imagery intelligence (IMINT) platforms, to include unmanned aerial vehicles (UAVs). The collective effect of mountain weather and terrain diminishes a commander's ability to achieve shared situational understanding among his subordinates. However, increased use of human intelligence (HUMI NT), clear orders and intents, and leaders capable of exercising initiative, allow commanders to dominate the harsh environment of a mountain area of operations.

As in any environment, mountain operations pose both tactical and accident risks. However, since most units do not routinely train for or operate in the mountains, the level of uncertainty, ambiguity, and friction is often higher than in less rugged environments. Commanders must be able to identify and assess hazards that may be encountered in executing their missions, develop and implement control measures to eliminate unnecessary risk, and continuously supervise and assess to ensure measures are properly executed and remain appropriate as the situation changes. Although risk decisions are the commanders' business, staffs, subordinate leaders, and individual soldiers must also understand the risk management process and must continuously look for hazards at their level or within their area of expertise. Any risks identified (with recommended risk reduction measures) must be quickly elevated to the chain of command (see **FM** 3-100.14).

SECTION I - ASSESSMENT OF THE SITUATION

2-1, Although higher-elevation terrain is not always key, the structure of a mountain area of operations (AO) often forms a stairway of key terrain features Identification and control of dominant terrain at each operational terrain level form the basis for successful mountain maneuver. Key terrain features at higher elevations often take on added significance due to their inaccessibility and ease of defense. Tomainlain freedom of maneuver, commanders must apply combat power so that the terrain at Levels II and III can be exploited in the conduct of operations. Successful appl i cat ion of this concept requires commanders to think, plan, and maneuver vertical l y as well as horizontal ly.

2-2. Mountain operations usually focus on lines of communication, choke points, and dominating heights. Maneuver generally attempts to avoid strengths, envelop the enemy, and limit his ability to effectively use the high ground. Major difficulties are establishing boundaries, establishing and maintaining communications, providing logistics, and evacuating wounded. Throughout the plan, prepare, and execute cycle, commanders must continuously assess the vertical impact on the mission, enemy, terrain and weather, troops and support available, time available, civil considerations (METT-TC),

HISTORICAL PERSPECTIVE

Importance of Controlling Key Terrain: Mustafa Kemal at Gallipoli (April 1915)

On 25 April 1915. the Allies launched their Gallipoli campaign. However, LTC Mustafa Kemal's understanding of the decisive importance of the hilly terrain, his grasp of the enemy's overall intent, and his own resolute leadership preserved the Ottoman defenses. His troops seized the initiative from superior forces and pushed the Allied invasion force back to its bridgehead. The result was nine months of trench warfare, followed by the Allies' withdrawal from Gallipoli.

German Fifth Army Commander General von Sanders expected a major Allied landing in the north, at Bulair. The British, however, were conducting a feint there; two ANZAC divisions were already landing in the south at Ari Burnu (now known as "ANZAC cove") as the main effort. The landing beaches here were hemmed by precipitous cliffs culminating in the high ground of the Sari Bair ridge, a fact of great importance to the defense. Only one Ottoman infantry company was guarding the area. Although prewar plans had established contingencies for using 19th ID, Kemal, the division commander, had received no word from his superiors regarding the developing scenario. Nevertheless, understanding that a major Allied landing could easily split the peninsula, he decided that time was critical and set off for Ari Burnu without waiting for his senior commander's approval. In his march toward Ari Burnu that morning, he recognized that the hilly terrain in general and the Sari Bair ridge in particular were of vital strategic importance: if the enemy captured this high ground they would be in an excellent position to cut the peninsula in half.

Kemal now engaged the enemy. He impressed upon his men the importance of controlling the hilltops at all costs, issuing his famous order: "I am not ordering you to attack. I am ordering you to die. In the time it takes us to die, other forces and commanders can come and take our place." Despite being outnumbered three-to-one, the Turkish counterattack stabilized their position and prevented the Allies from capturing the Sari Bair ridge. Nightfall brought about a lull in the fighting. There was some sniping and a few local encounters on 26 April, and on 27 April Kemal finally received major reinforcements. The front stabilized and the opposing armies settled down into trench warfare. On 16 January 1926, the Allies admitted defeat and withdrew.

The 19th ID's counterattack, which prevented the ANZAC from establishing themselves on the Sari Bair ridge, may well have decided the outcome of the entire Gallipoli campaign. Despite his lack of situational knowledge, Kemal instinctively understood the enemy's intent and, recognizing the importance of controlling the hilltops and ridgelines, was committed to concentrating his combat power to seize and hold this key terrain.

Compiled from "The Rock of Gallipoli," *Studies of Battle Command*, George W. Gawrych

MISSION

2-3. Mission analysis must include the spatial and vertical characteristics of the AO. Although defeating the enemy continues to be the basic objective of tactical operations, the task of controlling specific operational terrain levels will be paramount. At brigade level and below, major tactical objectives are normally translated into tasks pertaining to seizing, retaining, or controlling specific dominating heights at either Level II or Level III. Therefore, it is imperative to identify the tasks and assets necessary to access each operational terrain level.

2-4. At any operational terrain level, defending and delaying are easier at defiles, while attacking is more difficult. Due to the compartmented terrain, units usually execute offensive missions by conducting several simultaneous

smaller-scale attacks, utilizingthefull height, width, and depth of their area of operations. Consequentiy, commanders must always consider the impact of decentralization on security.

2-5. One method of maintaining freedom of action is to seize or hotd key terrain. In the mountains, key terrain is frequently identified as terrain that is higher than that held by the enemy. Seizing this terrain often depends on long and difficult envelopments or turning movements. Therefore, the specified and implied tasks associated with mobility and sustainment, as well as command and control, must be considered in terms of their vertical difficulty.

ENEMY

2-6. An enemy will normally position forces in depth and height along likely avenues of approach. Mountain terrain facilitates wide dispersal, allowing relatively smait units to hold dominant terrain in a connected system of strong points. To prevent bypassing and envelopment attempts, the enemy may adopt a many-tiered, perimeter defense. Aside from the relative size of forces, the type of enemy units and their equipment must be compared with those of friendly forces, to include a comparison of the suitability of forces, tactics, and training. When considering the enemy's ability to operate in mountainous terrain, commanders should consider how well the enemy can accomplish the tasks and actions listed in Figure **2-1.** Again, in analyzing both enemy and friendly factors during mountain operations, the vertical, as well as the horizontal, perspective should be fully integrated into all aspects of the assessment.

- **Utilize the environment to his advantage**
- **Conduct air operations**
- **Conduct decentralized operations**
- **Utilize the terrain in Levels II and III**
- **Employ obstacles or barriers to restrict maneuverability**
- **Conduct limited-visibility operations**
- **Sustain his maneuver elements**

Figure 2-1. Factors Affecting Assessment of the Enemy Situation

TERRAIN AND WEATHER

2-7. As in all military operations, terrain analysis involves observation and fields of fire, cover and concealment, obstacles, key terrain, and avenues of approach (OCOKA). Terrain often influences the conduct of operations more in the mountains than on flatter terrain. The mountains form the nonlinear and vertical structure of the battlefield, and the influences of geography and climate dictate the extent to which commanders modify tactics. Examples of these difficulties are often encountered in the concentration of forces, as well as in the maintenance of command and control.

2-8. In themountains, as elsewhere, surprise is easier to achieve for the force that knows the terrain better and has the skills and equipment necessary to achieve greater mobility. The appropriate use of vertical terrain improves the element of surpriseiftheterrain has been analyzed properly to determine the best means to counter the enemy's reactions. Once the commander decides on a preliminary course of action, he should immediately initiate a detailed terrain reconnaissance,

2-9. In a mountainous environment, the terrain normally favors the defender and necessitates the conduct of limited visibility operations. Highly trained units can achieve significant tactical gains and decisive victories by exploiting limited visibility. However, limited visibility operations in mountainous terrain require precise planning, careful daylight reconnaissance, exceptionally good command and control, and a high degree of training. Imaginative and bold limited visibility operations can minimize the advantage of terrain for the defender and shift the balance of combat power to the side that can best cope with or exploit limited visibility.

OBSERVATION AND FIELDSOF FIRE

2-10. Although mountainous terrain generally permits excellent long-range observation and fields of fire, steep slopes and rugged terrain affect a soldier's ability to accurately estimate range and frequently cause large areas to be hidden from observation. The existence of sharp relief and dead space facilitates covert approaches, making surveillance difficult despite such long-range observation. Four factors that influence what can be seen and hit in mountainous terrain are listed in Figure 2-2.

1. The ability to observe and identify targets in conditions of bright sunlight
2. The ability to estimate range in clear air
3. The ability to apply wind corrections
4. The ability to shoot accurately up and down vertical slopes

Figure 2-2. Factors Affecting Observation and Fields of Fire

COVER AND CONCEALMENT

2-11.The identification and proper use of the cover and concealment provided by mountainous terrain are fundamental to all aspects of mountain operations. The ridge systems found in Level II may provide covert approaches through many areas that are hidden from observation by the vegetation and relief. The difficulties a force encounters in finding available cover and concealment along ridges are fewer than those on the peaks, especially above the timberline. Uncovered portions of an approach leave a force exposed to observation and fire for long periods. The enemy can easily detect movement in this region, leaving commanders with three primary options to improve cover and concealment:

1, Identify and exploit avenues of approach the enemy would consider unlikely, due to their difficult ascent or descent.

2, Negotiate routes during periods of limited visibility.

3, Provide overwhelming route security.

OBSTACLES

2-12, Obvious natural obstacles indude deep defiles, diffs, rivers, landslides, avalanches, crevices, and scree slopes, as well as the physical terrain of the mountains themselves. Obstades vary in their effect on different forces Commanders must evaluate the terrain from both the enemy and friendly force perspective. They must look spedfreal ly at the degree to which obstades restrid operations, and at the ability of each force to exploit the tactical opportunities that exist when obstades are employed. M an-made obstades used in conjundion with restrictive terrain are extremely effedive in the mountains; however, their construction is very costly in terms of time, materiel, transportation assets, and labor. Commanders must know the location, extent, and strength of obstades so that they can be incorporated into their scheme of maneuver.

KEY TERRAIN

2-13. Key terrain generally increases in importance with an increase in elevation and a decrease in accessibility. In the mountains, however, terrain that is higher than that held by the opposing force is often key, but only if the force is capable of fighting there. A well-prepared force capable of maneuver in rugged terrain can gain an even greater advantage over an ill-prepared enemy at higher elevation levels.

2-14. The vast majority of operations in the mountains requires that the commander designate dedsive terrain in his concept of operations to communicate its importance to his staff and subordinate commanders. In operations over mountainouster rain, the analysis of key and ded si ve terrain is based on the identification of these features at each of the three operational terrain levels. There are few truly impassable areas in the mountains. The commander must recognize that what may be key terrain to one force may be an obstacle to another force. He must also recognize that properly trained combatants can use high obstruding terrain as a means to achieve dedsive vidories with comparatively small-sized combat elements.

AVENUES OF APPROACH

2-15. I n mountainous terrain, there are few easily accessible avenues of approach, and they usually run along valleys, defiles, or the crests and spurs of ridges. This type of geography allows the defender to economize in difficult terrain and to concentrate on dangerous avenues of approach A typical offensive tactic is to conduct a coordinated assault with the main effort along accessible avenues of approach, and supporting efforts by one or more maneuver elements on difficult and unexpected avenues of approach. Normally, high rates of advance and heavy concentration of forces are difficult or impossible to achieve along mountainous avenues of approach. Relief features may create large areas of dead space that facilitate covert movement. Units may use difficult and unlikely avenues of approach to achieve surprise: however, these are extremely high-risk operations and are prone to failure unless forces are well trained and experienced in mountaineering techniques. In mountainous terrain, the analysis of avenues of approach should be based on a thorough reconnaissance and evaluated in terms of the factors listed in Figure 2-3 on page 2-8.

WEATHER

2-16. As discussed in Chapter 1, weather and visibility conditions in the mountainous regions of the world may create unprecedented advantages and disadvantages for combatants. To fight effectively, commanders must acquire accurate weather information about their AO. Terrain has a dominant effect on local climate and weather patterns in the mountains. Mountainous areas are subject to frequent and rapid changes of weather, including fog, strong winds, extreme heat and cold, and heavy rain or snow. Thus, many forecasts that describe weather over large areas of terrain are inherently inaccurate. Commanders must be able to develop local, terrain-based forecasts by combining available forecasts with field observations (local temperature, wind, precipitation, cloud patterns, barometric pressure, and surrounding terrain). Forecasting mountain weather from the field improves accuracy and enhances the ability to exploit opportunities offered by the weather, while minimizing its adverse effects (see Appendix B).

- **Ability to achieve surprise**
- **Vulnerability to attack from surrounding heights**
- **Ability to provide mutual support to forces on other avenues of approach**
- **Effect on rates of advance**
- **Effect on command and control**
- **Potential to accommodate deception operations**
- **Ability to support necessary CS and CSS operations**
- **Access to secure rest and halt sites**
- **Potential to fix the enemy and reduce the possibility of retreat**

Figure 2-3. Factors Affecting Analysis of Avenues of Approach

TROOPS AND SUPPORT AVAILABLE

2-17. Commanders must assess the operational and tactical implications of the restrictive environment on mobility, protection, firepower, and logistics. The complex task of arranging activities in time, space, and purpose requires commanders to fully understand the impact of elevation, weather, and visibility on the capabilities of his subordinate elements and relative combat power. Mountainous terrain and weather can greatly enhance the relative combat power of defending forces and, conversely, it can drastically reduce those of the attacking forces. For example, an infantry battalion may be inadequate to defeat a defending infantry company in the mountains. Instead, an infantry battalion may only be capable of defeating a well-positioned infantry platoon. However, commanders must carefully consider each unique situation and weigh all tangible and intangible aspects of combat power (maneuver, firepower, leadership, protection, and information) when comparing strengths and determining the forces necessary to accomplish the mission.

2-18. Commanders must also assess the proper mix of heavy and light forces that capitalizes on the unique strengths that each type of force can bring to mountain operations while minimizing their limitations. While generally complicating command and control, an appropriate mix allows commanders more flexibility in the synchronization of their operations. Additionally, the difficulty providing combat support and combat service support for mountain operations must be evaluated to determine if the proportion of support troops to combat troops is sufficient.

2-19. Prior to and throughout an operation, commanders must continually assess the effect that the rugged mountain environment and sustained combat operations has on the ability of their soldiers to accomplish the mission.

Commanders may need to slow the pace of their operation, transition to the defense for short periods, or rotate units to ensure that their soldiers are physically capable of striking effectively at decisive times and locations. Too often, commanders consider only the operational readiness (OR) rate of equipment and logistics levels when determining their overall ability to continue offensive actions. Failure to consider this intangible human aspect may result in increased loss of lives and mission failure.

2-20. Vertical operations are an integral part of mountain operations and are one means to improve the success of decisive engagements. Commanders must review the state of training of their units to ensure they are adequately prepared to maneuver and fight at various elevations. Increased requirements for aviation support require aviation units to be capable of operating in the specific mountain environment. Units must also have sufficient numbers of pathfinders and trained air assault personnel to select and mark landing zones (LZs) and prepare sling loads.

TIME AVAILABLE

2-21. In the mountains, proper timing is fundamental to creating opportunities to fight the enemy on favorable terms. Restrictive terrain, weather, the accumulation of chance errors, unexpected difficulties, and the confusion of battle increase the time necessary to assemble, deploy, move, converge, and mass combat power, effectively decreasing the amount of time available to plan and prepare. To optimize the time available, commanders must continuously evaluate the impact of reduced mobility caused by the weather and terrain. At times, commanders may need to conduct a tactical pause to facilitate the concentration of combat power at a decisive point. However, they must consider time with respect to the enemy as time available is always related to the enemy's ability to execute his own plan, prepare, and execute cycle. Figure 2-4 summarizes the time considerations that are different from or greater than those encountered on flatter terrain.

- **Adaptability of plans to the terrain and varying weather**
- **Increased time needed to conduct reconnaissance, execute movements, and synchronize events on the battlefield**
- **Significant variance in the number of hours of visibility with season and elevation**

Figure 2-4. Factors Affecting Time Available

CIVIL CONSIDERATIONS

2-22. Generally, civilian population centers will be located at the lower elevations of Level I close to sources of water and along major lines of communications. Refugees and displaced civilians may increase congestion on the already limited road and trail networks normally found in mountainous environments, further complicating maneuver and sustaining operations.

2-23. Commanders must also consider the impact of operations on the often-limited civilian resources available in the mountains. The wisdom of using local resources to lighten in-theater supply requirements must be balanced against the impact on civilians and their local economy. While the purchase of goods and services from the local economy is generally welcomed, it may serve to inflate prices and make it impossible for local civilians to purchase their own scarce and needed supplies.

2-24. In mountainous regions, commanders often encounter a populace of diverse political and ethnic orientation that may support, oppose, or be ambivalent to US operations or the presence of US forces. Depending on friendly force objectives, commanders may conduct public relations, civil affairs, humanitarian assistance, and psychological operations (PSYOP) to influence perceptions and attitudes of neutral or uncommitted parties. Even if commanders cfioose not to commit resources to enlist civilian sympathy and support, they must still adjust their operations to minimize damage and loss of life to innocent civilians.

SECTION II-LEADERSHIP

2-25. To help ease their anxiety in combat, soldiers must have confidence in their leaders. This confidence may diminish rapidly unless leaders demonstrate the ability to lead over formidable terrain and under the most difficult weather conditions. Superficial knowledge of mountain warfare and ignorance or underestimation of mountain hazards and environmental effects may result in mission failureand the unnecessary loss of soldiers' lives.

2-26. Effective leadership in mountain operations combines sound judgment with a thorough understanding of the characteristics of the mountain environment. Commanders must first develop flexible and adaptable leadership throughout the chain of command. They must then be able to understand and exploit the operational and tactical implications of the mountain environment, as well as its effects on personnel, equipment, and weapons. The keys to meeting this challenge are proper training and operational experience in the mountains. To fight effectively, leaders creatively exploit the opportunities offered by the mountain environment while minimizing the adverse effectsit can have on their operations.

2-27. Leadership rapidly becomes the primary element of combat power on the mountain battlefield. Commanders must recognize the distinctive effects created by decentralization of command, develop a depth of leadership that forms the vital link to unity of effort, and organize and direct operations that require minimum intervention. While specific situations require different leadership styles and techniques, the nature of mountain warfare generally necessitates that commanders embrace the philosophy of command and control known as mission command (see FM 6-0). This type of command and control requires subordinates to make decisions rapidly within the framework of the commander's concept and intent. Commanders must be able to accept some measure of uncertainty, delegate, and trust and encourage subordinate leaders at all levels to use initiative and act alone to achieve the desired results, particularly when the situation changes and they lose contact with higher headquarters.

SECTION III - COMMUNICATIONS

2-28. The communications means available to support operations in mountainous regions are the same as those to support operations in other regions of the world. However, rapid and reliable communications are especially difficult to achieve and maintain in mountainous areas. The mountainous environment requires electronic equipment that is light, rugged, portable and able to exploit the advantages of higher terrain. The combined effects of irregular terrain patterns, magnetic and ionospheric disturbances, cold, ice, and dampness on communications equipment increase operating, maintenance, and supply problems and require precise planning and extensive coordination.

COMBAT NET RADIO

SINGLE-CHANNEL GROUND AND AIRBORNE RADIO SYSTEMS (SINCGARS)

2-29. The Single-channel Ground and Airborne Radio System (SINCGARS) family of frequency modulation (FM) radios is good for the control of battalion and smaller-sized units operating in a mountainous environment (see FM 6-02.32 and FM 6-02.18). If available, hands-free radios, such as helmet-mounted radios, are an excellent means of communication for small unit tactics and close-in distances, particularly while negotiating rugged terrain. In colder environments, shortened battery life greatly reduces the reliability of manpacked systems that rely on constant voltage input to maintain maximum accuracy.

2-30. Since even a small unit may be spread over a large area, retransmission sites may be needed to maintain communications and increase range. These sites require extensive preparation and support to ensure the survival of personnel and the continued maintenance of equipment. Retransmission systems are often placed on the highest accessible terrain to afford them the best line-of-sight; however, through simple analysis, these locations are often predictable and make them more vulnerable to enemy interdiction. The importance and difficulty of maintaining adequate communications in mountainous terrain requires commanders to devote additional resources for the protection of these limited assets and operators skilled in the proper use of cover and concealment, noise and light discipline, and other operations security (OPSEC) measures.

2-31. Physical range limitations, difficulties in establishing line-of-sight paths due to intervening terrain, and limited retransmission capabilities often make it difficult to establish a brigade and larger-sized radio net. However, commanders can, if within range, enter subordinate nets and establish a temporary net for various contingencies. In the mountains or if the mobile subscriber equipment network is not yet fully developed, commanders should consider the increased need for the improved high frequency radio (IHFR) family of amplitude modulation (AM) radios and single-channel tactical satellite communications terminals for extended distances.

SATELLITE COMMUNICATIONS (SATCOM)

2-32. Satellite communications (SATCOM) terminals are light, small, portable ground terminals that are able to communicate in spite of rugged terrain. During operations in mountainous areas having little or no infrastructure to support command and control, satellite communications become the primary means of communications. Single channel SATCOM are currently transmitted over the ultrahigh frequency (UHF) band and readily support forces operating in the mountains, while providing worldwide tactical communications, in-theater communications, combat net radio (CNR) range extension, and linkage between elements of long-range surveillance units (LRSUs) and Army special operations foroes (ARSOF). SATCOM can network with multiple users, communicate while enroute, penetrate foliage while on the ground, and has several other advantages making it an ideal system for mountain communications (see Figure 2-5). However, limitations include restricted access, low-rate data communications, and lack of antijam capability Commanders should review FM 6-02.11 for further information on the employment of SATCOM.

- Greater freedom from siting restrictions
- Extended range, capacity, and coverage
- Mobility and rapid employment
- Extremely high circuit reliability

Figure 2-5. SATCOM Advantages

COMMAND AND CONTROL (C^2) AIRCRAFT

2-33 Using C^2 aircraft can assist the commander in overcoming ground mobility restrictions and may improve communications that would otherwise limit his ability to direct the battle. In the mountains, terrain masking, while making flight routing more difficult, may provide the degree of protection needed to allow an increased use of aircraft. To avoid radar or visual acquisition and to survive, C^2 aircraft must use the same terrain flight techniques employed by other tactical aviation units. This flight method often degrades FM communications and reinforces the requirement for radio relay or retransmission sites

ANTENNAS AND GROUNDS

2-34. Directional antennas, both bidirectional and unidirectional, may be needed to increase range and maintain radio communications. Although easy to fabricate, directional antennas are less flexible and more time-consuming to set up. Positioning of all antennas is also crucial in the mountains because moving an antenna even a small distance can significantly affect reception

2-35. Antenna icing, a common occurrence at high elevations, significantly degrades communications. Ice may also make it difficult to extend or lower antennas, and the weight of ice buildup, combined with increased brittleness, may cause them to break. Antennas should have extra guy wires, supports, and anchor stakes to strengthen them to withstand heavy ice and wind loading. All large horizontal antennas should be equipped with a system of counterweights arranged to slacken before wire or poles break from the excess pressures of ice or wind. Soldiers may be able to remove wet snow and sleet that freezes to antennas by jarring their supports, or by attaching a hose to the exhaust pipe of a vehicle and directing the hot air on the ice until it melts. However, soldiers must exercise great care to ensure that the antenna is not damaged in their attempts to dislodge the ice.

2-36. Ground rods and guy wires are often difficult to drive into rocky and frozen earth. Mountain pitons are excellent anchors for antenna guys in this type of soil. In extreme cold, ropes can be frozen to the ground and guys tied to these anchor ropes. Adequate grounding is also difficult to obtain on frozen or rocky surfaces due to high electrical resistance. Where it is possible to install a grounding rod, it should be driven into the earth as deep as possible or through the ice on frozen lakes or rivers. Grounding in rocky soil may be improved by adding salt solutions to improve electrical flow.

MOBILE SUBSCRIBER EQUIPMENT

2-37. Like FM. mobile subscriber equipment [MSE) requires a line-of-sight transmission path and a tactical satellite or several relay sites to overcome mountainous terrain and maintain MSE connectivity (FM 6-02.55 contains in-depth information concerning the deployment and employment of MSE).

WIRE AND FIELD PHONES

2-38. Wire is normally one of the most reliable means of communication Unfortunately, in rugged mountains and particularly during the winter months, wire is more difficult and time consuming to install, maintain, and protect. Wire may be dispensed in mountain areas by tracked or wheeled vehicle, foot, skis, snowshoes, or oversnow vehicles. As in any environment, units must periodically patrol their wire lines to ensure that they have remained camouflaged and that the enemy has not tapped into them.

2-39. Snow-covered cables and wire can cause the loss of many man-days in recovering or maintaining circuits. This can be avoided by pulling the cable from under the snow after each snowfall and letting it rest just below the surface of the snow. Trees or poles can be used to support wire. Allowances must be made for drifting snow when determining the height above ground at which to support the lines. However, when crossing roads, it is preferable to run the wire through culverts and under bridges rather than bury or raise wire overhead. In addition to ease, this technique reduces maintenance requirements associated with vehicles severing lines, particularly with higher volumes of traffic on limited road networks. If long-distance wire communications are required, the integration of radio relay systems must be considered.

2-40. Great care must be taken in handling wire and cables in extreme cold weather. Condensation and ice on connectors make connecting cables difficult and can degrade the signal path. When rubber jackets become hard, the cables must be protected from stretching and bending to prevent short circuits caused by breaks in the covering. Therefore, all tactical cable and wire should be stored in heated areas or warmed prior to installation. TC 24-20 provides more detailed information on the installation and maintenance of wire and cable.

2-41. Field phones are usually set up in a stationary position, such as a mountain patrol base or an ambush site, although leaders must consider the weight and difficulties encountered in laying and maintaining wire in these sites of limited duration. The batteries that are used to operate field telephones and switchboards are subject to the same temperature limitations as those used to power tactical radiosets.

2-42. When used with a hands-free phone, commercially available rope with a communication wire in it is ideally suited for mountain operations. This system is lightweight and easy to manage, and provides an added measure of security during limited visibility operations. In addition to the standard uses, since it functions as both a rope and a wire, it can be used to control movement on all types of installations, and it can serve as a primary means of communication for climbing teams.

AUDIO, VISUAL, AND PHYSICAL SIGNALS

2-43. Leaders can use simple audio signals, such as voice or whistles, to locally alert and warn. Sound travels farther in mountain air. Although this effect may increase the possibility of enemy detection, interrupting terrain, wind conditions, and echoes can restrict voice and whistle commands to certain directionsand uses.

2-44. Like audio signals, visual signals such as pyrotechnics and mirrors have limited use dueto enemy detection, but may work for routine and emergency traffic at the right time and place. Blowing sand or snow, haze, fog, and other atmospheric conditions may periodically affect range and reliability.

2-45. Units should use hand and arm signals instead of the radio or voice when ever possi bl e, especi al I y wh en cl ose tothe enemy. Luminous tape on t he camouflage band, luminous marks on a compass, or flashlights may be used as signals at night over short distances. Infrared sources and receiving equipment, such as night vision goggles, aiming lights, and infrared filters for flashlights, can be used to send and receive signals at night However, an enemy outfitted with similar equipment can also detect active devices.

2-46. A tug system is a common method of signaling between members of a roped climbing team. However, tug systems are often unreliable when climbers are moving on a rope or when the distance is so great that the friction of the rope on the rock absorbs the signals. Separate tug lines can be installed in Static positions by tying a String, cord, or wire from one position to the next. Soldiers can pass signals quietly and quickly between positions by pulling on thetug line in a prearranged code.

MESSENGER

2-47. Although slew, communication by messenger is frequently the only means available to units operating in the mountains. Messengers should be trained climbers, resourceful, familiar with mountain peculiarities, and able to carry their own existence load. During the winter, advanced skiing skills may also be required. Messengers should always be dispatched in pairs. Air messenger service should be scheduled between units and integrated with the aerial resupply missions. Vehicles may also be employed to maintain messenger communications when conditions of time, terrain, and distance permit.

SECTION IV - TRAINING

2-48. Because US forces do not routinely train in a mountain environment, they must make extensive preparations to ensure individual and unit effectiveness. Ultimate success in the mountains depends largely on developing cohesive, combat-ready teams consisting of well-trained soldiers. To be successful, commanders must understand the stratification of mountain warfare, recognize the unique aspects of leadership required, and implement training programs that prepare soldiers for the rigors of mountain fighting.

2-49. In the mountains, commanders face the challenge of maintaining their units' combat effectiveness and efficiency. To meet this challenge, commanders conduct training that provides soldiers with the mountaineering skills necessary to apply combat power in a rugged mountain environment, and they develop leaders capable of applying doctrine to the distinct characteristics of mountain warfare.

- **Mountaineering skills**
- **Air assault and air movement operations**
- **Deception**
- **Stealth and infiltration**
- **Limited visibility operations**
- **Patrolling**
- **Reconnaissance**
- **Communications**
- **CS and CSS operations**

Figure 2-6. Training Areas of Emphasis

2-50. The ability to apply doctrine and tactics in mountainous environments is not as easy to develop as technical proficiency. Training, study, and garrison experimentation may provide the basis for competence. However, only through experience gained by practical application in the mountains will leaders become skilled in mountain warfare. Proficiency in the areas listed in Figure 2-6 will provide commanders with a degree of flexibility in the application of doctrine to a mountain area of operations.

2-51. The best combat and combat support plans cannot ensure victory unless commanders concentrate on developing a leadership climate that is derived from the human dimension of mountain warfare. The complexities of mountain combat make it extremely important to establish training programs that modify the traditional application of tactics so that units can reach their full potential. Training must simulate the tempo, scope, and uncertainty of mountain combat to create the versatility required to capitalize on the harsh environment as a force multiplier,

2-52. Competent units operate effectively in mountains and focus on the battle. Unprepared units, however, may become distracted by the environment and end up expending as much effort fighting the environment as they do fighting the enemy. Soldiers cannot be fully effective unless they have the proper clothing and equipment, and are trained to protect themselves against theeffectsof terrain and frequent and sudden changes in weather.

INITIAL TRAINING ASSESSMENT

2-53. In addition to the questions applicable to every mission, commanders must consider the following when preparing for operations in a mountainous environment:

- What kind of mountains will the unit be operating in?
  - What elevations will the unit be operating at?
  - What are the climatic and terrain conditions of the AO?
  - » Are at least two years of accurate weather reports available (see Appendix B)?
- When must the unit be ready to move?
- What training resources are needed and available?
- D Are local training areas and ranges available?
  - If not. what alternative arrangements can be made?
  - What available training areas most closely resemble the AO?
- D What special equipment does the unit require?
- D What training assistance is available?
  - Does the unit have former mountain warfare instructors, military mountaineers, or others with experience in a mountainous environment?
  - Are instructors available from outside the unit?
- P What special maintenance is required for weapons and equipment?
- What is the level of physical fitness?
- 0 What additional combat, combat support, and combat service support units are necessary to accomplish the operational missions?
  - Can specific units be identified for possiblecoordinated training?
- D Will allied and multinational troops participate?

2-54, As commanders get answers to these and other questions, they must develop training programs to bring their units to a level where they will be fully capable of operating successfully in mountainous conditions. To do this, they must establish priorities for training. The training requirements listed in Figure 2-7 are only a guide. Commanders should add, delete, and modify the tasks as necessary, depending on the specific AO, the state of readiness of their units when they begin preparations for mountain operations, and the time and facilities available (see FM 7-10).

PHYSICAL CONDITIONING

2-55. Soldiers who have lived and trained mostly at lower elevations tend to develop a sense of insecurity and fear about higher elevations - many are simply afraid of heights in general. With this in mind, leaders must plan training that accustoms soldiers to the effects of the mountain environment. Physical conditioning must be strictly enforced, since "new musde" strain associated with balance and prolonged ascents/descents quickly exhausts even the most physically fit soldiers. Even breathing becomes strenuous, given the

TRAINING REQUIREMENTS	ALL	STAFF AND LEADERS	TEAMS AND CREW MEMEBERS	SPECIALISTS
Physical Conditioning and Acclimatization	✔	✔	✔	✔
Mountain Illnesses and Injuries	✔	✔	✔	✔
Mountain Living and Survival	✔	✔	✔	✔
Mountain Navigation Techniques	✔	✔	✔	✔
Mounted and Dismounted March Planning	✔	✔	✔	✔
Communications Techniques	✔	✔	✔	✔
Weapons/Equipment Training	✔	✔	✔	✔
Additional Maintenance Requirements	✔	✔	✔	✔
Camouflage and Concealment	✔	✔	✔	✔
Obstacles		✔	✔	
Above-ground Fortifications	✔	✔	✔	✔
Level 1 Mountaineering	✔	✔	✔	✔
Level 2 and 3 Mountaineering				✔
Driver and Pilot Training				✔
Air Assault/Air Movement Operations	✔	✔	✔	✔
NBC Operations	✔	✔	✔	✔

Figure **2-7.** Mountain Preparatory Training

thinner atmosphere at higher altitudes. Therefore, training must emphasize exercises designed to strengthen leg muscles and build cardiovascular (aerobic) endurance (see FM 3-25.20). Frequent marches and dimbswith normal equipment loads enhance conditioning and familiarize soldiers with mountain walking techniques,

MOUNTAIN LIVING

2-56. Successful mountain living requires that personnel adjust to special conditions, particularly terrain and weather. To develop confidence, soldiers should train in conditions that closely resemble those they will face. Lengthy exercises test support facilities and expose soldiers to the isolation common to mountain operations. The mountain area of operations can be harsh, and training should develop soldiers who possess the necessary field craft and psychological edge to operate effectively under mountainous conditions. Although FM 4-25.10 and FM 3-25.76 do not specifically address mountain environments, much of their information applies. Regardless of the level of technical mountaineering training required, all soldiers deploying to a mountainous region should be trained in the areas listed in Figure *2-8* on page 2-**18.**

2-57. Navigation in the mountains is made more difficult because of inaccurate mapping, magnetic attraction that affects compass accuracy, and the irregular pace of the soldiers It is easy to mistake large terrain features that are very far away for features that are much closer The increased necessity for limited-visibility operations restricts the use of terrain techniques as the primary means of determining and maintaining direction. Individuals must train to use a variety of equipment, such as a compass, an altimeter, global positioning system devices, and maps, as well as learn techniques pertaining to terrestrial navigation, terrain association, dead reckoning, resecfioning, and artillery marking (seeFM 3-25.26).

- Temperature extremes and clothing requirements
- Bivouac techniques and shelter construction
- Elevation and rarified air effects
- Hygiene, sanitation, and health hazards
- Locating and purifying water
- Food-gathering techniques

Figure 2-8. Mountain Living Training

MD EQUIPMENT

2-58. Nearly every weapon or piece of equipment familiar to the soldier is affected to some degree by the mountain environment. In addition to honing skills, training must focus on the specific operational area and ways to overcome anticipated environmental impacts when using weapons and equipment.

2-59. Individual marksmanship training must emphasize the effect of wind and include practical training in wind measurement techniques and adjusted aiming points (holdoff). Practical training in range estimation techniques, combined with using laser range finders, M19 binoculars, target reference points, and range cards, helps lo overcome difficulties in range estimation.

2-60. In the conduct of their preparations, commanders should strive to increase the number of qualified snipers within their units, as they are ideal in the mountains and can be used to adversely affect enemy mobility by delivering long range precision rifle fire on selected targets. They can inflict casualties, slow enemy movement, lower morale, and add confusion to enemy operations A single sniper team in well-concealed positions, such as mountain passes, can severely impede enemy movement (see FM 3-21.20 and FM 3-91.2 for further information on sniper employment).

;E AND CONCEALMENT

2-61. The basic principles of camouflage and concealment also apply in mountain operations (see FM 3-24 3). However, certain elements must be adjusted for snow. With snow on the ground, standard camouflage nets and paint patterns are unsuitable. In areas where snow cover is above 15 percent of the background, winter camouflage nets should take the place of standard nets and temporary white paint should be used over the green portions of vehicles. In terrain with more than 85 percent snow cover, the vehicles and equipment should be solid white. However, with less than 15 percent snow cover, standard patterns should be maintained.

2-62. Snow provides excellent conditions for threat thermal and ultraviolet sensor detection. To counter these types of sensors, soldiers must be trained to utilize the terrain to mask themselves and their weapons and equipment from enemy detection. The mountainous terrain often limits the access routes to and from selected positions. Commanders must take appropriate measures to conceal vehicle tracks and limit movement times to periods of limited visibility. Snow presents a significant problem, making movement discipline an absolute requirement. When moving, leaders should be trained to follow the shadows along windswept drift lines as much as possible. Drivers should learn to avoid sharp turns, which are easily recognizable in the snow, and follow existing track marks where possible.

FORTIFICATIONS

2-63. Fighting and protective positions in the mountains do not differ significantly from other environments, except in areas of snow and rock (see FM 3-34.112 for more information on common survivability positions and FM 3-97.11 for positions created in snow). Digging positions in rocky ground is difficult and often impossible. If demolitions, pneumatic drills, and jackhammers are available, positions may be blasted or drilled in the rock to afford some degree of protection. More often, it will be necessary to build above-ground positions by stacking boulders, stones, and gabions. If possible, existing rock formations should be used as structural wall components.

2-64. If above-ground positions are to be used, considerable care should be taken to avoid siting them in view of any likely enemy avenues of approach. Even a two-man position is difficult to conceal if it is above the timberline. Camouflage nets and the use of background rocks are necessary to break up theoutlineof theposition and hide straight edges.

2-65. Positions should be built of the largest rocks available, wedged securely together. Extreme care should be taken that the walls are stable and not leaning or sloping downhill. An unstable wall is more of a liability than an asset, as the first Impact may cause it to collapse onto the defenders. Rocks and gabions should be stacked to systematically overlap each joint or seam to help ensure stable construction. Larger rocks or stones can be used to help bond layers of rock beneath. If possible, a layer of sandbags should be placed on the top of and around the inside of the wall. Substantial overhead cover is normally required in rocky areas. The effects of artillery bursts within and above a protective position are greatly enhanced by rock and gravel displacement or avalanche Figure 2-9 on page 2-20 shows simple examples of the right and wrong way to build these positions.

MILITARY MOUNTAINEERING

2-66. The skills required for movement are often difficult to learn and usually very perishable. Commanders must understand the application and mechanics of technical mountaineering systems needed for mobility and movement of soldiers and equipment. In the mountains, a unit may be ineffective unless it has the prerequisite technical training. However, some mountains may feature terrain that is relatively benign, requiring minimal specialized techniques. Other areas will mandate the need for more advanced mountaineering skills. One key to quickly determining the type and extent of training required is to analyze and classify the level of individual movement required according to the dismounted mobility classification table introduced in Chapter 1. Once commanders have determined the specific level and tasks required, TC 90-6-1 will provide them with detailed information on specific mountaineering techniques and equipment (described below).

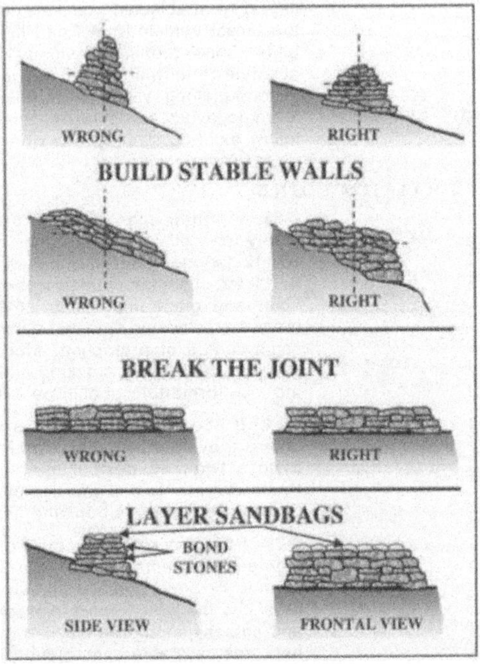

2-67. Military mountaineering training provides tactical mobility in mountainous terrain that would otherwise be inaccessible. Soldiers with specialized training who are skilled in using mountain climbing equipment and techniques can overcome the difficulties of obstructing terrain. Highly motivated soldiers who are in superior physical condition should be selected for more advanced military mountaineering training (Levels 2 and 3) conducted at appropriate facilities. Soldiers who have completed advanced mountaineering training should be used as trainers, guides, and lead climbers during collective training. They may also serve as supervisors of installation teams (see Chapter 4) and evacuation teams (see Chapter 5). Properly used, these soldiers can drastically improve mobility and have a positive impact disproportionate to their numbers. Units anticipating mountain operations should strive to achieve approximately ten percent of their force with advanced mountaineering skills.

LEVEL 1: BASIC MOUNTAINEER

2-68 The basic mountaineer, a graduate of a basic mountaineering course, should be trained in the fundamental travel and climbing skills necessary to move safely and efficiently in mountainous terrain. These soldiers should be comfortable functioning in this environment and, under the supervision of qualified mountain leaders or assault climbers, can assist in the rigging and use of all basic rope installations. On technically difficult terrain, the basic mountaineer should be capable of performing duties as the "follower" or "second" on a roped dimbing team, and should be well trained in using all basic rope systems. These soldiers may provide limited assistance to soldiers unskilled in mountaineering techniques. Particularly adept soldiers may be selected as members of special purpose teams led and supervised by advanced mountaineers. Figure 2-10 lists the minimum knowledge and skills requi red of basicmounlaineers

▯ Characteristics of the mountain environment (summer and winter)	▯ Rope management and knots
▯ Mountaineering safety	▯ Natural anchors
▯ Use, care, and packing of individual cold weather clothing and equipment	▯ Familiarization with artificial anchors
▯ Care and use of basic mountaineering equipment	▯ Belay and rappel techniques
▯ Mountain bivouac techniques	▯ Use of fixed ropes (lines)
▯ Mountain communications	▯ Rock climbing fundamentals
▯ Mountain travel and walking techniques	▯ Rope bridges and lowering systems
▯ Hazard recognition and route selection	▯ Individual movement on snow and ice
▯ Mountain navigation	▯ Mountain stream crossings (to include water survival techniques)
▯ Basic medical evacuation	▯ First aid for mountain illnesses and injuries

Figure 2-10. Level 1: Basic Mountaineer Tasks

2-59. In a unit training program, level 1 qualified soldiers should be identified and prepared to serve as assistant instructors to train unqualified soldiers in basic mount aineeri ng skills. **All** high-risk trai ning, however, must be conducted under the supervision of qual ified level 2 or 3 personnel.

LEVEL 2: ASSAULT CLIMBER

2-70, Assault climbers are responsible for the rigging, inspection, use, and operation of all basic rope systems. They are trained in additional rope management skills, knot tying, belay and rappel techniques, as well as using specialized mountaineering equipment. Assault dimbers are capable of rigging complex, multipoint anchors and high-angle raising/lowering systems. Level 2 qualification is required to supervise all high-risk training associated with Level 1. At *a* minimum, assault climbers should possess the additional knowledge and skillsshown in Figure2-11 on page 2-22.

LEVEL 3: MOUNTAIN LEADER

2-71, Mountain leaders possess all the skills of the assault climber and have extensive practical experience in a variety of mountain environments in bot h winter and summer conditions. Level 3 mountaineers should have well-developed hazard evaluation and safe route finding skills over nil types of mountainous terrain. Mountain leaders are best qualified to advise commanders on all aspects of mountain operations, particularly the preparation and leadership required to move units over technically diffiouft, hazardous, or

□ Use specialized mountaineering equipment □ Perform multipitch climbing: - Free climbing and aid climbing - Leading on class 4 and 5 terrain □ Conduct multipitch rappelling □ Establish and operate hauling systems □ Establish fixed ropes with intermediate anchors	□ Movement on moderate angle snow and ice □ Establish evacuation systems and perform high angle rescue □ Perform avalanche hazard evaluation and rescue techniques □ Familiarization with movement on glaciers

Figure 2-11. Level 2: Assault Climber Tasks

exposed terrain. The mountain leader is the highest level of qualification and is the principle trainer for conducting mountain operations. Instructor experience at a military mountaineering training center or as a member of a special operations forces (SOF) mountain team is critical to acquiring Level 3 qualification. Figure 2-12 outlines the additional knowledge and skills expected of mountain leaders. Depending on the specific AO, mountain leaders may need additional skills such as snowshoeing and all-terrain skiing

Figure 2-12. Level 3: Mountain Leader Tasks

DRIVER TRAINING

2-72. Driving in mountains is extremely difficult. To be successful, drivers must know their equipment's limitations and capabilities Training should center on practical exercises in mountainous terrain that gradually introduce drivers to more complex terrain and weather conditions. The exact nature of the mountainous terrain determines the training (see Figure 2-13).

□ Identification and recognition of potential dangers
□ Movement along steep grades combined with:
- Narrow roads and sharp curves
- Loose rock and gravel
- Ice and snow (to include using tire chains for wheeled vehicles)
- Towed loads
□ Increased cold weather maintenance requirements

Figure 2-13, Driver Training

ARMY AVIATION

2-73. The mountainous environment, particularly its severe and rapidly changing weather, affects aircraft performance capabilities, accelerates crew fatigue, and influences basic flight techniques. These techniques can be acquired only through a specific training program for the particular type of mountainous terrain. Additionally, limited visibility operations in the mountains are extremely hazardous and require extensive training for those aviation units involved. Common problems associated with mountain operations become much more complex at night, even when using night vision devices. Few Army aviation units regularly train for mountain operations, so it is critical to alert them as soon as possible to fadlitate the required training to ensure safe and successful mission execution.

RECONNAISSANCE AND SURVEILLANCE

2-74. Training in reconnaissance and surveillance should focus on trafficability (route, mobility, and bridge classification), potential drop zones or landing areas, likely defensive positions, and potential infiltration routes. Infiltration and exfiltration are relatively easy in mountainous terrain and constitute a significant threat tothe maneuver elements and their support units.

TEAM DEVELOPMENT

2-75. The decentralized nature of mountain combat and the need for the exercise of a mission command philosophy of command and control involve assigning missions to i ndependenlty operating small teams that may be i sotated from thei r higher headquarters. The disruptive i nfl uences of the environment and sustained physical stress further increase the perception of i sol at ion (see FM 4-02.22). The most important factor that sustains a soldier in combat is the powerful psychological support that he receives from his primary group, such as a buddy team, squad, or platoon. H e is less l i kely to feet the stress of loneliness under the isolated conditions of mountain warfare if his primary group maintains its integrity.

2-76, The soldier's ability to survive and operate in the mountains is the basis for the self-confidence needed to feel accepted by the team. Leaders must develop small-unit cohesion down to the buddy team. Each soldier must have a buddy to share both responsibilities and rewards. The leader must not simply assign two soldiers as a buddy team, but pair soldiers whose skills and attributes complement each other. Each soldier can then learn his buddy's specialized skills adding depth to the unit if one soldier becomes disabled. Soldiers work with their buddies, as well as function as part of the larger squad team. The combined strengths of buddies enhance both unit effectiveness and combat power. FM 6-22 has more information on team development.

Chapter 3

Firepower and Protection of the Force

Employing fire support systems, which are an integral part of maneuver, is included in this chapter. This arrangement, however, does not suggest any change in the close doctrinal relationship between fires and maneuver during mountain operations.

SECTION I – FIREPOWER

FIELD ARTILLERY

3-1. The basic tactical principles for artillery remain valid in mountains, subject to the limitations imposed by terrain and weather.

MOVEMENT AND POSITIONING

3-2, Rugged terrain and reduced mobility increase the reliance on field artillery fire support. However, the employment and positioning of field artillery systems may be severely impacted by the extreme difficulty of ground mobility in mountainous terrain Self-propelled artillery is often limited to traveling on the existing road and trail networks and positioning in their immediate vicinity. Towed field artillery is usually more maneuverable; it can be brought into position with the aid of trucks, tractors, and fixed or rotary-winged aircraft. Therefore, gun crews should be

CONTENTS

proficient in equipment-rigging techniques and air assault procedures, and possess ample sling-load equipment. Field artillery emplaced by helicopter normally requires continued airlift for subsequent displacement and ammunition resupply, and often necessitates substantial engineer support,

3-3. Light field artillery may require forward displacement of gun sections by helicopter to provide forward troops the necessary support. Medium field artillery may give the longer range required, but may be limited by high-terrain crest clearance. Normally, field artillery is employed far enough to the rear to take advantage of increased angles of fall. Flat areas, such as dry riverbeds, villages and towns, and farmland, can usually accommodate firing units, however, these positions present particular problems in the mountains for the following reasons:

- Dry riverbeds are hazardous because of the danger of flash flooding.
- Towns and villages usually have adequate flat areas such as parks, schoolyards, and playing fields but they are relatively scarce and often targeted by the enemy.
- Farmland is often difficult to negotiate from spring to fall. In the winter, if the ground is frozen, farmland may provide good firing positions; however, frozen ground may cause difficulty emplacing spades, base plates, and trails.

3-4, Good artillery positions, selected for cover, flash defilade, and accessibility to road nets and landing zones (LZs). are difficult to find, and their relative scarcity makes it easier for the enemy to target probable locations. In some instances, it may be necessary to by-pass the best position for one less suitable to reduce the enemy's counterfire effects. Commanders must ensure that positions on dominant terrain provide adequate defilade. Positions on commanding terrain are preferable to low ground positions because there is-

- A reduction in the number of missions requiring high-angle fire.
- A reduced amount of dead space in the target area.
- Less exposure to small arms fire from surrounding heights.
- Less chance of being struck by rockslides or avalanches.

3-5. Seme weapons may be moved forward to provide long-range interdiction fires or. in extreme cases, direct fires to engage a road-bound enemy in mountain passes or along valley floors. Because of rugged terrain, higher angles of fire, and reduced ranges, it is generally neoessary to displace artillery more frequently than on level terrain to provide continuous support. In the mountains, commanders must often employ field artillery in a decentralized manner because of the limited space for gun positions.

ACQUISITION AND OBSERVATION

3-6. Because of high angle fire requirements, radar can be effective against enemy indirect fire systems. In many instances, terrain masking and diminished line-of-sight may degrade its effectiveness. Sites should be selected on promi nent terrain to obtain the lowest possible screening crest. However, it is often difficult to obtain a low and consistent screening crest in mountainous terrain. Too low a screening crest drives the search beam into the ground. Too high a screening crest allows the enemy to fire under the beam and avoid detection. When positioning weapons locating radars, commanders should also consider thefollowing:

- Although time-consuming, visibility diagrams are extremely useful in determining the probability of acquiring targets within the sectors of search of the radar.
- To limit search areas, radars should focus on terrain that can be occupied by artillery and mortars.
- Accurate survey control is essential because of the extreme elevation variations in mountainous terrain. Helicopters may be useful in performing survey by use of the Position Azimuth Determining System (PADS) I f possible, digital radar maps may be used to minimize the time required for height correction of the weapon system. Digital maps allow the Firefinder systems to initially locate weapon systems to within 250 meters. This allows the radar operator to make only two to three visual elevation adjustments to accurately locate the weapon system.
- Impact predict is computed at the radar's elevation, therefore, excessive errors in the impact predict can be expected.
- Firefinder radars in the same area must not face one another and radiate at the same time. This causes interference and emissions burnout, resulting in equipment failure. If radars need to face one another to accomplish the mission, commanders must coordinate to ensure that they do not radiate at the same time.
- Computing track volume may become a critical task in determining a radar's effectiveness for a proposed position (see FM 3-09,12 for computation s).
- Units will use more shelling reports (SHELREPs) to determine enemy firing locations.

3-7. The majority of afl field artillery fires in mountains will be observed, especially dose support and defensive fires. Unobserved fires are frequently unreliable because of poor maps and rapidly changing meteorological conditions that cause registration corrections for high angle fire to be valid for only short periods of time.

3-8. Generally, field artillery observation posts should be emplaced on the highest available ground to increase observation. Low clouds or fog may require moving them to preplanned emplacements at lower elevations. Observers must be prepared to perform assault climbing to reach the most advantageous observation site. Commanders may use aerial observers or unmanned aerial vehicles (UAVs) to detect long-range targets and complement forward observers by adjusting fires beyond terrain masks, in deep defilade, and on reverse slopes. However, in extremely high mountains aerial observers may be confined to valleys and lower altitudes due to altitude limitations on different types of aircraft,

3-9. Laser weapons demand increased emphasis on observation techniques Laser target ranging and designation systems help to overcome difficulties in range estimation by providing accurate directional distance and vertical angle information for use in locating enemy targets. However, when positioning with a laser designator, an observer should consider line-of-sight with the target, as well as cloud height. Cloud ceilings that are too low will not allow laser guided munitions enough time to lock on and maneuver to the target.

TARGETING

3-10. Because of the decentralized nature of mountain operations, targets warranting massed fires may present themselves less often than in open terrain. However, narrow defiles used as routes of supply, advance, or withdrawal by the enemy are potentially high payoff targets for interdiction fires or large massed fires. Large masses of snow or rocks above enemy positions and along main supply routes are also good targets, because they can be converted into highly destructive rockslides and avalanches that may deny the enemy the use of roads and trails, and may destroy dements in ddilade. In the mountains, suppression of enemy air defenses takes on added importance because of the increased dependence on all types of aircraft. Commanders and thdr staffs should carefully review FM 3-60. A clear understanding of the targeting methodology combined with the knowledge of the capabilities and limitations of target acquisition and attack systems in a mountain environment is crucial to the synchronization of all available combat power.

3-11. To provide accurate and timely delivery of artillery fires in mountainous terrain, commanders must take into account the foil owing:

- High angles of elevation and increased time of flight for rounds to impact.
- Targets on reverse slopes, which are more difficult to engage than targets on flat ground or rising slopes, requiring more ammunition for the same oover age.
- I ncr eased amounts of dead space that cannot be hit by artillery fires.
- I ntervening crests that require detailed map analysis.

[] When the five requirements for accurate predicted fire (target location and size, firing unit location, weapons and ammunition information, meteorological information, and computational procedures) are not achievable, registration on numerous checkpoints becomes essential because of the large variance in elevation (see FM 3-09.40 for more detailed information).

MUNITIONS

3-12 Terrain and weather also affect the use of field artillery munitions Considerations for munitions employment in the mountains are discussed below.

- Impact fuze, high explosives (HE) shells and dual-purpose improved conventional munitions (DPI CM s) are very effective on rocky ground, scattering stones and splintering rocks, which themselves become missiles. However, deep snow reduoes their bursting radius, making them approximately 40 percent less effective. The rugged nature of the terrain may afford added protection for defending forces; therefore, large quantities of HE may be required to achieve the desired effects against enemy defensive positions.

- Variable time (VT) or time fuzes should be used in deep snow conditions and are particularly effective against troops on reverse slopes. There are some older fuzes that may prematurely detonate when fired during heavy precipitation (M557 and M572 impact fuzes and M564 and M548 time fuzes).

- Smoke, DPICM, and illuminating fires are hard to adjust and maintain due to swirling, variable winds and steep mountain slopes. Smoke (a base-ejecting round) may not dispense properly if the canisters become buried in deep snow. In forested mountains, DPI CMs may get hung up in the trees. These types of munitions are generally more effective along valley floorc.

- Using the artillery family of scatterable mines (FASCAM) and Copperhead is enhanced when fired into narrow defiles, valleys, and roads. FASCAM may lose their effectiveness on steep terrain and in deep sncw. Melting and shifting sncw may cause the anti-handling devices to detonate prematurely the munitions, however, very little settling normally occurs at temperatures lower than 5 degrees Fahrenheit. Remote antiarmor mine system (RAAMS) and area denial artillery munitions (ADAM) must come to rest and stabilize within 30 seconds of impact or the submunitions will not arm. and very uneven terrain may keep theADAM trip wires from deploying properly.

MORTARS

3-13. Mortars are essential during mountain operations. Their high angle of fire and high rate of fire is suited to supporting dispersed forces. They can deliver fires on reverse slopes, into dead space, and over intermediate crests, and, like field artillery, rock fragments caused by the impact of mortar rounds may cause additional casualties or damage.

3-14. The 60mm mortar is an ideal supporting weapon for mountain combat because of its portability, ease of concealment, and lightweight ammunition. The 81mm mortar provides longer range and delivers more explosives than the 60mm mortar. However, it is heavier and fewer rounds (usually no more than two per soldier) can be man-packed. The 120mm mortar may be more desirable in some situations, since they can fire either white phosphorous (WP) or HE at greater ranges than tighter mortars and have a significantly better illumination capability. However, because of the weight of these mortars and their ammunition, it may be necessary to transport fewer of them into mountainous terrain and use the remaining gun crews as ammunition bearers, or position them close to a trail network in a valley or at lower elevations. The second technique may be satisfactory if the movement of the unit can be covered and sufficient firing positions exist.

AIR SUPPORT

3-15, Air interdiction and dose air support operations can be particularly effective in mountains, since enemy mobility, like ours, is restricted by terrain. Airborne forward air controllers and close air support pilots can be used as valuable sources of information and can find and designate targets that may be masked from direct ground observation. Vehides and personnel are particularly vulnerable to effective air attack when moving along narrow mountain roads. Predsion-guided munitions, such as laser-guided bombs, can quickly destroy bridges and tunnels and, under proper conditions, cause landslides and avalanches to dose routes or collapse on both stationary and advandng enemy foroes. Moreover, air-delivered mines and long-delay bombs can be employed to seriously impede the enemy's ability to make critical route repairs. Predsion-guided munitions, as well as fuel air explosives, can also destroy or neutralize well-protected point targets, such as cave entrances and enemy forces in defilade.

3-16. Low ceilings, fog, and storms common to mountain regions may degrade air support operations. Although, global positioning system (GPS) capable aircraft and air delivered weapons can negate many of the previous limitations caused by weather. Terrain canalizes low altitude air avenues of approach, limiting ingress and egress routes and available attack options, and increasing aircraft vulnerability to enemy air defense systems. Potential targets can hide in the crevices of diffs and the niches of mountain slopes, and on gorge floors. Hence, pilots may be able to detect the enemy only at short distances, requiring them to swing around for a second run on the target and giving the enemy more time to disperse and seek better cover. Additionally, accuracy may be degraded due to the need for pilots to divert more of their attention to flying whi le simultaneously executing their attack.

ELECTRONIC WARFARE

3-17. The ability to use electromagnetic energy to deceive the enemy, locate his units and facilities, intercept his communications, and disrupt his command, control, and target acquisition systems remains as important in the mountains as elsewhere. Theeffectsof terrain and weather on electronic warfare (EW) systems are often a result of the effects on the components of those systems (particularly soldiers, communications, and aviation). Although a number of the effects are discussed in more detail elsewhere in this manual (and in applicable FMs and TMs), for ease some of the more common degrading effects of the mountainous environment on the components of electronic warfare systems are described in Figure 3-1 on page 3-8.

SECTION II - PROTECTION OF THE FORCE

AIR DEFENSE ARTILLERY

3-18. The severe mountain environment requires some modification of air defense employment techniques. Suitable positions are scarce and access roads are limited. In some instances, supporting air defense weapons may not be able to deploy to the most desirable locations. Consequently, the manportable air defense systems (MANPADS) may be the only air defense weapon capable of providing dose-in protection to maneuver elements.

3-19. Mountain terrain tends to degrade the electronic target acquisition capabilities of air defense systems. This degradation makes it more difficult for the air defense planner to locate and select position to provide adequate coverage for the force, and increases the importance of combined arms for air defense (CAFAD) and passive air defense measures (see FM 3-01.8). Individual and crew-served weapons can mass their fires against air threats. The massed use of guns in local air defense causes enemy air to increase their standoff range for surveillance and weapons delivery, and increase altitude in transiting to and from targets. These reactions may make the enemy air more vulnerable to air defense artillery (ADA).

3-20. Enemy aircraft will probably use defiles and valleys in mountainous terrain for low-altitude approaches to take advantage of terrain masking of radar. Congested roads and trails, and their junctions, may become lucrative targets for enemy air strikes. Enemy pilots may avoid early detection by using terrain-dearance or terrain-following techniques to approach a target. Rugged mountain terrain degrades air defense detection, but. at the same time, mountain ridges and peaks tend to canalize enemy aircraft. Detailed terrain analysis, coupled with predictive analysis to identify probable enemy air avenues of approach, aids in effective site selection.

EW COMPONENT	ENVIRONMENTAL FACTORS							REMARKS
	CLOUDS	FOG	RAIN	SNOW	WIND	TEMP	TERRAIN	
Soldiers[1]	✓	✓	✓	✓	✓	✓	✓	■ Clouds, fog, precipitation, and terrain affect visibility and observation. ■ Precipitation, temperature, and the rugged terrain affect soldier performance and ability to operate systems.
Electronics and-wire/cables[2]		✓	✓	✓		✓	✓	■ Extreme cold, combined with rugged terrain, increases fragility and breakage. ■ Precipitation and humidity affect electronic components.
Antennas[2]			✓	✓	✓		✓	■ Strong winds damage or prevent erection. ■ Precipitation and cold create ice, causing breakage (increased load and wind resistance) and reduce effectiveness. ■ Terrain affects masking and line-of-sight restrictions.
Aircraft[3]	✓	✓	✓	✓	✓	✓	✓	■ Clouds, fog, and precipitation degrade visibility and may prevent aircraft from flying under visual flight rules (VFR), precluding missions requiring aircraft landing at unimproved mountain LZs. ■ Cold and precipitation lead to icing, which impedes lift. ■ Compartmented terrain affects flight routes and target acquisition.
Vehicles			✓	✓		✓	✓	■ Rain, snow, and rugged terrain decrease mobility.
Radars/ Sensors		✓	✓	✓	✓		✓	■ Wind increases background noise, reducing efficiency. ■ Terrain affects masking and line-of-sight restrictions. ■ Fog and precipitation decrease infrared and electro-optical systems effectiveness.
Batteries						✓		■ Terrain reduces effectiveness and battery life – some systems may not even work under reduced power.

1 See Chapter 1 (Effects on Personnel)
2 See Chapter 2 (Communications)
3 See Chapter 4 (Helicopters) and the Previous Section (Air Support)

Figure 3-1. Effects of the Mountainous Environment on EW Systems

3-21 Movement to and occupation of positions in mountainous terrain require additional time. Planners must consider slope (pitch and roll), site preparation, and access route improvement prior to movement. Bradley Stinger fighting vehicle (BSFV) units often are unable to accompany small, lightly equipped maneuver elements, and may be restricted to supporting elements in more accessible areas of the battlefield. Avenger fire units can be sting-loaded by heavy lift aircraft and MANPADS airlifted into otherwise inaccessible positions. However, equipment emplaced by helicopters is resupplied and repositioned by the same means. When moving dismounted, MAIMPAO teams are limited to one missile per soldier, unless olher members of the unit are tasked to carry additional missiles.

3-22. Because of terrain masking of radars and the difficulty in establishing line-of-sigbt communications with lhe Sentinel or light and special division interim sensor (LSDIS) radar, early warning for short-range air defense (SHORAD) systems may be limited. Soldiers must maintain continuous visual observation, particularly along likely low-level air avenues of approach. Therefore, when possible. Sentinel or LSDIS radars should be emplaced on the highest accessible terrain that provides the best air picture for target detection and early warning, not necessarily peaks and summits.

ENGINEER OPERATIONS

3-23. Engineer combat support requirements increase in mountainous terrain because of the lack of adequate cover, the requirement for construction of field fortifications and obstades, and the need to breech or reduce enemy obstades. With such an enormous multitude of tasks, effective command and control of engineer assets is essential for the optimal utilization of these relatively scarce resources (see also the discussion of engineer augmentation and employment in themobility section of Chapter 4),

3-24, Digging fighting positions and creating temporary fortifications above the timberline is generally difficult because of thin soil with underlying bedrock. As described in Chapter 2, boulders and loose rocks may be used to build hasty, aboveground fortifications, Weil-assembled positions constructed in rock are strong and offer good protection, but they require considerable time and equipment to prepare,

3-25. Engineers assist maneuver units with light equipment and tools carried in or brought into position by ground vehicles or helicopters. Bulldozers, armored combat earthmovers (ACEs). and small emplacement excavators (SEEs) can be used in some situations to help prepare positions for command bunkers and crew-served weapons. They can also be used to prepare positions off existing roads for tanks, artillery, and air defense weapons. Conventional equipment and tools are often inadequate in rocky terrain, and extensive use of demolitions may be required- In the mountains, a greater number of engineer assets will be devoted to maintaining mobility and maneuver and unit commanders should assume that available engineer support will be limited to assist them with their survivability efforts. To enhance survivability and mobility a minimum of two soldiers per maneuver platoon should be capable of using standard demolitions.

NBC PROTECTION

3-26. Terrain and weather dictate a requirement for a high degree of nuclear, biological, and chemical (NBC) defense preparedness in mountainous areas. Due to limited mobility, viable tactical positions, and limited communication abilities, friendly units must be self-sufficient in protecting themselves against NBC weapon system effects.

3-27. Wearing mission-oriented protective posture (MOPP) gear at high elevations, when possibly combined with altitude sickness, increased dehydration, and increased physical exertion, degrades performance and increases the likelihood of heat casualties. Commanders should make every effort to keep soldiers out of MOPP gear until intelligence indicators reveal that an NBC attack is imminent or it is confirmed that a hazard actually exists (see FM 3-11.4 for a discussion on vulnerability analysis). When precautions must be taken against hazards, commanders must make decisions early and allow extra time for tactical tasks. Commanders should also refer to TC 3-10 for greater detail on tactics, techniques, and procedures necessary to operate under NBC conditions.

NUCLEAR

3-28 A mountainous environment can amplify or reduce the effects of and distort Ihe normal circular pattern associated with nuclear blasts. The irregular patterns reduce the accuracy of collateral damage prediction, damage estimation, and vulnerability analysis.

3-29. Air blast effects are amplified on the burst side of mountains (see Figure 3-2). Mountain walls reflect blast waves that can reinforce each other, as well as the shock front. Therefore, it is passible that both overpressure and dynamic pressure, and their duration will increase. An added danger is the creation of rockslides or avalanches A small yield nudear weapon detonated 30 kilometers or more from the friendly positions may still cause rockslides and avalanches, and easily close narrow roads and canalized passes. On the other hand, there may be little or no blast effects on the side of the mountain away from the burst.

3-30. Hills and mountains block thermal radiation, and trees and other foliage reduce it. Low douds, fog, and falling rain or snow can absorb or scatter up to 90 percent of a burst's thermal energy. During colder weather, the heavy dothing worn by soldiers in the mountains provides additional protection. However, the reflection from snow and the thin atmosphere of higher elevations may increase the effects of thermal radiation. Snow and ice melted by thermal radiation can result in flash flooding.

3-31. Frozen and rocky ground may make it difficult to construct shelters for protection from the effects of nuclear weapons. However, natural shelters such as caves, ravines, and diffs provide some protedion From nudear effects and contamination, in some instances, improvised shelters built of snow, ice, or rocks may be the only protedion avai lable. The dear mountain air extends the range of casualty-produdng thermal effeds. Within this range, however, the soldiers' added dothing reduoes casualties from these effects.

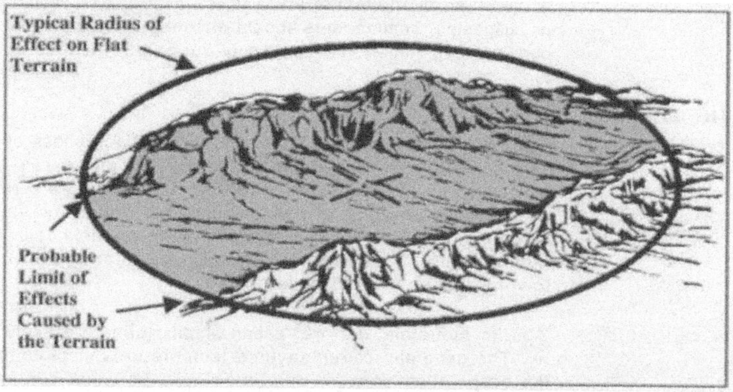

Figure 3-2. Effects of Mountains on Radiation and Blast

3-32. In mountainous regions, the deposit of radiological contamination is very erratic in speed and direction because of variable winds. Hot spots may occur far from the point of detonation, and low-intensity areas may occur very near it. Limited mobility makes radiological surveys on the ground difficult, and the difficulty of maintaining a constant flight altitude makes air surveys highly inaccurate. Additionally, melting snow contributes to the residual radiation pattern. After a nudear detonation, streams should be checked for radiation contamination before using them for drinking or bathing. As with the other effeds, the pattern of initial and induced nudear radiation may be modified by topography and the height of the burst,

BIOLOGICAL

3-33. Most biological pathogens and some toxins are killed or destroyed by the ultraviolet rays in sunlight. Above the timberline, there is little protedion from the sun; thus, the effectiveness of a biological attack may be reduced Downwind coverage may be greater because of the frequent occurrence of high winds over mountain peaks and ridges. Additionally, inversion conditions favor the downwind travel of biological agents through mountain valleys. Typically, winds flow down terrain slopes and valleys at night and up valieys and sunny slopes during the day. The effeds of mountainous terrain and rapidly changing wind conditions on the ability to predid and provide surveys of contamination for biological agents are similar to that for nudear radiation.

3-34 Temperatures and humidity also affect the survivability of biological agents. Generally, cool temperatures favor survival, and higher humidity increases the effectiveness of the agents. Extreme cold weather and snow deposited over a biologically contaminated area can lengthen the effective period of the hazard by allowing the agent to remain alive but dormant until it is disturbed or the temperature rises, if the use of biological agents is known or suspected, commanders should ensure that soldiers pay added attention to personal hygiene and consume only purified/treated water.

CHEMICAL

3-35. Wind and terrain can also cause the effectiveness of chemical agents to vary considerably. Depending on conditions, effects can be significantly enhanced or almost ineffective. High winds and rugged terrain cause chemical agent clouds to act in a manner similar to radioactive fallout. Inversions in mountain valleys may afso effectively cap an area, slowing the dissipation rate. Because of terrain and winds, accurate prediction of the downwind travel of toxic agent clouds is difficult

3-36. In mountain warfare, chemical munitions are likely to be delivered by air. The generally cooler daytime temperatures in mountainous terrain slow the evaporation process, thus allowing a potential contamination hazard to remain active longer Midday temperatures favor using persistent or blister-type agents, since nonpersistent agents dissipate too rapidly to cause any effect and unsupervised personnel are more likely to remove protective clothing for oomfort

3-37. The actions to protect against chemical agents in the mountains are not significantly different than from the requirements in less mountainous terrain. However, in extreme cold weather, survey and monitoring is often limited to the individual team mission, the FOX system may be limited to roads and trails, and the detection of vapor hazards is limited when the temperature falls below 32 degrees Fahrenheit. Decontamination may be more difficult due to freezing conditions, and the virulency period of contamination hazard for persistent agents may increase.

SMOKE AND OBSCURANTS

3-33. Smoke operations in mountainous areas are characterized by difficulties encountered due to terrain and wind Inadequate roads enhance the military value of existing roads, mountain valleys, and passes and add importance to the high ground that dominates the other terrain. Planners can use smoke and flame systems to deny the enemy observation of friendly positions, supply routes, and entrenchments, and degrade their ability to cross through tight, high passes and engage friendly forces with direct and indirect fires.

3-39. Thermally induced slope winds that occur throughout the day and night increase the difficulty of establishing and maintaining smoke operations, except in large and medium sized valleys. Wind currents, eddies, and turbulence in mountainous terrain must be continuously studied and observed, and their skillful exploitation may greatly enhance smoke operations rather than deter them. Smoke screens may be of limited use, due to enemy aerial observation, to include UAVs. and observation by enemy forces located on high ground. Smoke units may be required to operate for extended periods with limited resupply unless petroleum, oils, and lubricants (POL) supplies are emplaced in hide positions with easy access.

Chapter 4

Maneuver

The mountain environment requires the modification of tactics, techniques, and procedures. Mountains limit mobility and the use of large forces, and restrict the full use of sophisticated weapons and equipment. These limitations enable a well-trained and determined enemy to have a military effect disproportionate to his numbers and equipment. As such, mountain campaigns are normally characterized by a series of separately fought battles for the control of dominating ridges and heights that overlook roads, trails, and other potential avenues of approach. Operations generally focus on smaller-unit tactics of squad, platoon, company, and battalion size. Because access to positions is normally difficult, adjacent units often cannot provide mutual support and reserves cannot rapidly deploy. Attacks in extremely rugged

CONTENTS

terrain are often dismounted, with airborne and air assaults employed to seize high ground or key terrain and to encircle or block the enemy's retreat. While the mountainous terrain is usually thought to offer the greatest advantage to the defender, the attacker can often gain success with smaller forces by effectively using deception, bold surprise actions, and key terrain.

Although mountains often increase the need to employ light forces, commanders should not be misled into believing that this environment is the sole domain of dismounted units. On the contrary, the integrated use of mounted and dismounted forces in a mountainous environment, as elsewhere, increases a commander's capabilities while reducing his limitations. However, the employment of mixed forces must be based on sound mission, enemy, terrain and weather, troops and support available, time available, civil considerations (METT-TC) analysis of the specific mountain area of operations (AO), The infantry, armor, and combined arms series of field manuals, at both battalion and brigade level, provide the capabilities and limitations for each type force, planning and safety considerations, as well as, various concepts for employment. In all cases, commanders should assign complementary missions to each type force that capitalizes on their strengths and reduces their weaknesses, and takes into consideration the differential in operations tempo. Working together on the mountain battlefield, armored and dismounted forces can offset each other's weaknesses and provide much greater lethality than any one alone.

SECTION I - MOVEMENT AND MOBILITY

4-1. To move decisively in all directions without losing momentum in a mountain area of operations requires meticulous planning and careful preparation. In a mountainous environment, numerous conditions exist that affect mobility. The force that can maintain its momentum and agility under these conditions has the best chance of winning. Reduced mobility is a primary Mmitation to be considered during all phases of planning for mountain operations. Rugged terrain, the time of year, the weather, and the enemy have a decisive influence on movement in the mountains. Commanders must ensure that they have sufficient time and space to deploy their forces for battle by maintaining constant security and selecting proper routes and movement techniques. Additionally, they must closely manage limited off-road areas. Tactical operations centers, artillery units, aid stations, air defense artillery, battalion trains, and other supporting units will compete for limited space in restrictive mountainousterrain.

4-2. At any elevation level, movement is generally considered to be either movement across or along terrain compartments. When moving across terrain compartments from one ridge to another, elements should use bounding overwatch. Lead elements should secure the high ground and provide overwatching fires as the rest of the element crosses the low ground. When moving along a terrain compartment, forces should move on the high ground without silhouetting t hem selves or, at a minimum, place an element there to securetheir flanks

4-3. Maneuver forces should move by stealth and exploit the cover and concealment of terrain. Using rough, unlikely routes and movement during limited visibility helps avoid enemy detection. All movements must exploit known weaknesses in enemy detection capabilities. Whenever possible, movement should be planned to coincide with other operations that divert the enemy's attention,

4-4. Because of the narrow routes sometimes encountered, especially in the higher elevations, formations may be compressed to columns or files. To reduoe vulnerability to the enemy, forces should move separated from each other on multiple and unlikely routes. When moving dismounted along unlikely routes, special teams construct fixed ropes, hauling systems, traverse systems, and other mountaineering installations to provide access to higher elevation levels and increased mobility.

4-5. The danger of surprise attack is most acute in terrain that makes deployment from the march impossible. Even with well-thought-out movement plans, maneuver elements must take both active and passive security measures at all times. Restrictive terrain facilitates templating and determining the movement of forces, making the actions of an armored force more predictable. Elements may avoid detection by using planned fires to destroy known enemy sensors and observation posts or by pfadng fires to divert the enemy's attention away from an exposed area through which the element must move. However, the placement of fires in a particular area or along a route may compromise operational security.

4-6. When the danger of rockslides or avalanches exists, the distance between elements should be increased as much as four to six times more than required on flat terrain. The more the conditions vary for each unit, the more thorough the planning must be, especially if units must reach the objective simultaneously. Often, a reserve of time must be programmed if units move on multiple routes, over unfamiliar terrain, or during limited visibility, or if they face an uncertain enemy situation.

MOUNTED MOVEMENT

WHEELED AND TRACKED VEHICLES

4-7. Generally, the mountain terrain above the valley floor severely limits movement of wheeled vehides and is too restricted for tracked vehides. Trafficable terrain tends to run along features with steep slopes on either side, making mounted movement vulnerable to vehicular ambushes and attack aircraft. Recovery vehides must always accompany mounted forces in mountainous terrain to rapidly remove disabled vehides from the limited and narrow trail network.

4-8. Tanks and other armored vehides, such as infantry fighting vehides (IFVs), are generally limited to movement in valleys and existing trail networks at lower elevations. Even at these levels, the trails may require extensive engineer work to allow tracked vehides to pass over them. Tanks,

Bradley fighting vehicles (BFVs), and cavalry fighting vehicles (CFVs) can support by fire if accessible firing positions are available; however, it will rarely be possible for them to accompany dismounted infantry in the assault. In such cases, commanders may seek to use their increased firepower to isolate the objective for the dismounted assault. If employed above Level I, armored vehicles are forced to fight in smaller numbers, yet a single tank at a critical pant may have a decisive effect. Although antitank weapons employed from higher elevations can easily penetrate the top of armored vehides, in many situations, the inability to elevate the weapon system's main gun suffidently to return fire may further increase its vulnerability.

4-9. Low atmospheric pressure considerably increases the evaporation of water in storage batteries and vehide cooling systems, and impairs cylinder breathing. Consequently, vehicles expend more fuel and lubricant, and engine power is reduced by four to six percent for every 1,000-meter (3.300-foot) increase in elevation above sea level. This translates to a fuel and oil increase of approximately 30 to40 percent or more.

4-10. Figure 4-1 contains questions that are part of any mounted movement plan. Limited road networks and restricted off-road mobility significantly increase their importance. In the mountains, failure to address these questions in detail may seriously jeopardize the overall mission.

HELICOPTERS

4-11, Utility and cargo helicopters are key to the rapid movement of soldiers and equipment in the mountains. However, any operation that depends primarily on continuous aviation support to succeed is extremely risky. H rgh eievat ions and rapidly changing and severe weather common to mountainous regions is very restrictive to aviation operations and makes availability of aviation support very unpredictable. At high altitudes, weather that appears to be stable to the ground observer may significantly affect helicopters- The effects of fog, frontal systems, winds, and storms are readily discernible. Additionally, higher altitudes may restrict aircraft lift capabilities and decrease aircraft allowable gross weight in mission profile. Aircraft idng is common at high altitude and may occur suddenly. De-ice/anti-ice capabilities exist for rotor blades, however, idng may still decrease lift and. in severe cases, prevent flight altogether. Therefore, commanders must become intimately familiar with the conditions that may limit the full effectiveness of Army aviation when operating in a mountain environment (see FM 3-04.203).

- ☐ **How fast can the march be conducted?**
- ☐ **Will there be other traffic on the route?**
- ☐ **Are there potential areas that offer covered, off-road positions?**
- ☐ **Are there any locations along the route that could be used for resupply?**
- ☐ **Are there alternate routes?**

Figure 4-1. Mounted Movement Planning

4-12. Additionally, commanders must consider the effect of altitude on soldiers when planning air assault operations (see Chapter 1). If possible, commanders should use soldiers acdimatized at or above the elevation level planned for the air assault. Depending on the situation, it may be better to have troops walk in rather than fly them to the necessary elevation level.

4-13. Rugged, mountainous terrain complicates flight route selection and places an additional navigational load and strain on the entire crew, as they have little margin for error. Direct routes can seldom be flown without exposing aircraft to an unacceptable risk of detection and destruction by the enemy. Tactical flight routes follow valley corrida's, where it is possible to obtain cover and concealment while maintaining the highest possible terrain flight altitude. Terrain flight in the mountains may preclude using dosed formations. Multi-helicopter operations are normally flown in "loose" or "staggered trail" format ions with increased spadng between aircraft.

4-14. Terrain suitable for multiple helicopter landing zones (LZs) in mountainous regions is limited. Level areas that are suitable for mountain LZs frequently require little preparation beyond the dearance of loose material, since the ground is usually firm enough to support helicopters. Conversely, if LZs must be developed, dearing may be difficult due to the rocky ground. Stand-off space from rock wall faces must be deared and a level landing surface must be created. Demolitions may be required to dear large rocks but care must be used to prevent rockslides or avalanches started by the explosive shock. During the winter, snow must be packed to prevent whiteouts. Similarly, sandy or dusty LZs should be dampened with water to prevent brownouts.

4-15. When only single aircraft landing zones are available, in-flight spadng between helicopters must be significantly increased. Although helicopter LZs should be located on the windward side of ridges or peaks to take advantage of the more stablewinds, conceal ment from enemy observation and the mission are the most i m port ant factors in site selection in forward areas. When it is impossible for helicopters to land, personnel may rappel and light equipment may be sling-loaded into a LZ or. in some situations, lowered by rope while the helicopter hovers However, this may increase turnaround time and aircraft vulnerability Since available landing sites are often limited, the enemy can be expeded to target all likely locations. Personnel should secure terrain that dominates a landing site before using it. They must extensively suppress enemy air defense weapons during air assault or supply operations

4-16, Attack helicopters can be well suited for a mountain environment; however, commanders must be continuously mindful of weather and elevation effects on their employment. They can be the commander's most mobile maneuver forces in mountain warfare, enabling him to concentrate combat power quickly and exploit enemy weaknesses. During stable weather conditions, attack helicopters equipped with a variety of ordnance can rapidly engage targets beyond the range of other weapons or those masked by intervening crests. As discussed earlier, higher altitudes and icing conditions affect lift and subsequently armament loads. Ice can also prevent attack helicopters from firing their weapons altogether.

4-17, Employment time and fuel consumption increases because of the few direct routes. Terrain compartments provide excellent terrain masking and radar and visual acquisition avoidance, and allow for rapid movements to the flanks and rear of an isolated enemy force. However, these same compartments may limit aircraft maneuverability and necessitate smaller flight formations, which, in turn, may affect target engagement techniques. The compartmented terrain, combined with extended distances, may require engagement without the support of other combined arms. If terrain precludes placement of fuel and arming points in the forward area, turnaround time increases and on-station lime decreases. Since ground-to-air communication is often degraded by intervening terrain, in-flight operational control over extended ranges may be difficult.

4-18. Enemy motorized and mechanized forces may be slowed and canalized as they move up steep grades, down narrow valleys, and along mountain trails. These types of conditions allow attack helicopters to engage slower moving targets that have little room to maneuver or hide. However, these same conditions also make it difficult for pilots to select positions that allow line-of-sight to the target, sufficient tracking distance, acceptable standoff range, and adequate cover and concealment. Positions located high on a ridgeline may support successful target acquisition, tracking, and standoff, but create dangerous silhouettes and look-down angles (the angle from the aircraft to the target) that exceed aircraft weapon constraints. Lower positions, possibly in draws or saddles, may provide concealment to the flanks and an extensive backdrop to help conceal positions, but can decrease a pilot's ability to locate and track targets. Intervisibility lines may mask targets, and extreme terrain relief within the aircraft's optics field of view may inhibit tracking.

4-19. Remote Hellfire engagements avoid most of these problems (see Figure 4-2), but they may increase the time of flight of the missile. A remote engagement limits the number of aircraft exposed for tracking and lasing targets. When engaging with remote fires, a designating team is plaoed in a position overwatching the enemy's route of march. The remaining aircraft position themselves in covered and concealed battle positions oriented on the engagement area. If possible, these battle positions should be located below the reverse military

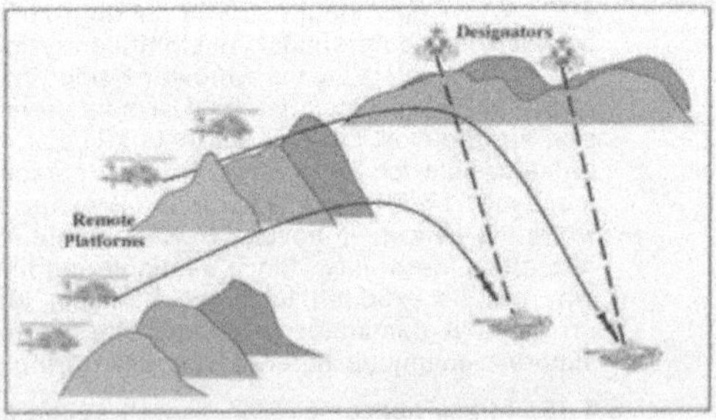

Figure 4-2. Remote Hellfire Engagements

crest or the counterslope of the ridgeline paralleling the engagement area. Aircraft in these battle positions will act as remote platforms by providing missiles to the designators without unmasking. This tactic prevents the enemy from achieving lineof sight on the firing aircraft.

4-20. The AH-64D Longbow Apache will provide an even greater killing capability for the mountain commander. It is able to detect, classify, prioritize, and engage targets with Longbow H el lf ire missiles without visually acquiring the target. The commander of the AH-64D Longbow Apache utilizes the its fire control radar and mast-mounted site to target all of the vehicles in the enemy's march formation. A data transfer handover is then executed to all other AH-64D Longbow Apaches within the company. After the data transfer has been completed, the company can then engage the enemy march column without exposing the remote platforms. When employing the AH-64D Longbow Apache with radar guided Heflfire missiles, engagement times are greatly reduced and aircraft exposure to the enemy is limited. Regardless of the type of helicopter, a thorough terrain analysis and early involvement of aviation operational planners are key to successful application of Army aviation assets in mountainous terrain.

DISMOUNTED MOVEMENT

4-21. Dismounted movement is often extremely slow and arduous, and may require the skills of technical mountaineering teams to secure the advance. For example, movement in Level 11 may dictate that elements secure the high ground in Level II I. As with any type of movement, proper movement techniques and formations and conslant security to avoid unplanned enemy contact are some of the keys to successful dismounted movement.

4-22. Foot marches in the mountains are measured in time rather than distance. When making a map reconnaissance, map distance plus one-third is a good estimate of actual ground distance. One hour is added for each 300 meters of ascent or 600 meters of descent to the time required for marching a map distance. Figure 4-3 on page 4-8 shows dismountcd movement calculations for an estimated 10-kilometer march on flat and mountainous terrain. Although not included in this example, commanders must also consider acclimatization, fatigue, soldiers' loads, limited visibility, and other factors that affect movement times (see FM 3-25.18 for additional factors influencing dismounted march rates).

4-23. Commanders cannot permit straggling or deviations from the selected route. Every aspect of march discipline must be rigorously enforced to keep a column closed with the knowledge that the interval between individuals depends on terrain and visibility.

4-24. To prevent an accordion effect, soldiers must allow enough distance between themselves todimb without causingthe following individual to change pace. In mountainous terrain, a slow, steady pace is preferred to more rapid movement with frequent halts. Commanders must incorporate scheduled rest halts into their movement plans based on distance, availability of covered and concealed positions, and other fadors described above. If possible, commanders should not condud rest halts during steep ascents or descents. At the start of a march, soldiers should dress lightly so that they begin slightly chilly. However, a short halt should be taken to adjust dothing and equipment after the first 15 minutes of movement. In addition, soldiers must put on special mountaineering equipment before reaching steep terrain.

Flat Terrain	Mountain Terrain
Total 16 kilometers / 4 kilometer per hour = 4 hours	Normal Time 16 kilometers / 4 kilometer per hour = 4 hours
	Ascent 600 meters / 300 meters per hour = 2 hours
	Descent 600 meters / 600 meters per hour = 1 hour
	Total 4 + 2 + 1 = 7 hours

Figure 4-3. Example Dismounted Movement Calculations

4-25. In glacial areas, the principal dangers and obstacles to movement are crevices, snow and ice avalanches. Exposure to the hazards of glaciated mountainsis increased at company-level and above, and movement should be limited to separate platoon and lower levels. When moving on glaciers, an advance element should be used- This group identifies the best routes of advance, marks the trail, and provides directions and distances to follow-on units. A marked trail is especially important during inclement weather and low visibility, and provides a route for retrograde. Commanders must carefully weigh the advantages of a marked route against the possibility of ambush and the loss of surprise.

MOBILITY

4-26. During mountain operations where limited mobility exists, it is critical that units maintain security and control of available road/transportation networks. This includes securing key bridges, fords, crossing sites, intersections, and other vulnerable choke points. These locations must be protected against enemy air. obstacle, and ground threats. However, commanders must carefully balance their available combat power between protecting their freedom of mobility/maneuver and allocating forces to critical close combat operations. Effective risk analysis and decisions are essential. Route clearance operations, patrols, traffic control points (TCPs), and other security operations aid commanders in securing routes. During offensive operations, commanders may need to commit forces to seize key terrain and routes that afford their forces greater mobility and tactical options against the enemy.

4-27. Engineer support in front of convoys and combat formations is often necessary to dear and reduce obstades, such as washouts, craters, mines, landslides, and avalanches, as well as, snow and ios in colder regions. Redudng obstades is more difficult in mountainous areas because of reduced maneuver space, lack of heavy equipment, and an increased competition for engineer support. Minefields should normally be breached, since bypassing properly sited obstades is often impossible. In the mountains, using mechanical mine plows and rollers is frequently impossible due to the lack of roads and trails, and removal of mines by hand or through demolitions is often required. Commanders must exerdse extreme caution when employi ng demolitions in the vidnity of snow and rock covered slopes because they can cause dangerous rockslides, avalanches, and secondary fragmentation FM 3-34 2 has information on breaching operations and synchronization required.

4-2S. Creating new road systems in mountainous regions is usually impractical because of the large amount of rock excavation required. Therefore, roadwork is generally limited to the existing roads and trails often requiring extensive construction, improvement, maintenance, and repair towithstand the increased military traffic and severe weather conditions. In certain mountainous areas, materials may be difficult to obtain locally and impossible to make full use of conventional heavy engineer equipment for road and bridge construction or repair. In such cases, large numbers of engineers are required and units must rely heavily on hand labor, light equipment, and demolitions.

4-29. Secondary roads and trails should be steadily improved to accommodate trucks and infantry fighting vehides, and. eventually, heavier vehicles. Their seledion depends on necessity and the speed with which the routes can be put into service. Abnormal gradients on roads may be necessary to ensure that construction keeps pace with lactical operations. Sidehill cuts are the rule, and the same contour line is followed to avoid excessive fills or bridging. Turnouts should be installed approximately every 500 meters to reduce traffic congestion on single-lane roads or trails. Drainage requirements must be considered in detail because of the effeds of abnormally steep slopes, damaging thaws, and heavy rains.

4-30. Stream and river crossing operations are difficult and must usually be accomplished by expedient means. Bridging operations in mountainous terrain are normally limited to spanning short gaps and reinfordng existing bridges by using prefabricated materials and fixed spans from floati ng bridge equipment. However, standard design or improvised suspension bridges may still be needed for longer spans. Because exist i ng bridges may have low vehide load classifications, standard fixed tactical bridges and bridging materials should be on hand to quickly ranforce or repl ace them. In extremely rough terrai n, cableways and tramways may be construded to move l ighl loads and personnel across gorges, and up and down steep slopes.

COUNTERMOBI LITY

4-31. Obstades become more important because of the compartmented terrain and already limited road and trail networks. It is easy to create effective obstructions in mountains by cratering roads, fully or partially destroying bridges, or inducing rockslides and avalanches. Units can use antitank minefields effectively to canalize the enemy, deny terrain, or support defensive positions. Commanders should remember that dearing or redudng these same obstades may be extremely difficult and a hindrance to future operations. Using reserve and situational obstades. lanes and gaps, and plans to rapidly reduce friendly obstades must be an integral part of all defensive operations. Commanders must also consider the enemy's ability to create similar obstades and minefields when developing courses of action that hinge on speed of movement or a particular avenue of approach.

4-32. Reinforcing obstacles can be used effectively with the natural ruggedness of mountains to deny the enemy terrain and to delay and impede his movement. As in all environments, the engineer and maneuver force commander must site obstacles based on terrain and the availability of weapon systems.

4-33. Antitank mines are laid along the comparatively narrow approaches suitable for mounted attacks. Flash floods and excessive runoff may dislodge mines from their original location; however, they normally remain armed. Family of scatterable mines (FASCAM), particularly artillery-delivered and helicopter-delivered mines, increases the flexibility of the maneuver unit commander, reduces the engineer effort, and is a valuable resource in protecting rear areas from enemy envelopment and breakthroughs. Using FASCAM should be weighed against thetime in delivery, displacement of the artillery, and the additional logistics burden that may be involved

ENGINEER AUGMENTATION AND EMPLOYMENT

4-34. Because mountain terrain requires small-unit decentralized operations, an engineer platoon or company should be allocated to each maneuver battalion, light or heavy. Allocation in this manner may leave division and brigaderear areas short of engineer support.

4-35. An additional corps engineer battalion (wheeled) and an engineer light equipment company may be needed to augment an infantry division. Platoons from the engineer light equipment company may be tasked to assist divisional platoons with the engi neer effort in each maneuver battalion area. The corps combat engineer battalion (wheeled) provides heavy equipment and dump trucksrequired to support road improvement and maintenance in division and brigade rear areas. Also, this corps combat engineer battalion (wheeled) can accomplish such tasks as constructing or reinforcing bridges. To operate efficiently, additional items, such as compressors, jackhammers. power drills, chain saws, and bulldozers, may be necessary, as well as large amounts of explosives and obstacle materials.

SPECIAL PURPOSE TEAMS

4-36. On steep, exposed, or technically difficult terrain, soldiers with advanced mountaineering skills may be required to maintain or improve mobility. Advanced climbers may deploy ahead of maneuver forces during limited visibility or inclement weather to erect aids that will help maneuver elements move in difficult terrain. They may also be committed to lead forces or to operate independently to strike the enemy suddenly over unlikely routes and to occupy certain key heights that can be defended easily because of their position. The specific employment of special purpose teams is based on the mission, tasks, and requirements of the commander.

4-37. Commanders must analyze operational terrain levels and identify the mobility requirements necessary to obtain and maintain freedom of both tactical maneuver and operational movement (see Figure 4-4). It is critical that special purpose teams are properly organized before a mission begins. Once movement is underway, unplanned deviations have little chance of success. Bypassing obstacles in mountainous terrain is almost always difficult or impossi ble. In many instances, the best available bypass will channel friendly forces into enemy kill zones or ambushes.

4-38. To enable a force to move personnel, equipment, and supplies on the mountain battlefield with limited delays due to terrain, visibility, or obstacles, commanders should organize soldiers with advanced mountaineering training as guides, lead climbing teams, installation teams, and evacuation teams (evacuation teams are covered in the combat health support portion of Chapter 5).

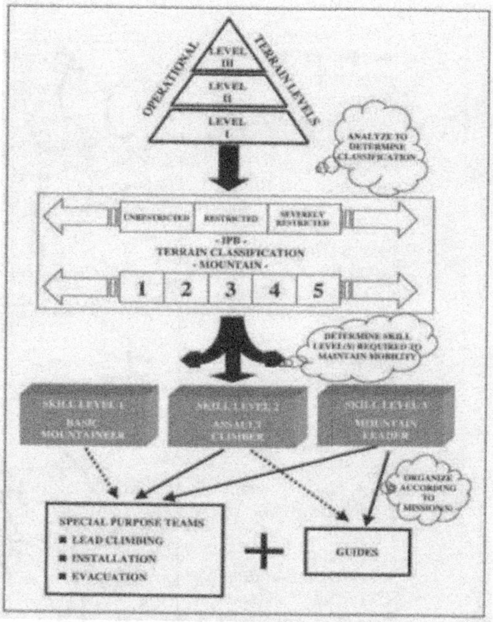

Figure 4-4. Relationship of Terrain to Skills and Special Purpose Teams

HISTORICAL PERSPECTIVE

The 10th Mountain Division and Riva Ridge (February 1945)

After attempts to capture Bologna in Italy's Po Valley during the winter of 1944-1945 failed. Allied commanders focused on the possibility of wide sweeping movements aimed at encircling Bologna and all the German armies in the region. The plan called for using the 10th Mountain Division to capture a series of mountain peaks and ridges dominating a 10-mile section of Highway 64, one of two main routes leading from Pistoia through the northern Apennines to Bologna. This would provide the Allies with a better position before starting the spring offensive, since seizure of this terrain would result not only in the Germans' being unable to protect their vital lines of communications, but also in the Americans' being able to observe German activity almost all the way to the Po Valley, approximately 40 miles away. The Division's baptism-by-fire was to be in a region dominated by two ridges whose highest peaks rose 3,000 to 5,000 feet. One of these ridges, Pizzo di Campiano-Monte Mancinello, became known as Riva Ridge (see Figure 4-5).

These heights appeared to be impregnable, as it was doubtful that any force large enough to overwhelm the Germans could be massed unobserved for an assault. The Germans had all the advantages of the commanding heights, and there was little cover for troops crossing the barren, snow-covered ground. It was clear that Riva Ridge would have to be cleared before the decisive attack could advance up Mount Belvedere and along the ridge towards Monte della Torraccia. Therefore, the plan was for the mountain troops to climb the 1500-foot cliff and surprise the Germans, who would not be expecting the attack up the face of the cliff.

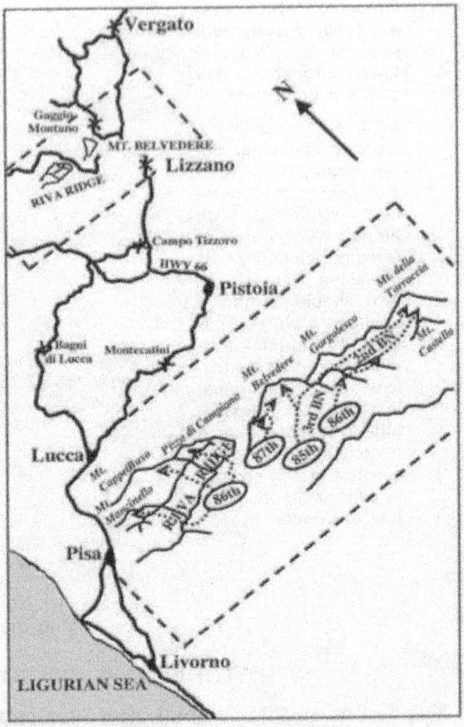

Figure 4-5. Riva Ridge

The 86th Mountain Infantry Regiment's 1st Battalion and one company from its 2nd Battalion were assigned to negotiate the cliff and capture Riva Ridge, thereby setting the conditions for other assault elements to achieve their objectives. Select teams of climbers assembled their ropes, pitons, and other gear. The teams climbed in the dead of night, hammering pitons into the rock, attaching links to them, and fastening ropes to the links. These served as fixed lines to assist those who followed in their ascent of the vertical face of the ridge. The lead climbing teams reached the top around midnight, and battalion units began their ascent in force. By 0400 on 20 February, all three of 1st Battalion's companies and the company from the 2nd Battalion had reached their individual objectives on top of the ridge without being seen. They attacked the defending Germans and completely surprised them. German daylight counterattacks were repulsed and the division's left flank was secured on Riva Ridge opening the way for the rest of the division to accomplish its mission.

The capture of Riva Ridge was the only significant action in which the division put to use its specialized stateside training. Nevertheless, no one would deny that this single mountain operation justified all the demanding training the 10th Mountain Division had undergone.

Adapted from *See Naples and Die*, Robert B. Ellis.

GUIDES

4-39. Mountain guides are useful for any operation, particularly *on* terrain above Class 3, which requires accurate judgment and extensive technical knowledge. They act primarily as advisors to unit commanders and assist with planning when technical mountaineering problems affect the tactical scheme of maneuver. They must be experienced in all aspects of mountaineering (usually a Level 3 mountaineer) and capable of ensuring that the force is never beyond the margin of its operational capabilities while operating on mountainous terrain. Mountain guides must possess the skills, knowledge, and experience necessary to develop a true perspective of the obstades and conditions that must be overcome.

4-40. Mountain guides should perform the following functions for the commander:

1. Conduct a terrain analysis of the assigned area.
2. Select the best march routes
3. Identify danger areas, obstades, and hazards.
4. Estimate the effects of elevation on force capabilities.
5. Determine the technical means necessary to employ the force.
6. Estimate tho time of unit movements.
7. Develop a movement sketch indicating the azimuth, time, and mountaineering requirements
8. Lead units, usually company-level and above, on difficult terrain.
9. Supervise al l aspects of mountaineering safety.

LEAD CLIMBING TEAMS

4-41. As with all spedal purpose teams, lead dimbing teams have highly skilled soldiers qualified in advanced mountaineering techniques (normally Level 2 mountaineers). To a lesser degree, lead dimbing teams perform many of the same functions as guides. They also accompany a unit over unprepared routes and assist with actually conducting its mission by maintaining and improving mobility. In some instances, lead dimbing teams may operate independently of other forces to accomplish spedfic missions.

4-42. These teams are capable of climbing at night with the aid of night-vision goggies and can conduct operations over any type of terrain. Lead climbing teams should be utilized to install fixed ropes to assist personnel over exposed terrain. They make the most difficult climbs and act as rope leaders. Members of the lead climbing teams must be extremely proficient in the technical aspects of mountaineering, since they select the specific routes to be climbed. This responsibility emphasizes the importance of accurate judgment, since a single mistake could jeopardize the success of the unit's mission. Commanders should consider assigning lead dimbing teams to, or developing an organic capability within:

- Ground reconnaissance elements.
- Forward observer parties.
- Air defense sections.
- Communications sections.
- Security elements.
- Assault elements.
- Sniper sections.

4-43. The number of lead dimbing teams required is dependent on the mission and difficulty of the elevation level or dismounted mobility classification. If more than one point of attack is to be used or a more mobile patrol is needed, it may be necessary to employ additional lead climbing teams. Lead dimbing teams may also assist in the attack of very steep objectives by negotiating the most unlikely avenues of approach. They can be dispatched ahead of the attacking elements to secure the advance at night and during periods of inclement weather. The commander's imagination and their limited availability are the only restraints toward using lead dimbing teams.

4-44. In predpilous terrain, lead dimbing learns can be used alone to conduct small reconnaissance patrols or to form the nudeus of larger patrols. They may also be used to expedite movement of flank security elements over difficult terrain and during poor visibility. At least one team should be attached to each element, depending on the size of flank security and the operational terrain level in which operations take place. When the AO dictates a rate of march on the flanks that is slower than that of the main body, flank security elements should be located adjacent to the advance echelon. Lead dimbing teams are detached as necessary to reconnoiter and hold dominant terrain features on the flanks of the line of march. As the trail element passes and the security position is no longer required, the lead dimbing team's flank outposts joi n the mai n body or proceed forward to another security location.

INSTALLATION TEAMS

4-45. Installation team members are qualified in the construction and maintenance of technical mountaineering systems, referred to as installations, which fadlitate unit movement. These teams deploy throughout the AO. in any weather or visibility conditions, to erect mountaineering installations that overcome obstacles to the movement of friendly forces and supplies.

4-46. In most situations, installation teams should consist of six qualified members, enough to build most installations. Installation teams deploy early and prepare the AO for safe, rapid movement by constructing various types of mountaineering installations (see Figure 4-6). Following construction of an installation, theteam, or part of it, remains on site to monitor the system, assist with the control of forces across it, and make adjustments or repairs during its use. After passage of the unit, the installation team may then disassemble the system and deploy to another area as needed.

- Fixed ropes
- Pulley systems
- Hauling lines
- Rope ladders
- Rope bridges
- Handrails
- Traverse systems

Figure 4-6. Types of Installations

4-47. Although installation teams can assist, commanders remain responsible for establishing a traffic control system and the rate of negotiation to expedite tactical movement and prevent unnecessary massing of personnel and equipment on either side of t he obstacle- Plans for the traffic control system includearrangements for staging and holding areas, as well as movement. A prearranged order of movement, dissemi nated to all elements of the force, facilitates rapid movement. The traffic control plan also includes instruct ions for dispersing forces on the far side once they negotiate the obstacle. The successful negotiation of systems depends on how quickly a force can consolidate on the far side and continue its mission.

4-48. Commanders must carefully consider the proportion of installation team assets allocated to maneuver with those assigned to sustain essential logistics activities. When considering a unit's scheme of maneuver, the number and types of installations depend on METT-TC, emphasizing factors listed in Figure4-7.

- Number of routes required by the maneuver force
- Size of the maneuver force
- Amount and type of maneuver force equipment
- Number of installations required to extend logistics support
- Weather and visibility conditions
- Number of installation teams available

Figure 4-7. Factors Influencing Mountain Installations

4-49. Once operations stabilize, installation teams can direct and assist engineers in the construction of fixed alpine paths. Fixed alpine paths consist of permanent or semi-permanent mountain aids that assist troops in traversing rugged mountain terrain and facilitate the movement of equipment and supplies to and from forward areas.

SECTION II - OFFENSIVE OPERATIONS

4-50. Offensive operations in the mountains vary depending on the degree of restrictions dictated by mountains of different heights and character, but are normally planned and conducted utilizing the movement to contact and the attack. Exploitation and pursuit are conducted, but less frequently than in other environments.

4-51. Mountain operations will most likely be fought to gain control of key or decisive terrain. The goal is to seize objectives that are important for continuing the battle, such as lines of communication, passes, ridges, and choke points. Every attempt must be made to fight from the heights down. Frontal attacks against defended heights have little chance of success and attacks are usually made along the flanks and to the rear of the enemy. Consequently, envelopment becomes the preferred form of maneuver.

4-52. The missions assigned to units operating in a mountainous region remain the same as those in lowland regions. Commanders should place an increased emphasison;

- Limited-visibility operations,
- Mobility and survivability.
- Securing friendly lines of communications while severing those of the enemy.
- Reconnaissance and security,
- Centralized planning and decentralized small-unit actions.
- Environmental factors favoring operations of short duration and violent execution.

PLANNING CONSIDERATIONS

4-53. When conducting offensive operations in the mountains, surprise is often a dominant consideration, rather than a supporting one. Units achieve surprise by achieving superior situational understanding, especially of the terrain, and by using that knowledge to do the unexpected. Friendly forces should conduct bold and imaginative operations to exploit enemy weaknesses or inability to operate in a mountainous environment. A well-trained force can achieve surprise by infiltrating and attacking the enemy's rear or attacking during periods of limited visibility, such as night, rain, or snow. They can use helicopters and their technical mountaineering skills to conduct decisive operations anywhere in the AO. The effects of surprise can be dramatically increased if commanders select objectives in restrictive terrain that decrease the enemy's mobility and ability to react effectively.

4-54, The mountainous terrain increases the threat to concentrated formations. Usually, it is difficult to coordinate all forces by time and location so that they can rapidly support each other and achieve massed effects. The compartmented terrain separates adjacent units, precluding mutual support, and may adversely affect supporting distances. Therefore, it is critical to anticipate the concentration of forces and fires before the battle begins to achieve effective synchronization.

4-55. Commanders usually select decisive points for the attack based, in part, on their ability to seize and control key terrain. Each of these objectives often necessitates the seizure of one or more intermediate objectives. The compartmented terrain and resulting dispersion make it difficult for commanders to visualize the entire AO and complicates command and control, while the terrain often affords the defender excellent observation and decreases the attacker's ability to concentrate forces undetected.

4-56. Once a battle is joined, helicopters are the only rapid means for massing forces on terrain higher than Level I. When weather conditions permit, attack helicopters and tactical air assets are essential because they can move and strike over a large AO in a short time *Therefore,* the suppression of enemy air defense may become a high priority during offensive operations.

4-57. Commanders must personally acquaint themselves with the terrain to the fullest extent possible They combine this knowledge with other factors of METT-TC to develop simple, yet precise plans and orders. As previously discussed, compartmented terrain tends to divide the battle into many isolated engagements that are difficult to control. A complex and inflexible plan wilt not withstand the changing situations associated with this type of decentralized combat. An uncomplicated plan with a well-thought-out intent that is clearly communicated ensures understanding at the lowest level and allows subordinates to exploit battlefield opportunities even if communications fail.

PREPARATION

4-58. The length of the preparatory phase is typically longer in a mountainous environment. An offensive action against an enemy defensive position must be based on thorough reconnaissance and orderly preparation. The primary factor in determining the technique to be used in destroying the enemy position is the strength of the enemy's defenses. The stronger the defenses, the more deliberate the attack must be. In mountainous terrain, evaluation and exploitation of the ground are essential. Commanders must prepare plans of attack that take advantage of the weaknesses found in the enemy's defensive system. In the mountains, a larger number of reconnaissance assets and additional time may be needed to determine the strength of enemy positions on the objective and all surrounding heights, and favorable routes to and past the objective.

4-59. Difficult approach routes should be marked and prepared for safe passage. Easily traversed slopes, broad hills, plateaus, and valley floors, as well as mountainous terrain with well-developed road and transportation nets, permit deployment in breadth. High ranges with ridges and crests leading to the objective require organization in depth with extended lines of communications

4-60. In trackless mountain terrain, company-sized teams usually conduct attacks, tfthearea assigned toa battalion permits, companies should approach the objective separately on multiple routes. In restrictive terrain, adequate maneuver space may not always be available and several units may be required to move along the same avenue of approach. It may even be necessary to conduct shaping operations to seize sections of terrain from which the enemy can dominate the forces' movement and approach.

4-61. The preparatory phase may also indude feints and demonstrations to shape lhe battlefield. In mountainous lerrain, the defending force has a number of advantages that allow it to defeat an attacking force much larger than its own, such as long-range observation, rugged natural cover, concealment, advance siting of weapons, and operations on familiar terrain. Feints and demonstrations mask friendly operations, expose enemy vulnerabilities, disorganize the defense, and allow the attacking force to engage the enemy discriminately. In a mountain area of operations, the benefits achieved from a successful deception effort may well outweigh the difficulties involved in mounting it.

FORMS OF MANEUVER

4-62 The forms of offensive maneuver are common to all environments, to indude mountainous terrain. While frequently used in combination, each form of maneuver attacks the enemy in a different way, and some pose different challenges to the commander attacking in the mountains.

INFILTRATION

4-63. Infiltration is frequently used in the mountains. The difficult terrain and recurring periods of limited visibility allow for undetected movement. Infiltration in a mountainous environment is used to shape the battlefield by attacking enemy positions from the flank or the rear, securing key terrain in support of the decisive operations, or disrupting enemy sustaining operations. Infiltration is normally conducted using one of threetechniques: movement in small groups along one axis, movement in one group, or movement in small groups along several routes at the same time. Regardless of the technique used, units must move in a covert manner to reduce the chance of enemy contact.

4-64 With movement in small groups along one axis, all members of the force use the best route. Small groups are harder to deled and easier to control, and do not compromise the total force if detected. This technique may require an excessive amount of time and an increased number of guides and lead dimbing teams, and does require an assembly area or linkup point prior to oondud of the dedsive action. If the lead group is detected, other groups may be ambushed.

4-65. The enemy can more easily deted movement in one group. If the force is deteded, the overall mission may be endangered. However, this technique has no reassembly problems, since reassembly is not required. Everyone uses the same route, easing navigation and redudng the number of guides and lead dimbing teams required. A large force can fight out of a dangerous situation more easily than a small one This technique minimizes coordination problems with other infiltrating units.

4-66. Movement in small groups along several routes at the same time has several advantages. It avoids putting the total force in danger and is less likely to be seen. Il forces the enemy to react in many locations and makes il harder for him to determine the size of the force or its mission. Groups travelling over severely restrictive terrain may have significant assembly, control, and sustainment difficulties, and may require more guides, lead climbing teams, and installation teams.

ENVELOPMENT AND TURNING MOVEMENT

4-67. The envelopment and its variant, the turning movement, are used extensively in mountain operations. Both forms of maneuver seek to avoid the enemy's strength and attack the enemy at a decisive point or points where the enemy is weakest or unprepared, and both rely on superior agility. In the mountains, the ability to react faster than the enemy may be achieved through greater mountaineering skills, using airborne and air assaults, and, depending on the specific AO, amphibious assaults.

PENETRATION

4-68. Mountainous terrain normally makes penetration extremely dangerous or impossible because of the difficulty in concentrating overwhelming combat power in the area of penetration. Due to mobility restrictions, it is also difficult to develop and maintain the momentum necessary to move quickly through a gap and on toward the objective. The area of penet ration is always vulnerable to flank attack, and this vulnerability significantly increases in mountainous terrain A penetration may be useful when attacking an enemy that is widely dispersed or overextended in his defense. If a penetration must be conducted, flank defensive positions must be eliminated before the initial breach of enemy positions. Successful penetration of a defensive positron in restrictive mountainous terrain requires using limited visibility, stealth, and covered and concealed terrain at selected breach points.

FRONTAL ATTACK

4-69. Frontal attacks in hilly or mountainous areas, even when supported by heavy direct and indirect fires, have a limited chance of success. Mountain terrain adds to the relative combat power of the defender. The frontal attack exposes the attacker to the concentrated fire of the defender while simultaneously limiting the effectiveness of the attacker's own fires. In the mountains, the frontal attack is an appropriate form of maneuver to conduct as a shaping operation designed to fix a force, while the decisive operation uses another form, such asthe envelopment, to defeat the enemy.

MOVEMENT TO CONTACT

4-70. The fundamentals regarding movement to contact also apply in a mountainous environment with the added likelihood of surprise attack and ambush. Limited mobility and dependence on restrictive terrain make it difficult to rapidly deploy from the movement formation. Plans and movement formations and techniques should be based on maintaining fiexibility and providing continuous security.

4-71. During a movement to contact, the advance guard normally advances in column, moving continuously or by bounds, until it makes contact. While requiring less physical exertion, movement along the topographical crest of a ridgeline increases the possibility of enemy observation and should normally be avoided. Given adequate concealment, this exposure may be reduced by moving along the military crest. Ridgelines and crests can often provide a tactical advantage to the force that controls them. Their control may allow rapid movement from one terrain compartment to another and afford excellent observation into lower terrain levels. In all cases, commanders must address the control or clearance of ridgelines that dominate their planned avenues of approach.

4-72. The main body should never be committed to canalizing terrain before forward elements have advanced far enough to ensure that the main body will not become endrded. This is a critical fador when employing mixed heavy and light forces that have sharp differences in operational tempo. Combat service support must be decentralized and readily available to sustain the combat elements. Major terrain compartments may physically separate maneuver units moving as part of a larger force. Continuous reconnaissance to the front and flank security is essential to prevent the enemy from infiltrating the gaps between units.

4-73. As the enemy situation becomes better known, commanders may shorten the distance between elements to decrease reaction time, or they may begin to deploy in preparation for the attack. Lateral movement between adjacent columns is frequently difficult or impossible. However, every attempt should be made to maintain at least visual contad. Commanders must emphasize the use of checkpoint reporting, contact patrols, and phased operations to coordinate and control the movement of the overall force. Control measures should not be so numerous as to impede operations and stifle initiative. Proper control ensures that units and fires are mutually supporting, objectives arecorrectly identified, and unitsare in position to attack.

ATTACK

4-74. Speed, flexibility, and surprise, normally advantages enjoyed by the attacker, are limited by restridive terrain and the defender's increased ability to see and acquire targets at greater distances. These limitations make it difficult for units above the company team level to conduct a hasty attack against prepared positions. In the mountains, commanders usually need more time to coordinate fire support, pick routes to prevent enemy observation and detedion. and select control measures to coordinate and control the operation. Therefore, deliberate attacks requiring a detailed scheme of maneuver and well-developed fire support plan become the norm at battalion-level and above. Since daylight contributes to the defender's ability to see and, thereby, reduces the attacker's chances of success, commanders should seek opportunities to exploit the advantages of limited visibility. Although these conditions slow movement even more and make coordinating forces more difficult, they decrease the enemy's ability to accurately sense what is happening and react effectively.

4-75. In planning and conduding the attack, commanders should recognize that the enemy will generally seek to control the valleys and trail networks, induding adjacent slopes and high ground. Defenses normally be anchored around obstades, and long-range, di rect fire weapons employed in poorly t rafficable terrain, often on slopes and protruding high ground. The enemy will attempt to engage the attacker in the valleys and low ground with flanking fires and artillery, often in a direct fire mode. Commanders must analyze the terrain to determine not only how the enemy will organize his defensive positions, but also how the terrain might contribute to the enemy's ability to counterattack As friendly forces attempt to deploy for the attack, the enemy, using his advance knowledge of the terrain and prepared routes, may maneuver forces to counterattack from the flank or rear.

4-76, All terrain features that can be occupied by even a small enemy force should be secured. I n many instances, overwatch positions may not be readily available within the range capability of organic weapons. Infiltration, technical dimbing, and extensive breaching may be required to position weapons to support the assault. On many occasions artillery support, espedally in high mountains, may not be available. In other instances, commanders may need to identify intermediate objectives for maneuver forces based on the need to ensure that artillery units have suitable, secure firing positions to range the enemy and support the attack. As in all environments, commanders must identify fire support requirements and allocate fires based on the ability to support and available ammunition. Because resupply may be limited and extremely difficult, they may need to place restrictions on the amount of ammunition expended on spedfictargets

4-77. Fire and movement during the assault are extremely difficult. In situations where machine guns can be positioned effectively, a rifle platoon can provide itself with support from a flank or from a height. However, during an assault up a slope, supporting fire cannot come from an overwatch position and must originate from the flanks or through gaps between the assaulting soldiers. Control is difficult to maintain when the assault is in steeply rising terrain. Commanders must pay speda! attention to the dangers of fratricide.

4-78. Fire and movement are easier in an assault over a downward slope Down-slope assaults often have the advantage of good observation, but dead spaces and intervening terrain may reduce the effectiveness of supporting fires. Defensive positions laid out by a skillful enemy on a reverse slope significantly increase the effect of unfavorable down-slope conditions. This type of defense compels tho attacking force to position its supporting weapons and observation posts on exposed crests. In this sit u at ion, support dements must be positioned to avoid terrain masking and crest dearancs problems.

4-79. Breaching obstades and preparing bypass routes that allow the assault force to move into the defensive position must bean integral part of the commander's plan. In rugged terrain, man-made obstades that are covered by fire create a particularly dangerous and formidable barrier. Command and control of a covert, in-stride, deliberate, or assault breaches is more difficult than in open terrain, and mobility support is extensive if the obstade cannot be reduced. Assaults in mountainous terrain almost always involve preparing routes that allow the assault force to rapidly move over difficult natural obstades and into the objectives.

4-80. Commanders should maintain a strong reserve, if possible. In the mountains, as elsewhere, commanders can use their reserves to restore the momentum of a stalled attack, defeat enemy counterattacks, and exploit success. Reserves must becarefulty positioned and organized so difficult terrain, limited road networks, or unpredictable weather does not delay their arrival.

Once committed, commanders make every effort to reconstitute another reserve from available units.

4-81. An attack should not be halted on a summit or on a ridgeline objective, which enemy artillery and mortar fire will likely target. Reorganization is generally best conducted well forward of a crest line on the next suitable slope. Commanders must ensure that the enemy is not allowed the opportunity to counterattack to recapture key terrain. Rapid adjustment of positions and coordination with flanking units are essential. Support weapons, especially mortars, should be brought forward as quickly as possible. Helicopters are useful for this purpose and may be used for backhaul of casualties

COUNTERATTACK

4-82. Counterattacks in the mountains must exploit the aspects of terrain that impair enemy momentum and make it difficult for him to mass and maneuver. Obstructing terrain that canalizes movement and restricts mobility significantly increases the potential for counterattacks. In planning a counterattack, the commander must carefully consider the enemy's weaknesses or inability to operate in a mountainous environmenl. A counterattack, even on a very small scale, can have a decisive impact in mountainous terrain.

RAID AND AMBUSH

4-83. The restrictive terrain also affords increased opportunities to conduct raids and ambushes. These operations should take advantage of limited visibility and terrain that the enemy may consider impassable. In steep terrain, movement time increases significantly, and only light equipment can be taken. The force should use special climbing techniques to negotiate the difficult routes during limited visibility. Commanders must carefully consider the routes and methods used for extraction to ensure that the combat force does not become isolated after executing the mission. They can ensure a successful operation by avoiding detection through proper movement techniques and by skillfully using natural cover and concealment. It may be necessary to reposition some indirect fire support assets to cover dead space or use attack helicopters and close air support. The ambush or raid commander must know in advance if supporting fires can not cover his routes to and from the objective.

DEMONSTRATIONS AND FEINTS

4-84 Because maneuver space is usually limited or confined and restricts the number of avenues of approach for heavier forces, deception plays an important part in the mountain battle To mislead the enemy regarding friendly intentions, capabilities, and objectives, commanders should plan systematic measures of deception.

EXPLOITATION AND PURSUIT

4-85. In a mountainous environment, exploitation and pursuit operations must be conducted discriminately and the mountain commander must always prepare for success. A battalion may exploit its own success to a limited extent, but it normally participates in the exploitation as part of a larger force. Air assault and attack helicopter units can be used to augment exploitation and pursuit operations. The exploiting commander must compensate for the ground mobility restrictions imposed by terrain and weather. Speed can best be achieved by isolating enemy positions with the smallest force possible Engineer support should be well forward with the necessary equipment to allow combat troops to maintain momentum and avoid delay by enemy obstacles. The commander must be careful to prevent overextending either the exploiting force or its sustaining logistics. A withdrawing force is capable of establishing numerous strong points and firing positions on heights that allow it to quickly dissipate the combat power of the exploiting force.

MOTTI TACTICS

4-86. Motti tactics are presented here to demonstrate how forces can exploit superior mobility skills and knowledge of the mountainous terrain and environment to defeat the enemy. The Finns developed these tactics during the Finnish-Russian War in 1939-1940. They are characterized by attacks on rear areas, bivouac sites, and command posts.

4-87. The Finnish word "motti" means a pile of logs ready to be sawed into lumbo; in effect, setting the conditions so that a larger force can be defeated in detail. These tactics were most successful in the forested areas of Finland during the arctic winters. During the Finnish-Russian War, the Soviets were neither prepared for, nor trained for, warfare under such conditions. They were almost totally trail-bound, with few ski troops. In the 1980s, the Soviet Union experienced similar difficulties in the mountains of Afghanistan, In both instances, the road and trail-bound nature of their forces and their basic tactics left them vulnerable to motti tactics in mountainous terrain.

4-88. Generally, the force utilizing motti tactics never becomes decisively engaged. It disrupts the enemy's supply lines, denies him warmth and shelter, infiltrates his bivouacs, and destroys his rear areas to the point where he must remain in a high state of alert. These attacks, in combination with the environment, help to destroy the enemy's will to fight Commanders should not only develop a thorough understanding of how to apply these tactics, but also understand the conditions that may leave their own forces vulnerable to its use (see Figure 4-B on page 4-24).

Motti tactics generally follow the sequence of:

1. Locating and fixing the enemy.
2. Isolating the enemy.
3. Attacking to defeat or destroy the enemy

4-89 Reconnaissance is conducted to locate an enemy force moving in or toward an area that will restrict his movements to roads, trails, or linear terrain. Once identified, the force must be fixed so that it presents a linear target along the axis of advance to which it is bound. This is accomplished using obstacles and a series of squad and platoon sized ambushes and raids Obstacles may be natural, such as snow, crevices, deep mud, steep terrain, and water obstacles, or man-made, for example mines, landslides, avalanches, or destroyed bridges.

4-90. The ambushes and raids not only fix the enemy, they also disturb his composure, create an air of uncertainty, and prevent uninterrupted sleep and rest. Friendly units attack the enemy from the high ground. They make maximum use of night vision devices, as well as the difficult restrictive terrain. They avoid enemy security and interdict his operations. As a further result of these actions, the enemy is compelled to use more forces on security tasks. Unless the enemy can be easily defeated or destroyed, the attacking force rapidly withdraws after forcing the enemy to deploy. In general, this series of attacks confuses the enemy as to the friendly unit's exact location and intent, and slows his decision-making cyde so that he reads ineffectively to subsequent operations.

FORCES CAN USE MOTTI TACTICS WHEN THEY:

- Have superior technical mobility skills necessary to negotiate Class 4 and 5 terrain
- Are able to operate effectively in a noncontiguous area of operations with limited support, and despite temperature extremes and inclement mountain weather
- Are able to navigate in high mountainous terrain, dense vegetation, darkness, storms, and fog while making good use of available cover and concealment
- Maintain the element of surprise

FORCES ARE VULNERABLE TO MOTTI TACTICS WHEN THEY:

- Operate within noncontiguous areas of operations
- Have limited mobility skills restricting their movements to roads, trails, and Class 1 and 2 terrain
- Have inadequate reconnaissance and security

Figure 4-8. Conditions Affecting the Use of Motti Tactics

4-91. The attacking force then isolates the enemy into smaller groups. Once isolated, the friendly force maneuvers to envelop and attacks to defeat or destroy the isolated elements. As the enemy exhausts himself in an effort to break out, the attacking force may regroup and repeat the sequence. It is imperative that the attacking force seal off the enemy and keep avenues of approach closed, and not ignore the threat to its flanks, which may increase as the attack progresses.

4-92. Overall, motti tactics wear the enemy down to a point where he is vulnerable to more direct attacks or to the point where it is no longer beneficial or feasible to continue operations in the area. Motti tactics employed alone only prove decisive over a long period of time, depending on the enemy's capabilities, strength, and resolve. Based on METT-TC, friendly forces normally must increase the operation's tempo to gain a quick, decisive outcome. Still, these type tactics may complement other more direct offensive operations in support of the overall pfan.

SECTION III - DEFENSIVE OPERATIONS

4-93. In the mountains, more so than in the lowlands, the strength of the defense depends on its selection and use of key and decisive terrain. Key and decisive terrain provides the defender-and usually denies the attacker-excellent observation and fighting positions. Reinforcing obstades significantly enhance the natural obstacles of rugged mountainous terrain.

4-94. The immediate objective of a mountainous defense is to deny the enemy access to key terrain that helps him conduct further operations Therefore, it is necessary to defend in terrain that restricts and contains the enemy, as well as control the high ground that dominates this terrain. The effects of rapidiy changing weather, visibility, and mountain hazards must be continually assessed. The terrain provides the defender with cover, concealment, and camouflage that can deceive the enemy regarding the strength and dispositions of friendly forces. The advantages of knowing the terrain, having fortified positions, siting weapons in advance, stockpiling supplies, and identifying or preparing lateral trail networks favor the defense. They allow the defender to shift forces on the ground more rapidly than the attacker. Delaying operations are particularly effective in the mountains and can be accomplished by a smaller force. These advantages combine to make the mountai ns an ideal place for defensive operations. Regardless of the scale of defensive operations, key factors in achievi ng success in the mountai ns are having good observation and aggressive reconnaissance, white denying the same to the enemy.

PLANN*t*NG CONSi DERATIONS

4-95. Defending commanders must develop flexible pEans for control of fire, maneuver, communications, and logistics. Initially, the attacker has the Initiative and decides whorc and when combat will take place. The defender must be agile enough to maintain control of the heights, strike effectively, and shift his effort quickly without losing momentum and flexibility. Tactical flexibility depends on planning in detail, organizing in depth, and retaining an adequate, mobile reserve

4-96. Although the mountains generally allow observation at greater dislances, intervening terrain features and weather often prevenl commanders from seeing the area of operations beyond the area to their immediate front and flanks. Consequently, commanders normally allocate more assets for reconnaissance and security, echeloned in depth and in height, to ensure that they are able to sense all aspects of the AO and gain the time needed to decisively apply combal power.

4-97. Commanders must prevent the enemy from concentrating overwhelming combat power against isolated sections of their defense. The restrictive terrain is one of the primary advantages of the mountain defender, as it interferes with the attacker's synchronization, canalizes his movement, and impedes his ability to maneuver. However, unless oommanders carefully analyze the terrain from both the friendly and enemy viewpoints, to indude the horizontal and vertical perspectives, they leave themselves vulnerable to infiltration and possible attack from the flanks and rear along difficult and unexpected routes.

4-98. In the mountains, commanders usually organize for a perimeter defensetobe prepared to defeat the enemy from any approach, to indude those that may appear impassable. Although preparing for an all-around defense, they should weight a portion of the perimeter to cover the most probable direction of enemy attack. Rocky terrain may make it more difficult to prepare defensive positions and rapidly changing weather may halt preparations altogether. If suffident forces are not available, the commander must economize in some areas and rely moreheavily on prepared positions, to indude alternate and supplementary positions, obstades, and well-planned indired fires to cover gaps and dead space.

4-99. The width of an area to be defended depends mainly on the degree to which terrain is an obstade. Terrain that significantly restricts enemy movement tends to lavor a larger AO. Normally, an area should be approximately as deep as it is wide, and may indude the entire length and surrounding heights of a valley. Ridges that run at right angles to the enemy's direction of attack also permit increased width with less depth. A defense in depth is required in valleys that run in the direction of the enemy's attack. In either case, it is essential to have forces deploy on the dominating heights that control approach routes.

4-100. Ideally, reserves should be mobile enough to read to enemy adion in any portion of the perimeter. Less mobile reserves are positioned to block the most dangerous avenues of approach and assigned on-order positions on other critical avenues. Sharply compartmented terrain may require the creation of more than one reserve. Helicopters may be used to deploy reserves, but their use depends on the availability of suitable, secure LZs and favorable weather conditions.

PREPARATION

4-101. The process of preparing the defense must begin with a thorough reconnaissance. Preparations for a mountain defense require more time than in other terrain, and as units arrive they must begin immediate preparation of their defensive positions. In some instances, technical mountaineering skills may be needed to establish effective security and to emplace crew-served weapons properly. However, commanders must weigh the advantages gained from these inaccessible positions against difficulties in repositioning and resupply. Preparations for the defense must alsoindudeinstalling communications, stocking forward supply points with particular attention to Class IV. empladng medical elements, adjusting air defense coverage, and arranging for security of installations in the rear area. Commanders must ensure that time is available to develop alternate routes and positions, rehearse and time movements between positions and along routes, and rehearse counterattacks.

4-102. Commanders must seek every opportunity to recapture the initiative from the attacker and transition to offensive operations. Preparations for a counterattack in the mountains must include caching ammunition, preparing counterattack positions and routes to attack downhi 11, identifying crew-served weapon positions, and establishing rally pointsthat are usually on the reverse slope.

ORGANIZATION OF THE DEFENSE

4-103. Defensive operations in the mountains derive their strength, balance, and freedom of action from the effective use of terrain. The mobility restrictionsfound in mountainous areas, combined with the necessity to hold dominating ground, dictates that an area defense be used. Mountain defenses use security forces, continuous reconnaissance and combat patrols, as well as numerous observation posts. The mountain AO is usually organized into a security area, main battle area (MBA), and rear area.

SECURITY OPERATIONS

4-104. While a screening force is often thought to be the most preferable form of security in extremely rugged mountainous terrain, all forms of security operations, to include guard, cover, and area, may be employed effectively in a mountain AO based on the factors of METT-TC with particular emphasis on:

- Forces available for security operations.
- Abilitytomaintainamobility advantage.
- Size of the security area and the number of avenues of approach.
- Likelihood of enemy action,
- Size of theexpected enemy force.
- Amount of earlywarningand reaction time needed.

4-105 A screen primarily provides early warning to the protected force and is usually an econorny-of-force measure. The compartmented nature of mountainous terrain often serves to create multiple gaps and exposed flanks The rugged terrain may also serve to restrict the movement of not only advancing enemy forces, but also the movement and mobility of largor friendly security forces In thooc instances, commanders may choose to use minimum combat power to observe, identify, and report enemy actions at these locations, and engage and destroy enemy reconnaissance within the screening force's capability. The screening force may be able to avoid decisive contact by withdrawing into restrictive terrain that forces the enemy to utilize difficult climbing techniques if he continues the pursuit.

4-106. In the mountains as elsewhere, the screening force should adjust to the enemy advance and continue to screen as far forward as possible, even though elements of the force may have to withdraw. Retention of selected forward positions may allow surveillance and targeting forward of the MBA, upsetting the enemy's coordination. By allowing the enemy to bypass advance positions, the screening force can facilitate counterattack to the front of the forward edge of the battle area (FEBA) by providing observation of. and access to. the flanks and rear of attacking forces.

4-107. If a significant enemy force is expected or a significant amount of time and space is required to provide the required degree of protection, commanders usually resource a guard or cover mission instead of a screen. As long as trie security force can maintain a mobility advantage over the enemy, it can effectively delay and attack the enemy force by using obstacles and the restrictive terrain to its advantage. Although utilizing a greater proportion of his combat power, the appropriate use of a guard or cover force should provide the mountain commander greater depth in his security area and the ability to defeat, repel, or fix lead elements of an enemy ground force before they can engage the main body with direct and indirect fires.

4-108. No matter the type of security used, defending forces must prevent enemy infiltration by carefully positioning observation posts (OPs) and conducting continuous patrols and ambushes. Combat reconnaissance patrols and other intelligence gathering assets observe the enemy as far ahead of friendly positions as possible and report his strength and composition, as well as his route of movement. To accomplish this, reconnaissance patrols may need to rely heavily on technical climbing skills. Ground surveillance radar and unattended ground sensors can be used effectively, but the defender must besureto cover all gaps and dead spaces. The defender must make best use of his time to study the ground and determine ail possible infiltration routes.

MAIN BATTLE AREA

4-109. In rugged mountainous terrain, it may be difficult to maintain mutual support and overlapping observation. Elements should be employed to man observation posts, assist the passage of security forces into the MBA. cover obstacles and avenues of approach by fire, screen gaps between defensive positions, and ambush enemy infiltrators.

4-110. Defensive positions along ridges or domi nating heights should include as much of the forward and reverse slopes as possible to add depth and all-around security. The actual size of unit positions is terrain-dependent. At a minimum, fighting positions and observation posts should be echeloned vertically, as well as in depth. When defending a mountain valley, forces should establish fighting positions that are located on adjacent heights and in depth to permit coveringthevalleywith interlocking fire. These positions must also be anchored to restrictive terrain or adjacent defensive forces to prevent enemy envelopment. In wooded terrain, defensive positions may be organized on theforwardedgeofthe woods, as well as on commanding heights. Obstacles should be widely employed to slow or stop enemy movement throughout.

4-111, Mountain warfare demands that forces conduct an aggressive defense. Defending units must infiltrate enemy units and attack headquarters, supply lines, and rear areas. Smaller patrols and OPs should be deployed well forward to direct artillery fire and attack aircraft on targets of opportunity, and to conduct personnel and antiarmor ambushes. Disruption operations should be conducted to force the enemy to deploy additional assets to protect lines of communication and delay and upset preparations for the attack. In the mountains, enemy forces can frequently be isolated if they are discovered in time and reserves are effectively placed and highly mobile.

4-112. In the defense, the reserves' primary purpose must preserve the commander's flexibility In mountainous terrain, the commander may need to rely on the reserves as his principal means of restoring his defense's integrity or exploiting opportunities through offensive action. Because of the difficulties of movement, small reserves may be located near primary defensive positions, ready for immediate counterattack. This type of small, responsive counterattack may be much more effective than a large-scale, major counterattack. It can catch the enemy exhausted after an uphill assault and before position consolidation. Large, centrally placed counterattack forces are normally unableto intervene in time unless the terrain permits mounted movement, or sufficient helicopter lift assets are committed to the reserve force or made rapidly available.

REAR AREA

4-113. To minimize the vulnerability of sustaining operations and extended lines of communication, command and control, as well as support operations, in the rear area must be dispersed, redundant, and as far from potential enemy approaches as possible. Because of limited space available in the rear area, the commander must be careful in selecting and locating positions for combat service support activities. These positions are likely to be confined to small valleys. Therefore, they are high-priority targets for enemy artillery and air attack or raids by small combat patrols, particularly at night or in bad weather. When possible, combat service support elements must avoid the most obvious positions and occupy atypical sites. However, they should be in the vicinity of a defined road network and an air loading area, even if the network or area is within Level 11 Locating base defenses at Level II elevations may allow more access to supply bases for air resupply during inclement weather, such as the heavy fog often encountered in valleys and at lower elevations

4-114. A perimeter defense is planned for each combat service support unit within the defensive area. Defensive positions should be selected on the dominating high ground. Sensors, OPs. and radars are used to cover avenues of approach and gaps between positions. Rear area forces must routinely conduct patrols and ambushes around the perimeter, especially at night and during other periods of limited visibility. Air defense assets should be located to protect rear area facilities. Tactical combat forces (TCP) must be prepared to respond rapidly to rear area threats and should be prepared to move to any of their objectives by multiple routes. However, units within the rear area must not fall into the trap of relying solely on a TCP for their security. No matter how well-organized or mobile the TCP, rear area units must provide their own well thought-out and active security measures, even at the cost of a reduced ability tosustain the force.

REVERSE SLOPE DEFENSE

4-115. Reverse slope defenses apply particularly well to mountain operations and pursue offensive opportunities through surprise and deceptive operations by defending in a manner for which the enemy is unprepared. This defense seeks to reduce the effects of massed indirect fire from mortar, artillery, and dose air support, and draws the battle into the small arms range of infantry weapons. The overall goal is to make the enemy commit his forces against the forward slope of the defense, causing his forces to attack in an uncoordinated fashion across the exposed topographical crest. Once this type of defense is employed, subsequent use may be of limited value, due to the loss of the key element of surprise.

4-116. All or partsof thedefending force may use reverse slope techniques. I n many instances, mountainous terrai n favors a defense that employs combined forward and reverse slope positions to permit fires on enemy approaches around and over the crest and on the forward slope of adjacent terrain features. Key factors to t his type of defense are mutually supporting covered and concealed positions, numerous natural and man-made obstacles, theability to bring fire from all availableweapons onto the crest, and a strong and mobile counterattack force.

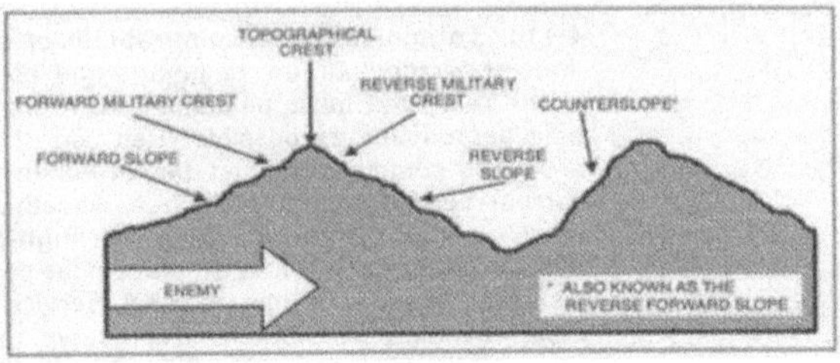

Figure 4-9. Reverse Slope Defense

4-117. The reverse slope defense is organized so that the main defensive positions are masked from enemy observation and direct fire by the topographical crest (see Figure 4-9). It extends rearward from the crest only to the maximum effective range of small arms fire. Observation and fires are maintained over the entire forward slope as long as possible to continue to destroy advancing enemy forces and prevent him from effectively massing for a final assault. A successful reverse slope defense is based on denying the topographical crest to the enemy, either by fire or by physical occupation. Although the crest may not be occupied in strength, control of the crest by fire is essential for success. For more detailed discussions of the reverse slope defense, see FM 3-100.40 and FM 3-21.30,

RETROGRADE OPERATIONS

4-118. Retrograde operations of delay and withdrawal can be conducted in mountainous terrain with fewer assets because of the mobility difficulties of an advancing enemy. Delaying operations are particularly effective in the mountains. Numerous positions may exist where elements as small as a machine-gun or sniper team can significantly delay a large force. When conducting retrograde operations in mountainous terrain, the friendly force must accomplish several tasks.

- it must make maximum use of existing obstacles- However, the addition of relatively few reinforci ng obstacles, such as the antitank mining of a route with very steep sides, often increases the value of existing obstacles.
- The force must conduct detailed reconnaissance of routes to rearward positions. Routes of withdrawal are not as numerous in mountainous terrain and often do not intersect as they do on flat terrain. These

factors complicate subsequent link-up operations and necessitate meticulous planning.

- It musl protect the flanks and rear to prevent encirclement, particularly by air assault. There are only a few LZs and they can significantly influence the outcome of a battle. At a minimum, they must be covered by fire.

STAY-BEHIND OPERATIONS

4-119. The compartmented terrain in the mountains lends itself tothe employment of stay-behind forces as a tool for offensive action. Stay-behind operations involve the positioning of friendly elements within operational areas before the enemy advances through the area. Stay-behind forces conceal their location and allow themselves to be bypassed as the enemy advances. (Figure 4-10 outlines the important tasks that stay-behind forces can accomplish for the mountain commander.)

- **Attack the enemy throughout the depth of his formations**
- **Disrupt the cohesion of the enemy offense by interrupting lines of communication and logistics**
- **Detract from the enemy's main effort by forcing him to allocate combat forces to rear areas**
- **Provide immediate intelligence**
- **Call for and control indirect fire and close air support**

Figure 4-10. Tasks for Mountain Stay-Behind Forces

4-120. Stay-behind forces may be positioned forward of the FEBA, in the MBA, and not participate in the initial fight, or, under certain conditions, in the MBA after the fighting has started. When planning for stay-behind operations in the mountains, commanders must consider the following:

- Stay-behi nd forces should be a combined arms force that indudesengineers.
- Indirect fire support or dose air support should be available throughout their operations.
- Return routes must be well planned and reconnoitered in advance. Exfiltration, regardless of element size, should follow covered and concealed routes, and rally points should be designated forward of and behind the lines of friendly forces. Reentry must be carefully coordinated to prevent fratridde.
- After an attack, stay-behind forces may be unable to reach a hide position to subsequently return to friendly lines by exfiltration. If this occurs, they must be prepared to condud a breakout,
- The tactical situation and logistics supplies that were stockpiled or cached in the AO have an impad on the length of time stay-behind forces remain in enemy territory.

Chapter 5

Logistics and Combat Service Support

Mountainous terrain poses great challenges to combat service support (CSS) forces and complicates sustaining operations. Existing roads and trails are normally few and primitive, and cross-country movement is particularly demanding. Highways usually run along features that have steep slopes on either side, making them vulnerable to disruption and attack. Rivers become major obstades because of rapid currents, broken banks, rocky bottoms, and the lack of bridges. Landslides and avalanches, natural as we! l as man-made, may also pose serious obstacles to CSS operations. Mountainous areas have wide variations in climate and are subject to frequent and sudden changes of weather that may preclude reliance on continuous aviation support. Together, these conditions compound the obstacle-producing effects of mountainous terrain and create major challenges for the CSS planner. Therefore, the forward distribution of supplies may depend upon the knowledge, skill, and proficiency of CSS personnel in both basic mountaineering and aerial resupply operations.

CONTENTS

HISTORICAL PERSPECTIVE

The Importance of Lines of Communications: The Satukandav Pass (Soviet-Afghan War, November-December 1987)

One of the characteristics of the Soviet-Afghan War (December 1979 - February 1989) was the attempt by both sides to control the other's lines of communications (LOCs). In an effort to deprive the guerrillas of their source of sustainment, the Soviets used various methods to drive the rural population into exile or into cities. For their part, the Mujahideen regularly interdicted supply routes through the establishment of blocking positions and vehicular ambushes. In some regions, they were able to effectively interdict supply routes for weeks, months, and even years at a time. The Soviet main supply route was a double-lane highway network winding through the rugged and inhospitable Hindu Kush Mountains. The continued Soviet presence in Afghanistan depended, in large part, on their ability to keep the roads open. Therefore, much of heavy Soviet combat was a fight for control of this road network, with this control often changing hands during the course of the war.

In the fall of 1987, the Mujahideen had established a series of blocking positions that severely limited the supply of weapons, ammunition, and food to Soviet forces in the Khost district. In response, the Soviets planned and

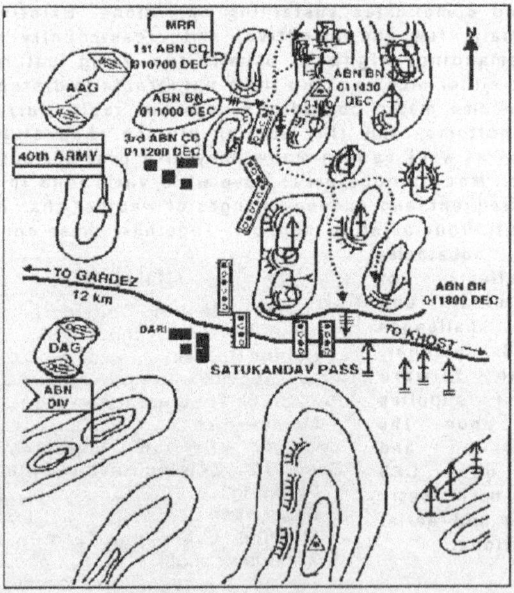

Figure 5-1. Satukandav Pass

conducted Operation Magistral, "main highway," to open the LOCs (see Figure 5-1). The guerrilla forces had established strong positions in the Satukandav Pass, virtually the only way through the mountains between Gardez and Khost. For the operation, the Soviets massed a motorized rifle division, airborne division, separate motorized rifle regiment (MRR), separate airborne regiment, various 40th Army regiments, special forces, and other subordinate units, and regiments from the Afghan Armed Forces. On 28 November, in order to determine the location of Mujahideen positions, particularly air defense systems, the Soviets conducted a ruse in the form of an airborne assault using dummy paratroopers. When the Mujahideen fired at the dummies, Soviet artillery reconnaissance was able to pinpoint enemy strong points and firing positions. The Soviets hit these positions with air strikes and a four-hour artillery barrage. The next day, however, an MRR failed to make its way up the foothills to seize the dominant terrain along the crest, and suffered heavy casualties. The 40th Army Commander, General Gromov, nevertheless, decided to continue to press his advance using the 1st Airborne Battalion and a battalion of Afghan commandos. On 1 December, two airborne companies captured key terrain and used this to support the decisive operation against the dominant peak to the south. This flanking attack took the Mujahideen by surprise and they began to withdraw. While calling in artillery fire on the retreating guerrilla forces, primarily on the reverse slope and along the probable avenue of approach for the commitment of enemy reserves, the Soviet battalion commander used this hard-won, key terrain to support a simultaneous, two-prong attack to the south toward the Satukandav Pass. Now it was the Soviets who were in a position to cut off supplies, especially fresh drinking water, from the Mujahideen. The latter were forced to withdraw, and the two battalions captured the pass. However, while the operation itself was a success, Soviet and Afghan Army forces could keep the road open for only 12 days, after which the Mujahideen once again cut off the supply route to Khost.

Both sides recognized the vital importance of LOCs, and this shift of LOC control was a constant feature throughout the entire duration of the Soviet-Afghan War. The Mujahideen's ability to interdict the LOCs prevented the Soviets from maintaining a larger occupation force there, a key factor in the eventual Soviet defeat.

Compiled from *The Other Side of the Mountain* and *The Bear Went Over the Mountain*.

SECTION I - PLANNING CONSIDERATIONS

5-1. Mountainous areas of operations dictate that commanders foresee needs before demands are placed upon CSS personnel The main logistical differences between mountain operations and operations in other terrain are a result of the problems of transporting and securing material along difficult and extended lines of support Logistics support must emphasize a continuous flow of supplies to specific locations, rather than the build-up of stocks at supply points along the main supply routes. Supply point operations alone are insufficient; the proposed support structure must plan for redundancy in the ability to distribute supplies directly to units operating from predetermined supply routes. If possible, commanders should plan to use multiple supply routes designed to support maneuver elements moving on separate axis.

5-2. Commanders must be concerned not only with the sustainment of current operations but also with the support of future operations. A detailed logistics preparation ol the theater (LPT) to identify the potential lines of communication plays a major part in determining the conduct of CSS operations. A detailed reconnaissance should be conducted to determine:

- The type and maximum number of vehicles that the road network can support in the area. New roads may need to be constructed or improvements made to existing ones to support protracted operations in isolated areas.
- Classification of bridges.
- Suitable sites for drop zones (DZs), loading zones (LZs), and short, tactical airstrips.
- Availability of water sources.
- Availability of local resources, facilities, and service and support activities.

5-3. Because of terrain constraints, it may be necessary to disperse support units over a wider area and ensure that supplies are positioned closer to supported units. Dispersion reduces vulnerability of CSS assets, which also creates problems with command, control, and security CSS units are often high-priority targets, and must ensure adequate protection against ground and air attacks.

5-4, In mountainous terrain, battalion CSS elements are normally echeloned into combat and field trains to increase responsiveness, provide adequate space, and decrease the logistics footprint. Combat trains are routinely located in ravines or valleys on the rear slope of the terrain occupied by the unit. This permits the personnel officer (S1) and logistics officer (S4) to operate in close proximity to the tactical operations center (TOC), and allows them to keep abreast of unit requirements,

5-5. In the mountains, unresolved logistical problems can quickly lead to mission failure. Ground operations may increase fuel consumption rates of individual vehicles by 30 to 40 percent, requiring more frequent resupply operations. The operation of equipment in mountainous terrain has proven that maintenance failures far exceed losses due to combat, and most breakdowns can be attributed to operator training. Air operations are characterized by a significant increase in lift requirements; however, increased elevations decrease aircraft lift capabilities,

5-6. Commanders must carefully consider combat loads in the mountains, based upon a thorough mission analysis. Excess equipment and supplies reduce the efficiency of the individual soldier and seriously impede operations. In steep terrain above 1.500 meters (5,000 feet), soldier loads may need to be reduced by nearly 50 percent. Commanders must develop priorities, accept risk, and require the combat foroe to carry only the bare essentials needed for its own support. Nonessential equipment should be identified, collected, and stored until it is needed. In situations where there are conflicts between the weighl of ammunition and weapons, experience has shown that it is belter lo carry more ammunition and fewer weapons. In the mountains, commanders should strive to achieve the imperatives indicated in Figure 5-2

- Limit supplies to essentials.
- Lighten the individual soldier's combat load.
- Improvise methods and supply sources, to include utilizing captured enemy supplies and equipment.
- Use aviation assets to increase responsiveness.
- Anticipate maintenance requirements.
- Develop plans that place realistic demands on the CSS system.

Figure 5-2. Mountain Supply Imperatives

SECTION II - SUPPLY

5-7. Units operating in mountainous terrain transport supplies by a combination of wheeled vehicles, over snow vehicles, i ndigenous pack animals and personnel (see FM 3-05.27), assisted by Army and Air Force lift assets. These combinations depend on equipment availability, location of combat units, type of terrain, and weather. However, any combination of resupply usually i ndudes combat soldiers man-packing supplies to their positions.

5-8. Since combat operations in the mountains are decentralized, CSS operations are correspondingly decentralized. This decentral ization serves to create heavier man-loads, while rough, steep terrain decreases the amount soldiers are able lo carry. Although most soldiers are eventually able to acdimate themselves to higher elevations, their pace and subsequently the overall pace of the entire operation slows down as elevation increases.

5-9. Mountain warfare is highly dependent on accurate logistical planning if supply operations are to function smoothly. To win in any area of operations (AO), commanders normally seek to move and strike as rapidly as possible. Rapidly changing tadical situations may cause long supply lines, resulting in delay or complete disruption of supply operations. To mitigate these risks, situational understanding, rapid dedsions, and continuous coordination between tactical and logistical planners are essential. Stockpiling and caching supplies may also help to decrease the risks to resuppply.

5-10. The total tonnage of supplies required by the force may also decrease. For example, while individual vehide petroleum, oils, and lubricants (POL) consumption may increase, overall consumption may decrease because of lower vehide movement. The quantity of supplies needed by the individual soldier normally increases. Soldiers consume more food because of increased energy expenditure, and need many additional items of equipment, such as extra clothing, sleeping bags, climbing equipment, tents, and stoves, all of which must be stored and transported.

SUPPLY ROUTES

5-11. Main supply routes are generally limited to the roads located along major valleys and, through necessity, to the smaller, more restrictive trails that follow or parallel the ridgelines. The limited number of routes increases the volume of traffic and places heavy demands on engineer units to maintain them. In most cases, engineer units require assistance in clearing and developing, as well as in securing, these routes. Travel times for ground transportation assets are significantly increased due the generally poor quality of mountain roads and trails, frequent switchbacks, and steep grades that require lower vehicle speeds. Traffic control assumes increased importance due to the limited number of routes in the mountains, and may require an increased number of military police dedicated to the task of battlefield circulation control. In particular-

- Existing roads should be rapidly analyzed for bottlenecks, deployment areas, passing places, and turnarounds for various vehicles.
- Routes should be classified as one- or two-way, and schedules developed for the use of one-way routes.
- Signs should be placed for both day and night moves on difficult and dangerous routes.
- Whenever possible, separate routes should be designated for vehicular and dismounted movement. Additionally, separate routes should be designated for wheeled and tracked vehicles, particularly if the latter are likely to damage road surfaces.

5-12, The enemy will emphasize destroying logistical units and interdicting supply activities. Enemy units will infiltrate and seize key terrain that dominates supply routes in an effort to disrupt and isolate units from their logistics support. Using mountain trails and roads without securing the high ground on both sides invites ambush Patrols must be continually conducted at irregular intervals to verify the status of roads and prevent enemy infiltration. Patrols must be continuously alert for ambush and they must be skilled at locating and identifying mines. However, a combination of patrols and aerial reconnaissance is the best means of providing route security-Observation posts on dominant terrain along supply routes are also essential for early warning of enemy infiltration into rear areas.

5-13. Most often, units have to use the narrrow ridge trails as alternate supply routes, i n some instances as main supply routes, to reduce the volume of traffic on the main supply routes located along valley floors. This involves movements in much more restrictive terrain and exposure to excellent observation and fire by the enemy. Supply columns moving along separate routes face the same problems as combat units; they face the difficulties of being able to provide mutual support due to compartmented terrain, should one column come under attack. Movement of supplies at night may reduce vulnerability to enemy attack, but night marches present other hazards due to the difficult terrain, and require daylight reconnaissance, careful route preparations, and using guides

CLASSES OF SUPPLY

CLASS I: RATIONS AND WATER

5-14. The strenuous activities required during mountain operations increase caloric requirement to 4,500 calories or more per day. Improper or too little food means soldiers will lack the stamina to accomplish the mission. Although combat rations are normally used, unitized group rations (UGRs) should be provided once a day if the siluation permits. Individual packages of oatmeal and dehydrated soup mixes should be issued if the UGR cycle cannot be maintained.

5-15. In abrupt ascents to high altitude, soldiers do not have time to acclimate themselves, so their entire circulatory system labors to supply oxygen to the body. In this situation, standard rations are hard to digest and special rations, such as the ration, cold weather (RCW), that allow soldiers to eat light and often should be procured. The totally self-contained operational ration consists of one full day's feeding in a flexible, white-camouflaged meal bag. It contains cooked, freeze-dried, or other low moisture entrees, as well as a variety of items such as oatmeal, a nut-raisin mix, and fruit-cookie bars. The RCW provides sufficient calories (approximately 4.500 kilocalories) to meet the increased energy expenditure during heavy exertion, while limiting sodium and protein content to reduce the risk of dehydration. Because of rapidly changing weather conditions and the difficulty of resupply, each soldier may need to carry two to three days' supply of rations. However, this increases the soldier's load by approximately 10 to 15 pounds.

5-16. Proper water production, resupply, and consumption are essential and a constant challenge during mountain operations. In low mountains, planners should count on at least four quarts of water per soldier per day when static and up to eight quarts per day when active. In high mountains, planner should increase those requirements by about two quarts per soldier. In the mountain environment, medical care often requires an increased water supply and must be considered as part of the original planning and contingency factors.

5-17. Units should always be prepared to use natural water sources to help reduce the logistics burden However, far above the timberline, water is extremely difficult to find. Special measures must be taken to protect it from freezing in cold weather, such as placing canteens in the chest pockets of the extended cold weather clothing system (ECWCS) coat, hanging a two-quart canteen on a strap under the coat, or utilizing a camel-back type, commercially available, canteen under overgarments Purification and chemical sterilization are always necessary no matter how clean mountain water may appear. Micro-organisms present in mountain water may cause serious illness and rapidly degrade the strength of a unit. If above ground water sources cannot be located or are not reasonably available, drilling for underground sources may become a critical engineer task. Once engineer units access the water, quartermaster units have responsibility for completing the water points and purifying the water.

CLASS II: GENERAL SUPPLIES

5-18. General supplies indude expendable administrative items, individual dothing and equipment, tentage, and other items authorized by common tables of allowance. All units must deploy with enough Class 11 items lo last until routine resupply can be established. Spedal items, such as extended cold weather dothing, gloves, dimbing equipment, extended cold-weather sleep systems, batteries, and one-burner cook stoves, will be in great demand. Due to the rugged nature of the terrain, mountain operations also increase requirements for replacement items of individual dothing and equipment. Combat boots, for example, may be expected to last approximately two weeks in harsh rocky terrain.

CLASS III: FUEL AND PACKAGED PETROLEUM PRODUCTS

5-19. Individual vehicles need much more fuel in mountainous terrain. However, limited road nets and steep slopes reduce the volume of vehide traffic and overall fuel consumption. The heavy reliance on aviation assets for resupply and movement increases aviation fuel requirements. A commander must routinely plan for the emplacement of a forward arming and refueling poinl (FARP) within their AO to support intensive aviation operations. Battalions should establish a fuel point in the field trains using collapsible fuel drums. These drums should be operational as soon as the field trains are established and prepared to receive fuel from the forward supply company as soon as it arrives. When terrain makes refueling operations vulnerable to attack, units should conduct forward refueling using supply point distribution, and dispense fuel to unit vehides using the tailgate technique. The lack of suitable terrain normally increases the percentage of forward refueling done by this met hod

5-20. At 600 meters (2,000 feet), multi-fuel cook staves operate at about 75 percent effidency. When soldiers refuel cook stoves, they must avoid using automotive fuel. Fuel points must supply units with refined or white gasoline that is spedfically produced for pressurized stoves. Relatively large quantities of this fuel will be used when procuring water and preparing food. Adequate quantities of five-gallon cans, nozzles, and one-quart fuel bottles musi be on-hand before deployment.

CLASS IV: CONSTRUCTION, BARRIER, AND FORTIFICATION MATERIALS

5-21. Soldiers should make maximum use of local materials to reduce Class IV requirements and demands on the transportation system. Gabion-type material is espedally versatile during mountain operations. Gabions are widely used in the mountains for constructing obstades, fighting positions, anchors, mountain installations, traverse platforms, and helicopter LZs; for creating landslides or rockfalls; and for repairing roads. Units should stock adequate quantities of easily transportable sizes of reinforcing mesh and other suitable materials for constructing gabions.

CLASS V: AMMUNITION

5-22. Because of terrain, ammunition resupply is difficult, making strict fire control and discipline an absolute necessity. Ammunition transfer poi nts need to be as far forward as possible without revealing friendly unit locations or pfad ng ammunition stocks at risk of capture or destruction. Di rect delivery to the user may be required using aeri al resupply. Innovation and flexibility are critical, in the mountains, the traditional mixes of tank ammunition may be less effedive. Depending upon the spedfic threat, more rounds may be needed to attack light vehicles and fortified positions and less may be needed to engage tanks. Ammunition consumption for direct fire weapons may be low, however, consumption of indirect fire munitions, such as grenades, mortars, and artillery, may be high because of the dead space common to mountainous terrain. Planners must ensure that increased consumption of indirect fire munitions is induded in computating required supply rates.

CLASS VII AND IX: MAJOR END ITEMSAND REPAIR PARTS

5-23. Rugged terrain and dim ate extremes cause an increase in repair parts consumption. However, overall vehide utilization decreases in many situations Because it is difficult to transport large end items to toward units, the commander must place additional emphasis on preventive maintenance and repair.

CLASS VIII: MEDICAL SUPPLIES

5-24. The medical platoon obtains medical supplies for the battalion from the supporting forward support medical company (FSMC) or similar task organized medical element Medical supply organizations may distribute supply by various means: supply point distribution, unit distribution, or a combination of both. Mountainous terrain necessitates using suppty point distribution to a great extent. Medical supply activities must maximize use of empty evacuation assets moving forward to execute unit distribution of supplies as often as possible. The terrain will severely constrain ground movement operations. Pre-planned unit distribution via air assets is a must for emergent situations, such as mass casualty scenarios. Medical supplies must have a high priority for movement. Distribution of Class V! II via air lines of communications (LOCs) should occur as often as tactically feasible.

SECTION III — TRANSPORTATION AND MAINTENANCE

5-25. Transportation assets For mountain operations are often limited, and their use requires sound planning. Although vehicles are used to move supplies as far forward as possible, they may not be able to reach deployed units. Usi ng smai ler cargo vehides with i mproved cross-country mobi lity and dedicated aircraft is paramount to sustaining units in the mountains. Locally obtained animals, indigenous personnel, or combat soldiers must often move supplies from roads and trails to unit positions. The poor quality of road networks requires increased engineer effort. The rugged mountain terrain aids in infiltration increasing security requirementsalong the route.

5-26. Air resupply should always be considered lo reduce the transportation burden on ground assets. Therefore, support personnel should be well-trained in aerial resupply and sling-load operations. Aerial resupply, either by parachute drop, free drop, or cargo helicopter, may be available for a variety of tactical situations. However, unpredictableweather and air currents, cloud cover, and lack of suitable landing zones make aerial delivery unreliable, and higher elevations decrease overall aircraft lift capabilities. The integrated use of available helicopter transport should be used for forward transport of mail, replacements, returnees, and personnel service support providers, such as chaplains and finance support teams.

5-27. Fixing equipment as far forward as possible takes on added importance during mountain operations. In low mountains, equipment recovery and maintenance teams are critical in keeping limited routes clear and returning damaged vehicles to the battle in the shortest possible time. In high mountains, these teams are generally less critical to units operating there because terrain often limits vehicle use. Helicopter repair teams are critical in all mountainous environments due to helicopters flying at or near the maximum limits of their operational capabilities to meet increased needs for helicopter support, 1 n all cases, maintenance turn-around time increases to compensate for fatigue and the other effects of the environment on maintenance personnel. Figure 5-3 lists some of the critical repair parts that often fail or require frequent replacement during mountain operations.

- **Tires**
- **Tracks**
- **NVG Batteries**
- **Communications Equipment**
- **Cooling Systems**

Figure 5-3. Key Repair Parts

5-28. Drivers well trained in proper maintenance and driving techniques eliminate a great deaf of unnecessary maintenance and reduce maintenance requirements and vulnerable bottlenecks. All soldiers must devote increased attention to applicable FMs and TMs for their weapons and equipment and must conduct preventative maintenance, to include the availability and use of suitable cleaning solvents and lubricants, appropriate for the weather and terrain conditions.

SECTION IV - PERSONNEL SUPPORT

5-29. Personnel support in the mountains is not unlike that provided toother lypesof operations except for the limitations on soldiers and equipment posed by the environment. Key personnel support missions are to provide manning and personnel service support to unit commanders, soldiers, and Army civilians.

5-30. Personnel units normally depend on large quantities of automation equipment to successfully accomplish their mission. Adverse weather and rugged terrain may decrease their reliability and commanders should take extra precautions to protect this equipment. Additionally, the difficulty in establishing and maintaining communications may require an increased reliance on manual strength reporting until communications and automated systems are firmly established.

5-31. Postal services establish the link between soldiers and their family and friends and assist in defeating the isolation caused by the compartmented terrain and the resulting dispersion of units. However, the limited lines of communications in mountainous terrain may adversely affect mail distribution. Inefficient distribution of mail can quickly undermine morale, regardless of the AO. The timely delivery or mail is especially important in countering the shock of entering a new environment. Commanders should consider devoting a high priority to the distribution of mail on arrival in the theater of operations. FM 1-0 describes in detail the critical personnel systems and functions essential to providing manning and personnel service support.

SECTION V - COMBAT HEALTH SUPPORT

PLANNING

5-32. Combat health support (CHS) in the mountains is characterized by–

- Difficulty in accessing casualties in rugged terrain.
- Increased need for technical mountaineering skills for casualty evacuation.
- Longer periods of time needed for casualties to be stabilized.

5-33. When planning CHS, commanders must consider the tactical situation, the nature of the terrain, and speed of movement along the chain of evacuation. Aid stations should be centrally located in relation to supported units. The exact location should be based on the ability to provide shelter from the elements, cover and concealment from the enemy, ease of evacuation, and expected casualty rates.

5-34. The decentralization in the mountain area of operations also forces the decentralization of CHS. Commanders may need to establish casualty collection points, operated by aidmen from the evacuation section, to support isolated elements. These points are designated at intermediate points along the routes of evacuation where casualties may be gathered. Additionally, multiple ambulance exchange points may be required to transfer casualties from one type of transportation to another.

EVACUATION

5-35. Aeromedical evacuation remains the preferred form of casualty evacuation in mountain operations. Aircraft provide the best capability of evacuating casualties from isolated locations and transporting them to treatment centers. However, in many instances, even lightly wounded personnel may not be able to move unassisted over rough terrain and LZs may not be available.

5-36, Medical evacuation teams must complete reconnaissance and install necessary evacuation systems along each evacuation route before the onset of casualties. Litter relay stations may be required at predetermined points to conserve the stamina of litter bearers and aocelerate evacuation. The evacuation plan must include measures to care for wounded soldiers at points along the route of evacuation where delays are possible. The plan must also depict all evacuation routes and provide for proper disposition of medical personnel along the lines of evacuation (see FM **4**-02,2). Evacuating the wounded from mountain combat zones normally requires a larger number of medical personnel and litter bearers than on flat terrain. The number and type of evacuation systems depend on mission, enemy, terrain and weather, troops and support available, time available, civil considerations (METT-TC) and the factors listed in Figure **5-4.**

- **The patient's condition**
- **Anticipated casualty rate**
- **Importance of rapid movement**
- **Number of available evacuation teams**
- **Number of evacuation routes available**
- **Transportation assets and equipment available**
- **Availability of suitable and secure LZs**

Figure 5-4. Factors Affecting Evacuation Systems

MOUTAIN EVACUATION TEAMS

5-37. As the battle to control the heights evolves, combatants attempt to exploit technical aspects of terrain. Consequently, commanders must anticipate the need for evacuation teams, normally Level 2 mountaineers, that have the capability to reach, stabilize, and evacuate casualties in rugged terrain. Ground evacuations are generally classified as steep slope (non-technical) or high angle (technical). The mission of trained mountain evacuation teams is to move casualties over cliffs, obstacles, and other technical terrain that would significantly impede the mobility of standard litter bearers. Using evacuation systems to negotiate obstacles shortens routes and i ncreases the speed of evacuation.

5-38. Because units normally deploy over a wide area and the availability of qualified technical evacuation teams is likely to be limited, all soldiers should be trained to conduct less technical, steep-slope evacuations. Two of the most qualified evacuation teams should be identified in each battal ion-si zed unit prior to planned operations. They should be designated as the battalion's technical evacuation assets and should undergo more advanced mountaineering training and rigorous evacuation training. These soldiers can also develop and teach a program of instruction that will increase the proficiency of the company evacuation teams.

5-39. Mountain evacuation teams must install the necessary evacuation systems before casualties occur. They must man the systems, move casualties over the obstacle until the evacuation route is no longer required, disassemble the system, and redeploy as necessary. Depending on the specific terrain, evacuation teams may demand extensive additional training in some of the areas listed in Figure 5-5.

- High-angle ascending and descending techniques.
- Anchor points and systems.
- Litter rigging.
- Lowering and raising systems.
- Avalanche search and rescue.

Figure 5-5. Mountain Evacuation Team Tasks

TREATMENT

5-40. Treatment of the wounded in forward areas by medical personnel is extremely difficult in restrictive terrain, since even a single company is often deployed over a wide area. Combat in the mountains demands a greater reliance on self-aid, buddy-aid, and unit combat lifesaver techniques. Emphasis must be placed on lifesaving and life-preserving measures to be performed before medical personnel arrive. Unit combat lifesavers must be identified and trained to perform in the absence of medics. Units operating in mountainous areas should strive to meet or exceed Army standards for the number of combat lifesavers required for their specific unit. See FM 4-02.92 for more information on combat lifesavers.

5-41. Soldiers in mountain regions are exposed to many and varied types of illnesses and injuries. Appendix A describes the cause, prevention, symptoms, and treatment of common mountain illnesses and injuries.

Appendix A

Mountain Illnesses and Injuries

Table A-1. Chronic Fatigue and Its Effects

CHRONIC FATIGUE (ENERGY DEPLETION)			
CAUSE	**PREVENTION**	**SYMPTOMS**	**TREATMENT**
Low blood sugar. Sources of energy are depleted. Insufficient caloric intake.	Provide adequate food (type and quantities). Monitor food intake and ensure soldiers eat 4,500 calories or more per day. Eat small, frequent meals rather than large, infrequent meals. Snack lightly and often. Increase amounts of fat in diet.	Difficulty sleeping. Fatigue, irritability, and headache. Difficulty thinking and acting coherently -- impaired judgement. Victims begin to stumble and become clumsy and careless. Energy depletion resembles and aggravates hypothermia. The body does not have enough fuel to maintain proper body temperature. As a result inadequate sources of energy, coupled with cold, create a compound or synergistic effect.	Proper diet and rest. Treat synergistic effects if required.

Table A-2. Dehydration and Its Effects

DEHYDRATION			
CAUSE	**PREVENTION**	**SYMPTOMS**	**TREATMENT**
Loss of too much fluid, salt, and minerals due to poor hydration. Contributing Factors: Water loss occurs through sweating, breathing, and urine output. In cold climates, sweat evaporates so rapidly or is absorbed so thoroughly by clothing layers that it is not readily apparent. In cold weather, drinking is inconvenient. Water is hard to resupply, heavy to carry, and freezes in colder climates. Lack of humidity in the dry mountain air. Diminished thirst sensation induced by hypoxia.	Drink 3 to 4 quarts of water per day when static and up to 8 quarts during increased activity. Adequate rest. Avoid caffeine (coffee, tea, soda) and alcohol, as they compound dehydration. Increase command supervision. Keep canteens full. Use flavored powdered drink mixes to encourage water consumption.	Generally tired and weak. Mouth, tongue, and throat become parched and dry, and swallowing becomes difficult. Darkening of urine. Constipation and painful urination. Loss of appetite. Rapid heartbeat. Headache, dizziness, and nausea with or without vomiting. Difficulty focusing eyes. Dehydration compounds the effects of cold and altitude.	Sufficient hydration to offset water loss. Rest. Severe cases may require an IV. Insulate as required and evacuate.

Table A-3. Giardiasis and Its Effects

GIARDIASIS (PARASITICAL ILLNESS)			
CAUSE	**PREVENTION**	**SYMPTOMS**	**TREATMENT**
Parasitical illness contracted from drinking unpurified water.	Drink only potable water. Boil water for 3 to 5 minutes. Use approved water purification tablets or filters. Keep water containers clean.	Abdominal pain. Weakness and nausea. Frequent diarrhea and intestinal gas. Loss of appetite.	Proper hydration with potable water. Evacuation and prescribed medications.

Table A-4. Hypoxia and Its Effects

HYPOXIA			
CAUSE	PREVENTION	SYMPTOMS	TREATMENT
Rapid ascent to high altitudes (above 3,000 to 4,000 meters or 10,000 to 13,000 feet).	Acclimatization. Slow ascent. Limited activities. Lonq rest periods.	Impaired judgment, perception, and higher mental functions increasing with altitude.	Evacuation to tower altitude.

Table A-5. Acute Mountain Sickness (AMS) and Its Effects

ACUTE MOUNTAIN SICKNESS (AMS)			
CAUSE	PREVENTION	SYMPTOMS	TREATMENT
Rapid ascent to high altitudes {2,400 meters or 8.000	Acclimatization. Staged and/or graded ascent. During stops, no strenuous activity and only mild activity with frequent rest periods. Increased carbohydrate intake (whole grains, vegetables. peas and beans, potatoes, fruits. honey, and refined sugar). Acetazol amide prescribed by a physician.	Headache and fatigue. Insomnia. irritability, and depression. Coughing and shortness of breath. Loss of appetite, nausea, and vomiting. Dizziness. Swelling ol the eyes and face.	Stop and rest Symptoms will normally subside in 3-7 days if soldiers do not continue to ascend. Observe tor the development of HAPE or HACE. If symptoms do not disappear, a rapid descent of 150 to 300 motors (500 to 1,000 feet) or greater is necessary. Re-ascent should take place only after symptoms are resolved.

Table A-6. High Altitude Pulmonary Edema (HAPE) and Its Effects

HIGH ALTITUDE PULMONARY EDEMA (HAPE)			
CAUSE	**PREVENTION**	**SYMPTOMS**	**TREATMENT**
Unacclimatized soldiers rapidly ascending to high altitudes (2,400 meters or 8,000 feet}*. Acclimatized soldiers ascending rapidly from a high to a higher altitude. Usually begins within the first 2 - 4 days after rapid ascent and generally appears during the second night of sleep at high or higher altitudes. Fluid accumulation in the lungs.	Acclimatization. Staged and/or graded ascent. Sleeping at the lowest altitude possible. Slow assumption of physical activity. Protection from the cofd.	Wheezing and coughing (possibly wkth pink sputum). Gurgling sound in chest. Difficulty breathing. Coma Death may occur if rapid descent is not initiated.	Rapid evacuation recommended. Observe for the development of HACE. Seek qualified medical assistance.
'HAI'fc most often <ioi.^ -KH ';ur uni: .ibc-'A' 3 -300 ivehv-'-v I'I? 000 U-<>;\			

Table A-7. High Altitude Cerebral Edema (HACE) and Its Effects

HIGH ALTITUDE CEREBRAL EDEMA (HACE)			
CAUSE	**PREVENTION**	**SYMPTOMS**	**TREATMENT**
Unacclimatized soldiers rapidly ascending to high altitudes (2,400 meters or 8,000 feet)\ Acclimatized soldiers ascending rapidly from a high to a higher altitude. Excessive accumulation of fluid in the brain.	Acclimatization Staged and/or graded ascent. Slow assumption ot physical activity. Protection from the cold.	Most severe high altitude illness. Severe headache, nausea, and vomiting. Staggering walk/sway. Contusion, disorientation, and drowsiness. Coma, usually followed by death.	1 mmediate evacuation; preferably by air evacuation Seek qualified medical assistance.
"HACE like haPE. most often does not occur unm above ^.000 meters [12.000 feel)			

Appendix B

Forecasting Weather in the Mountains

The Air Force provides the bulk of the weather support required by the Army; however, reports from other branches of the military service, our own National Weather Bureau, or a foreign country's weather service can also aid in developing accurate forecasts {see FM 2-33.2). Weather at different elevations and areas, even within the same general region, may differ significantly due to variations in cloud height, temperature, winds, and barometric pressure. Therefore, general reports and forecasts must be used in conjunction with the locally observed weather conditions to produce reliable weather forecasts for a particular mountain area of operations.

INDICATORS OF CHANGING WEATHER

MEASURABLE INDICATORS

B-1. In the mountains, a portable aneroid barometer, thermometer, wind meter, and hygrometer are useful to obtain measurements that will assist in forecasting the weather. Marked or abnormal changes within a 12-hour period in the indicators listed in Figure B-1 may suggest a potential change in the weather.

- Barometric Pressure
- Wind Velocity
- Wind Direction
- Temperature
- Moisture Content of the Air

Figure B-1. Measurable Weather Indicators

CLOUDS

B-2 Clouds are good indicators of approaching weather conditions. By reading doud shapes and patterns, observers can forecast weather even without additional equipment.

B-3. Shape and height are used to identify douds. Shape provides information about the stability of the atmosphere, and height above ground level provides an indication of the distance of an approaching storm. Taken together, both indicate the likelihood of predpitation (see Figure B-2). The heights shown in the figure are an estimate and may vary, based on geographical location.

Clouds by Shape

B-4. Clouds may be classified by shape as cumulus or stratus.

- Cumulus douds are often called "puffy" clouds, looking like tufts of cotton. Their thickness (bottom to top) is usually equal to or greater than their width. Cumulus douds are primarily composed of wator droplets that cause them to have sharp, distinct edges. These clouds usually indicate instability at the altitude of the atmosphere where they are found. The stormy weather associated with cumuius clouds is usually violent with heavy rains or snow and strong, gusty winds. Precipitating cumulus douds are called cumulonimbus.

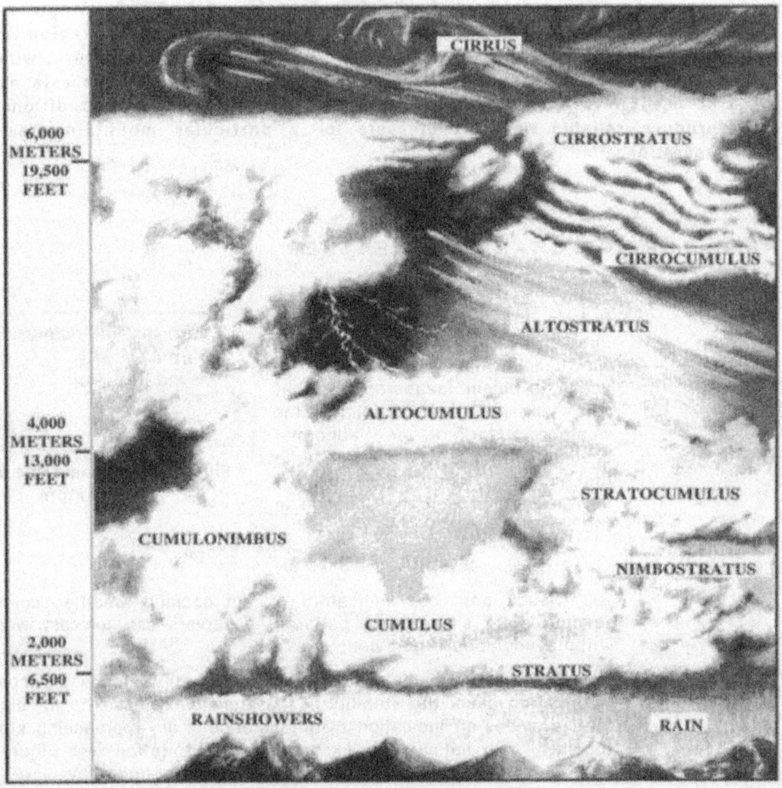

Figure B-2. Types of Clouds

- Stratus clouds are layered, often appearing flattened, with greater horizontal than vertical dimensions. They usually indicate a stable atmosphere, but can indicate the approach of a storm. Stormy weather assodated with stratus clouds usually does not normally include violent winds, and precipitation is usually light but steady, lasting up to 36 hours. Lightning is rarely associated with stratus douds, however, slee! may occur. Fog is also associated with the appearance of stratus clouds. Precipitating stratus clouds are called nimbostratus. and clouds that cannot be determined as stratus or cumulus are referred to as stratocumulus. These latter types may be evolving from one type to another, indicating a change in atmospheric stability.

Clouds by Height

B-5. Clouds are also classified by the height of their base above ground level into three categories- low. middle, and high.

- Low clouds, below 2,000 meters (6,500 feet), are either cumulus or stratus, or their predpitating counterparts. Low douds may be identified by their height above nearby surrounding relief of known elevation. Most precipitation originates from low douds because rain and snow from higher clouds usually evaporates before reaching the ground. As such, low douds usually indicate predpitation. especially if they are more than 1.000 meters (3.000 feet) thick (douds that appear dark at the base usually are at least that thick).
- Middledouds, between 2.000 and 6,000 meters (6,500 and 19,500 feet) above ground, have a prefix of "alto", and are called either altostratus or altocumulus. Middle douds appear less distinct than low douds because of their height. Warm "alto" douds have sharper edges and are composed mainly of water droplets. Colder douds, composed mainly of ice crystals, have distinct edges that grade gradually into the surrounding sky. Middle clouds indicate potential storms, though usually hours away. Altocumulus clouds that are scattered in a blue sky are called "fair weather" cumulus and suggest the arrival of high pressure and clear skies. Lowering altostratus douds with winds from the south indicate warm front conditions, decreasing air pressure, and an approaching storm system within 12 to 24 hours.
- High douds, higher than 6,000 meters (19,500 feet), are drrus. drrostratus, and drrocumulus. They are usually frozen douds with a fibrous sliudure and blurred outlines. The sky is often covered with a thin veil of drrus that partly obscures the sun or, at night, produces a ring of light around the moon. The arrival of drrus indicates moisture aloft and the approach of a storm system. Predpitation is often 24 to 36 hours away. As the storm approaches, the drrus thickens and lowers becoming altostratus and eventually stratus. Temperatures warm, humidity rises, and winds approach from the south or southeast.

Other Clouds

B-6 Somedoudsindicate serious weather ahead.

- Towering cumulus douds have bases below 2,000 meters (6,500 feet) and tops often over 6,000 meters (19.500 feet). They are the most dangerous of all types and usually do not occur when temperatures at the surface are below 32-degrees Fahrenheit. They indicate extreme instability in the atmosphere, with rapidly rising air currents caused by solar heating of the surface or air rising over a mountain barrier. Mature towering cumulus douds often exhibit frozen stratusdouds at their tops, producing an "anvil head" appearance. Towering cumulus clouds may be local in nature, or they may be associated with the cold front of an approaching storm. The iatter appears as an approaching line of thunderstorms or towering cumulus clouds. Towering cumulus clouds usually produce high, gusty winds, lightning, heavy showers, and occasionally hail and tornadoes (although tornadoes are rare in mountainous terrain). Such thunderstorms are usually short-lived and bring clear weather.

G Cloud caps often form above pinnacle and peaks, and usually indicate higher winds aloft. Cloud caps with a lens shape (similar to a "flying saucer") are called lenticular and indicate very high winds (over 40 knots). Cloud caps should always be watched for changes. If they grow and descend, bad weather can be expected.

APPLYING THE INDICATORS

B-7. Weather forecasts are simply educated estimations or deductions based on general scientific weather principles and meteorological evidence. Forecasts based on past results may or may not be accurate. However, even limited experience in a particular mountainous region and season may provide local indications of impending weather patterns and increased accuracy. Native weather lore, although sometimes greatly colored and surrounded in mystique, should not be discounted when developing forecasts, as it is normally based on the local inhabitants' long-term experience in the region.

BAD WEATHER

B-8. Signs of approaching bad weather (within 24 to48 hours) may include—

- A gradual lowering of the clouds. This may be the arrival or formation of new lower strata of clouds. It can also indicate the formation of a thunderhead.
- An increasinghaloaroundthesunorlhe moon.
- An increase in humidity and temperature.
- Cirrus clouds.
- A decrease in barometric pressure (registered as a gain in elevation on an altimeter).

STORM SYSTEMS

B-9. The approach of a storm system is indicated when—

- A thin veil of cirrus clouds spreads over the sky, thickening and lowering until altostratus clouds are formed. The same trend is shown at night when a halo forms around the moon and then darkens until only the glow of the moon is visible. When there is no moon, cirrus clouds only dim the stars, but altostratus clouds completely hide them.
- Low clouds, which have been persistent on lower slopes, begin to rise at the time upper clouds appear.

- Various layers of clouds move in at different heights and become abundant.

D Lenticular clouds accompanying strong winds lose their streamlined shape, and other doud types appear in increasing amounts,

0 Achangeinthedi rection ofthewindis accompanied by a rapid rise in temperature not caused by solar radiation. This may also indicate a warm, damp period.

OA light green haze is observed shortly after sunrise in mountain regions above the timber line.

THUNDERSTORMS

B-10. Indications of local thunderstorms or squally weather are—

- An increase in size and rapid thickening of scattered cumulus clouds during the afternoon.

Q The approach of a line of large cumulus or cumulonimbus clouds with an "advance guard" of altocumulus douds. At night, increasing lightning windward of the prevailing wind gives the same warning.

P Massive cumulus douds hanging over a ridge or summit (day or night).

STRONG WINDS

B-11. Indications of approaching strong winds may be—

[] Plumes of blowing snow from the crests of ridges and peaks or ragged shreds of doud moving rapidly.

0 Persistent lenticular douds, a band of douds over high peaks and ridges, or downwind from them.

- A turbulent and ragged banner doud that hangs to the lee of a peak.

PRECIPITATION

B-12. When there is predpitation and the sky cannot bc seen—

fj Small snowflakes or ice crystals i ndicate that thedouds above are thi n, and fair weather exists at high elevations.

- A steady fall of snowflakes or raindrops indicates that the predpitation has begun at high levels, and bad weather is likely to be encountered on ridges and peaks.

FAIR WEATHER

B-13. Continued fair weather may be assodated with—

[] A doudless sky and shallow fog, or layers of haze at valley bottoms in early morning.

- A doudless sky that is blue down to the horizon or down to where a haze layer forms a secondary horizon.
- Conditions under which small cumulus douds appearing before noon do not increase, but instead decrease or vanish during the day.

- Clear skies except for a law cloud deck that does not rise or t hick en during the day.

B-14. Signs of approaching fair weather include—

- A gradual rising and diminishing of clouds.
- A decreasing halo around the sun or moon.
- Dew on theground in the morning.
- Small snowfiakes, ice crystals, or drizzle, which indicate that the clouds are thin and fair weather may exist at higher elevations.
- An increase in barometric pressure (registered as a loss in elevation on an altimeter).

GLOSSARY

ABN	airborne
acclimatization	the physiological changes that allow the body to adapt or get used to the effects of a new environment, especially low oxygen saturation at higher elevations
ACE	armored combat earth mover
acetazolamide	a pharmaceutical drug used to accelerate acclimatization
ADA	air defense artillery
ADAM	area denial artillery munitions
AH-64	attack helicopter also called the Apacbe
aid	in mountaineering, a climbing device, such as pitons, bolts, chocks, and stirrups, used for body support and upward progress; also used for artificial height in the absence of handholds and footholds
ALOC	air lines of communications
AM	amplitude modulation
ambient temperature	encompassing atmosphere
AMS	acute mountain sickness
anchor	a secure point (natural or artificial} to which a person or rope can be safely attached
aneroid	using no liquid
ANZAC	Australia and New Zealand Corps
AO	area of operations
apnea	temporary suspension of respiration
ARSOF	Army special operations forces
ART	Army tactical task
ARTEP	Army training and evaluation program
assault climber	military mountaineer possessing advanced (Level 2) skills, capable of leading small teams over class 4 and 5 terrain and supervising rigging/operation of all basic rope systems
AT4	a man-portable, lightweight, self-contained, antiarmor weapon
ATGM	anti-tank guided missile
basic mountaineer	a military mountaineer trained in fundamental (Level 1) travel/climbing skills necessary to move safely and efficiently in mountainous terrain

belay	a rope management technique used to ensure that a fall taken by a climber can be quickly arrested: belay techniques are also used for additional safetyfcontrol in rappelling. raising and lowering systems, and for mountain stream crossings
BFV	Bradley fighting vehicle
BSFV	Bradley Stinger fighting vehicle
BN	battalion
&.	command and control
CAFAD	combined arms for air defense
CFV	cavalry fighting vehicle
CHS	combat health support
CNR	combat net radio
continental climate	bitterly cold winters, extremely hot summers; annual rain and snowfall is minimal and often quite scarce for long periods
Cordillera	principal mountain ranges of the world, named after the Spanish word for rope
crampons	climbing irons, attached to the bottom of boots, used on ice or snow in mountaineering
crevice	a narrow opening resulting from a split or crack as in a cliff
CS	combat support
CSS	combat service support
DA	Department of the Army
defile	a narrow passage or gorge
DPI CM	dual-purpose improved conventional munition
DZ	drop zone
ECWCS	extended cold weather clothing system
edema	a local or general condition in which the body tissues contain an excessive amount of tissue fluid
evacuation team	a team trained to move casualties over steep slopes, cliffs, and other obstacles that would significantly impede the mobility of standard litter bearers
EW	electronic warfare
F	Fahrenheit
FARP	forward arming and refueling point
FASCAM	family of scatterable mines
FEBA	forward edge of the battle area

fixed alpine path	a mountain path created by any combination of aids, to include steps, stanchions, standoff ladders, suspended walkways, cableways, or other improvements made of materials available; normally an engineering task.
fixed rope	a rope, or series of ropes, anchored to the mountain at one or more points to aid soldiers over steep, exposed terrain; usually installed by lead climbing teams (normally Level 2 qualification)
flash defilade	to arrange fortifications to protect from fire
FM	field manual; frequency modulation
FSMC	forward support medical company
FOX system	a lightly-armored, wheeled laboratory that takes air, water, and ground samples and immediately analyzes them for signs of weapons of mass destruction
gabion	a wicker basket filled with earth and stones often use in building fortifications; can also be created out of similar materials, such as wire mesh/fence, lumber, plywood, or any suitable material that forms a stackable container for rocks, gravel, and soil
giardiasis	parasitical illness
glaciated	covered with glacial ice
GPS	global positioning system
GTA	graphictraining aid
guide	a soldier experienced in all aspects of mountaineering who has the skills and knowledge to identify obstacles and ways to overoome them; commander's advisor on technical mountaineering matters that could affect the tactical scheme of maneuver; primary function of mountain leaders (Level 3 qualification)
HACE	high altitude cerebral edema
HAPE	high altitude pulmonary edema
HE	high explosives
Hellfire	tank-killing missile carried by the Apache attack helicopter
high mountains	mountains that have a local relief usually exceeding 900 meters (3,000 feet)
HUMINT	human intelligence
HWY	highway
hygrometer	an instrument used to measure humidity or moisture content in the air
hypoxia	a deficiency of oxygen reaching the tissues of the body
ice fog trails	steam/smoke trails created by firing weapons
ID	infantry division

IFV	infantry fighting vehicle
IHFR	improved high frequency radio
IMINT	imagery intelligence
installation team	a team organized to construct and maintain rope installations used to facilitate unit movement; usually comprised of Level 1 and 2 mountaineers
interdiction	to stop or hamper
ionospheric	a part of the earth's atmosphere of which ionization of atmospheric gases affects the propagation of radio waves; starts at about 30 miles above ground
IPB	intelligence preparation of the battlefield
IV	intravenous
JSTARS	joint surveillance, target attack radar system
km	kilometer
lead climbing team	a roped dimbing team (usually Level 2 qualification) trained to lead on dass 4 and 5 terrain; establishes/prepares the entire route for the remainder of the unit
leeward	the side sheltered from the wind
lenticutar	having the shape of a double-convex lens
LOC	line of communication
local relief	the difference in elevation between valley floors and the surrounding summits
look-down angles	the angle from the aircraft to the target
low mountains	mountains that have a local relief of 300 to 900 meters (1,000 ta 3,000 feet)
LPT	logistics preparation of the theater
LRS	long-range surveillance
LRSU	long-range surveillance unit
LSDIS	light and spedal division interim sensor
LTC	lieutenant colonel
I 7	landing zone
MANPADS	man-portable air defense system
maritime climate	mild temperatures with large amountsof rain or snow
MBA	main battlearea
METT-TC	mission, enemy, terrain and weather, troops and support available, time available, dvil considerations
MK-19	40-mm grenade machine gun, MOD 3

MOPP	mission-oriented protective posture
motti	Finnish word meaning "a pile of logs ready to be sawed into lumber"; used in military terms to descri be setting the condit ions so a larger force can be defeated in detai l
mountain leader	a military mountaineer possessing the highest level (Level 3) of mountaineering skills with extensive experience in a variety of mountain environments in both winter and summer months
WISE	mobile subscriber equipment
MSRT	mobile subscriber radio terminal
MTF	manual terrain following
NBC	nuclear, biological, and chemical
OCOKA	observation and fields of fire, cover and concealment, obstacles, key terrain, and avenues of approach
OH-58D	a scout and attack helicopter known as the Kiowa Warrior
OP	observation post
OPORD	operation order
OPSEC	operations security
OR	operational readiness
orographic	pertaining to the physical geography of mountains and mountain ranges
PADS	Position Azimuth Determining System
POL	petroleum, oils, and lubricants
protection	in mountaineering, special anchor points established during a roped party dimb to limit potential fall distances, protecting dimbers from severe fall/ground-fall consequences
PSYOP	psychological operations
RAAMS	remote antiarmor mine system
rappel	method of controlled fridional descent down a rope
RCW	ration, cold weather
rockfai l	rockfall occurs on alt steep slopes. It is caused by other dimbers or by the continual erosion of the rock on a mountainside resulting from freezing, thawing, and heavy rain; grazing animals; or enemy action.
SATCOM	satellite communications
scree	small unconsolidated rocks or gravel, fist-si2e or smaller, located mostly below rock ridges and diffs
screen i ng crest	a hill or ridge located in front of a radar set to mask it from unwanted returns (dutter) at dose range, and to provide security

	against electronic detection or jamming; screening crest also prevents visual observation and attack by direct fires
SEE	small emplacement excavator
SHELREP	shelling report
SHORAD	short-range air defense
SINCGARS	Single-channel Ground and Airborne Radio System
SOP	special operations forces
squall	a sudden, violent wind
SR	special reconnaissance
talus	accumulated rock debris that is much larger than scree, usually basket ball-size or larger
TBP	to be published
TC	[raining circular
TCF	tactical combat force
TCP	trafficcontrd point
temperature inversion	when the temperature is warmer at higher elevations than lower elevations
TM	technical manual
TOC	tactical operations center
TOW	tube-launched, optically tracked, wire-guided, heavy antitank missile system
TRADOC	United States Army Training and Doctrine Command
tundra	treeless, black, mucky soil with permanently frozen subsoil; located in mountainous regions above the timberline
tussocks	grassy clumps
UAV	unmanned aerial vehicle
UGR	unitized group ration
UHF	ultrahigh freciuoncy
Venturi effect	as a fluid (such as air) flows through a constriction (like a mountain pass), the speed increases and the pressure drops
VFR	visual flight rules
VT	variabletime
wind chill	the rate at which a man or object cools to the ambient temperature; wind increases the rate of cooling and adds to the risk of frostbite, hypothermia, and other cold-weather injuries
windward	beingin or fadngthedirection from which thewind isblowing
WP	white phosphorous

Bibliography

The bibliography lists field manuals by new number followed by old number.

ARMY PUBLICATIONS

Most Army doctrinal publications are available online:
http://155.217.58.58/atdls.htm

AR 385-10. *The Army Safety Program*. 23 May 1988.

FM 1-0 (12-6). *Personnel Doctrine*. 09 September 1994.

FM 2-01.3 (34-130). *Intelligence Preparation of the Battlefield*. 08 July 1994.

FM 2-33.2 (34-81). *Weather Support for Army Tactical Operations*. 31 August 1989.

FM 2-33.201 (34-81-1). *Battlefield Weather Effects*. 23 December 1992.

FM 3-0 (100-5). *Operations*. 14 June 1993.

FM 3-01.8 (44-8). *Combined Arms for Air Defense*. 01 June 1999.

FM 3-01.43 (44-43). *Bradley Stinger Fighting Vehicle Platoon and Squad Operations*. 03 October 1995.

FM 3-01.44 (44-44). *Avenger Platoon, Section, and Squad Operations*. 03 October 1995.

FM 3-04.100 (1-100). *Army Aviation Operations*. 21 February 1997.

FM 3-04.203 (1-203). *Fundamentals of Flight*. 03 October 1988.

FM 3-05.27 (31-27). *Pack Animals in Support of Army Special Operations Forces*. 15 February 2000.

FM 3-09.4 (6-20-40). *Tactics, Techniques, and Procedures for Fire Support for Brigade Operations (Heavy)*. 05 January 1990.

FM 3-09.12 (6-121). *Tactics, Techniques, and Procedures for Field Artillery Target Acquisition*. 25 September 1990.

FM 3-09.40 (6-40). *Tactics, Techniques, and Procedures for Field Artillery Manual Cannon Gunnery*. 23 April 1996.

FM 3-09.42 (6-20-50). *Tactics, Techniques, and Procedures for Fire Support for Brigade Operations (Light)*. 05 January 1990.

FM 3-11.4 (3-4). *NBC Protection*. 29 May 1992.

FM 3-11.5 (3-5). *NBC Decontamination*. 28 July 2000.

FM 3-]] ,6 (3-6), *Field Behavior of NBC Agents (Including Snufke and Incendiaries).* 03 November (986.

FM 3-11.19 (3-19). *NBC Reconnaissance.* 19 November 1993.

FM 3-21.7 (7-7). *Tile Mechanized Infantry Platoon and Squad <APC).* 15 *Match* 1985.

FM 3-21,20(7-20), *The Infantry Battalion.* 06 April 1992,

FM 3-21.30 (7-30), *The Infantry Brigade.* 03 October 1995.

FM 3-21.38 (S7-38). *Pathfinder Operations.* 09 April 1993.

FM 3-23.10 (23-10). *SniperTraining.* 17 August 1994.

FM 3-23,30 (23-30), *Grenades and Pyrotechnic Signal).* 27 Decem be r 1988,

FM 3-24.3 (20-3). *Camouflage. Concealment, and Decoys.* 30 August 1999.

FM 3-24.32 (20-32). *Mine/Countermine Operations.* 29 May 1998.

FM 3-25,9(23-9), *MI6A) and M16A2 Rifle Marksmanship.* 03 July 1989.

FM 3-25.18(21-18). *Foot Marches.* 0 J June 1990.

FM 3-25.26 (21-26). *Map Reading and Land Navigation.* 07 May 1993.

FM 3-25,60(21-60), *Visual Signals.* 30 September 1987,

FM 3-25,76 (21-76). *Survival.* 05 June 1992.

FM 3-34.1 (90-7). *Combined Arms Obstacle Integration.* 29 September 1994.

FM i 34.2 (9(1 13 I 1. *Combined Arms Brea, lung Operations.* -N iilmi.iri I V) I .

FM 3-34.102 (5-102), *Cotmtermobility.* l4March 1985.

FM 3-34.112(5-103). *Survivability.* 10 June 1985.

FM 3-34.214 (5-250). *Explosives and Demolitions.* 30 July 1998.

FM 3-34,223 (5-7-30). *Brigade Engineer and Engineer Company Combat Operations I Airborne, Air Assault, Light}.* 28 December 1994.

FM 3-34.330 (5-33). *Terrain Analysis.* I 1 July 1990.

FM 3-60 (6-20-10). *Tactics. Techniques, and Procedures for the Targeting Process.* 08 May 1996.

FM 3-90.21 (6-20-1). *Tactics, Techniques, and Procedures for the Field Artillery Cannon Battalion.* 29 November 1990.

FM 3-91 -2 (7 I -2). *The Tank and Mechanized Infantry Battalion Task Force.* 21 September 1988.

FM 3-91,3 (71-3), *The Armored and Mechanized Infantry Brigade.* 08 January 1996,

FM 3-9], 323(71-123). *Tactics and Techniques for Combined Arms Heavy Forces: Armored Brigade, Battalion Task Force, and Company Team,* 30 September 1992,

F M 3-97.3 (90-3). *Desert Operations. 24* Augusl 1993.

FM 3-97.1 I (90-1 I). *Cold Weather Operations. TUP.*

FM 3-97,22 (90-22), *(Night) Multi-Service Night and Adverse Weather Combat Operations.* 31 January 1991.

FM 3-97,50 (3-50), *Smoke Operations,* 04 December 1990,

F M 3-100.14 (100-14). *Risk Management,* 23 April 1998.

FM 3-100.15 (100-15). *Corps Operations.* 29 Oelnber 1996.

FM 3-100,40(100-40), *Tactics,*TBP,

FM 3-100.55 (100-55). *Reconnaissance and Surveillance.* TBP,

FM 3-100.71 (71-100). *Division Operations.* 28 August 1996.

FM 4-02.2 (8-10-6). *Medictd Evacuation in a Theater of Operations Tactics. Techniques, and Procedures.* 14 April 2000.

FM 4-02,22 (22-51), *Leaders' Manual for Combat Stress Control.* 29 September 1994.

FM 4-02.92 (8-10-4). *Medical Planum Leaders' Handbook Tactics, Techniques, and Procedures.* I6November 1990,

F M 4-20.1 (10-27). *General Supply in Theaters of Operations.* 20 April 1993.

FM 4-20.2 (10-23). *Basic Dtfcrrine for Artnv Field Feeding and Class I Operations Management.* IK April 1996.

FM 4-25,10 (21-10), *Field Hygiene and Sanitation.* 21 June 2000

FM 4-25.11(21-11). *First Aid for Soldiers.* 27 October 1988,

FM 4-30.32 (9-207). *Operations and Maintenance of Ordnance Materiel in Cold Weather.* 20 March 1998.

FM 5-0 (101-5). *Sniff Organization and Operations.* 31 May 1997.

FM 6-0 (100-34). *Command and Control.* TBP.

FM 6-02.1 I (24-1 I), *Tactical Satellite Communications.* 20 September 1990.

FM 6-02.18 (24-1 8). *Tactical Single-Channel Radio Communications Techniques.* 30 September 1987.

F'M 6-02.32 (I 1-32). *Combat Net Radio Operations.* 15 Otlciber 1990.

FM 6-02,43(1 1-43), *The Signal Leader's Guide,* 12 June 1995,

FM 6-02,55 (11-55). *Mobile Subscriber Equipment < MSB) Operations. 22* June 1999.

FM 6-22 (22-100), *Army Leadership Be, Know, Do.* 31 August 1999,

FM 7-0 (25-100). *Training die Farce.* 15 November 1988.

FM 7-10(25-101). *Battle Focused Truining.* 30 September 1990.

FM 7-I5(25-XX). *Army Universal Task List.* TBP,

FM 21-305. *Manna! far the Wheeled Vehicle Driver.* 27 August 1993.

GTA 8-5-60. *A Soldier's (hade to Staying Healthy at High Elevations.* 02 September 1996.

GTA 8-6-12, *Adverse Effects of Cold.* 01 August 1985.

TC3-10. *Commander's Tactical NBC Handbook.* 29 September 1994.

TC 24-20. *Tactical Wire and Cable Techniques.* 03 October 1 988.

TC 24-21. *Tactical Multichannel Radio Communications Techniques.* 03 October 1988.

TC 90-6-1. *Military Mountaineering.* 26 April 1989.

NONMILITARY PUBLICATIONS

Jalali. Ali Ahmad and Lester W. Grau. *The Other Side of the Mountain.* Ouanlko, Va.: Marine Corps Combat Development Command, USMC Studies and Analysis Division, 1999.

Ellis, Robert B. *See Naples and Die.* London: McFarland & Co. 1996.

Gawryeh, George W. "The Rock of Gallipoli" in *Studies in Battle Command.* Kurt Leavenworth. Kans.: LIS Army Command and General Staff College, Combat Studies Institute. 1995.

Cirau, LesLcr W. *The Bear Went Over the Mountain: Soviet Combat Tactics in Afghanistan.* Washington. D.C: National Defense University Press, 1996.

DOCUMENTS NEEDED

These documents should be available to the intended users of this manual.

DA Form 2028, Recommended Changes to Publications and Blank Forms. I February 1974,

FM 7-0 (25-100). *Training die Force.* 15 November 1988.

FM 7-10 (25-101). *Battle Focused Training.* 3(1 September 1990.

FM 3-0 (100-5), *Operations.* 14 June 1993,

FM 3-40(100-40). *Tactics.* TBP.

TC 90-6-1. *Military Mountaineering.* 26 April 1989.

Index

Entries are by paragraph number unless stated otherwise.

Made in the USA
Monee, IL
05 November 2026

67807979R00203